MATHEMATICAL ECONOMICS TEXTS

2

UNCERTAINTY AND ESTIMATION IN ECONOMICS

VOLUME ONE

MATHEMATICAL ECONOMICS TEXTS

Students of mathematical economics and econometrics have two unnecessary difficulties. One is that much of what they have to read is spread over the journals, often written in different notations. The second is that the theoretical and the empirical writings often make little reference to each other, and the student is left to relate them.

The main object of this series is to overcome these difficulties. Most of the books are concerned with specific topics in economic theory, but they relate the theory to relevant empirical work. Others deal with the necessary mathematical apparatus of economic theory and econometrics. They are designed for third-year undergraduates and postgraduate students.

The editors are drawn from both sides of the Atlantic and are people who are known both for their contribution to economics and for the emphasis they place on clear exposition.

Titles in the series

1. Dynamic Programming: D. J. WHITE
2. Uncertainty and Estimation in Economics—Volume I: D. G. CHAMPERNOWNE
3. Uncertainty and Estimation in Economics—Volume II: D. G. CHAMPERNOWNE
4. Uncertainty and Estimation in Economics—Volume III: D. G. CHAMPERNOWNE

Other titles are in preparation

UNCERTAINTY AND ESTIMATION IN ECONOMICS

VOLUME ONE

D. G. CHAMPERNOWNE
M.A.
Reader in Economics in the University of Cambridge

OLIVER & BOYD
EDINBURGH
HOLDEN DAY
SAN FRANCISCO . CALIFORNIA

05 002006 4
Library of Congress Catalog Card No. 73-76889

First published 1969

Copublished by
OLIVER & BOYD LTD
Tweeddale Court Edinburgh 1
and HOLDEN DAY INC.
500 Sansome Street
San Francisco California

Printed in Great Britain by
Robert Cunningham and Sons Ltd
Longbank Works Alva

PREFACE

The student of economics is likely to encounter a very wide range of subjects in which he needs to understand the nature and influence of uncertainty and the methods available for proceeding in the face of it. He will require, for example, some insight into the basic problems of the *meaning* of uncertainty, probability, irrelevance, independence and the like: he will need at least some acquaintance with the techniques available for interpreting imperfect or incomplete statistical information and obtaining statistical estimates, or choosing decisions in the light of such evidence. Finally he will need to understand something of how importantly and in what ways the behaviour of men of affairs is altered because they have to make big decisions on the basis of very imperfect knowledge of the prospects.

The judgements and skills required for these different tasks are by no means all of the same kind. It is therefore unlikely that any reader will find all chapters in the three volumes of this book equally interesting, or that any will find them all equally easy reading. Those readers who are uninterested in statistical techniques may omit large sections of the second volume and still, it is hoped, find much to interest them in the philosophical and economic discussions in the first and third volumes. On the other hand, it is hoped that those mainly interested in statistics may find much that is novel and illuminating in the approaches in the first two volumes to the long-standing problems of statistical estimation from the point of view of an economist.

In writing this book I have been helped by many friends at Cambridge University, and I am particularly indebted to the following: to Frank Hahn for persuading me in the first place to undertake this work; to Peter Swinnerton-Dyer for teaching me how to construct the example of a non-numerical probability ordering which is discussed in the Appendix to Chapter 2; to Brian Reddaway, Phyllis Deane and Jim Mirrlees for reading and criticising early drafts of some chapters of the book and to Ken Arrow (who had visited Cambridge) for very detailed criticism of a later draft of Volume 1. Of course none of these is responsible for the imperfections that remain, especially since I did not always adopt their suggestions.

I am also much indebted to the Cambridge University Computing Laboratory for the use of their facilities and to Harry Burley for help in programming the computer. I thank the Council of the Senate for their kind permission to reproduce examination questions set in Part 2 of the Economics Tripos in recent years. Finally, I acknowledge with thanks the many facilities provided by the Faculty of Economics and the Department of Applied Economics at Cambridge University: without these facilities I doubt whether the work would have been completed.

D. G. C.

Faculty of Economics,
Cambridge, England

CONTENTS

Chapter 1

IMPERFECT KNOWLEDGE IN ECONOMIC THEORY

1.1. The Uses of Probability Theory in Science and in Economics: a Contrast

The applications of the theory of probability to the natural sciences largely arise from the fact that random variations are found between the results of one experiment and another similar to it—variations which are due to causes beyond the control or direct observation of the experimenter. The presence of these random variations introduces a lack of precision in his conclusions, and the theory of probability is a means for discussing the degree of error to which his estimates may be subject.

But the need to use a theory of probability in economics does not primarily arise from the existence of random variation in observations. It arises from the fact that most economic decisions are made against a background of uncertainty and imperfect knowledge. People often behave quite differently in their choices and their decisions when they feel uncertain from how they behave when they are sure of their facts. Even if errors of opinion cancelled out where large numbers of persons were concerned, the effect of their awareness of uncertainty would still remain; and the mere variety of economic opinions can have important effects even if no one is aware of the imperfection of his own knowledge and opinions. In the following paragraphs we shall discuss a number of the important economic effects of the existence of ignorance and uncertainty.

1.2. The Fear of Uncertainty

Exposure to the risks of unforeseeable adverse change in an economic climate has led many a man to attach an almost overriding importance to the need to reduce this uncertainty. Galbraith's bestseller, *The Affluent Society*, describes how paramount is the fear of facing unfamiliar situations in determining a citizen's choice of career and of policy.

This is one reason for the economists' concern to take account of the prevalence of uncertainty: a first effect of the existence of ignorance and uncertainty is a general desire to avoid economic risk and situations causing uncertainty. Alongside of the desire for profit, comfort and prestige, this desire provides a main motive which governs the economic behaviour of modern man.

1.3. Methods of Reducing Ignorance and Uncertainty

A second and related effect of the existence of ignorance and uncertainty is that a considerable amount of economic activity may be undertaken in order to reduce ignorance and uncertainty by

(*i*) acquiring more information;
(*ii*) influencing prospects and probabilities;
(*iii*) providing insurance against loss.

1.4. Effort Diverted to Influencing Opinion

A third effect of ignorance and uncertainty is that considerable effort may be exerted towards influencing the economic behaviour of consumers and other purchasers by providing them with suitably selected information. Ignorance and awareness of his ignorance has the effect of rendering a consumer's choices, at times, somewhat aimless so that they are more easily affected by trivial influences which are irrelevant to the aims which might have guided his choices, had he been more accurately informed. Such a customer's choices will depend largely on the packaging and placing of the goods on display, and on the suggestions of the retailer, and accordingly less on a consideration of the suitability of the contents of the package to fulfil their functions.

1.5. Imperfect Knowledge Reduces Stability

A fourth effect of imperfect knowledge is that there can be rapid shifts in opinion and expectation, causing widespread revisions of expenditure plans. Such changes may cause further shift of expectation in the same direction, so that from a small and fortuitous event there may develop a calamitous wave of pessimism which will in the end cause nation-wide economic loss. Such waves of pessimism and optimism are particularly associated with speculative markets and the decisions when and whether to embark on a risky economic venture with uncertain prospects of considerable gain.

1.6. Uncertainty and Oligopoly

A fifth effect of imperfect knowledge can be detected in the economic behaviour of oligopolists. When a firm has only a small number of rivals, its directors must be constantly aware of the risk that any success it has in gaining profits at the expense of rivals may lead to retaliation on their part, which may more than offset the gain: moreover, any success which is shared with the rival firms may attract competitors into the industry and again may lead to subsequent losses. The inability to foresee just what the ultimate effects of a firm's own individual policy will be causes a strong incentive to reduce the number of unpredictable and unmanageable factors in the situation, by arriving at some understanding, tacit or otherwise, with its rivals, as to the principles which will govern the competitive behaviour of each. Just how far such understanding will go will depend in part upon the legal and institutional background of the economy concerned: but in any case, it will be an influence in the direction of securing that different firms charge the same prices for products that are very similar, and that these prices change only as a result of events which are industry-wide in their effects.

1.7. Uncertainty and the Prescription of Policy

Most of the effects of ignorance and uncertainty which we have discussed so far are of interest to the economist because he is concerned with their influence on people's economic behaviour. But the economist is not only anxious to diagnose; he is also interested in the prescription. Much of economic theory has been concerned with 'optima' of various kinds, and with criteria for improving economic policy. Very much of this discussion has been based on the simplifying assumption that the effects of imperfect knowledge and imperfect foresight conveniently cancel out. If proper account could be taken of the uncertain elements in the economic prospect, this might cause very substantial changes in the policy implications of economic theory and prescription. An argument for increased State expenditure on the health services may be based on the fact that ill health is a hazard that cannot be accurately foreseen: people who are unexpectedly visited by serious illness can be saved much hardship by subsidised medical services. Ill health is in fact the kind of risk which ought to be insured against; but left to themselves, individuals fail to insure on a sufficient scale, either through easy-going ways and indolence, or because ill health strikes them before they have earned money with which to insure against it. Since individuals will not, or cannot, insure sufficiently of their own accord, the case for State medical insurance and State expenditure on health services is all the stronger. The case in favour of State expenditure on industrial research is strengthened by the fact that, whereas a single enterprise runs a considerable risk that through ill luck its research expenditure will pay only poor dividends, and the benefit may accrue largely to other firms, the

State may spread its research expenditure over so wide a field that a fairly steady and high rate of return on the expenditure may be relied upon. Moreover, the State can afford to wait longer for the dividends on research to flow in, and the State is not concerned whether the benefit accrues primarily to one firm rather than another, so long as much of the benefit remains within its own economy.

1.8. Uncertainty causes Haphazard Economic Behaviour

Although the questions outlined in the last few paragraphs are of immense importance to the economist, they are not ones on which the accepted theories of probability are well adapted to shed light. This is due not so much to any defect in the theories as to the fact that the solution to these problems depends so largely on acquiring an insight into the behaviour patterns of human beings. So long as one may assume that economic man behaves in a manner prescribed by some very simple overriding principle—such as the maximisation of the expected value of profits or of some ordinal utility-measure—then general theory may be informative. But if economic man is driven by vague fears of the unknown, and if his economic choices are ill-considered and haphazard, and sensitive to a host of miscellaneous influences, general theory will not take one very far: what would be required would be a huge series of case-studies illustrating the peculiarities of his economic-behaviour reactions to the more potent stimuli. Such research would be very expensive and academically unrewarding, although its results might enable some to make economic gain at the expense of others. It would perhaps be more sensible to use the same intellectual effort in formulating advice to people on how to develop more considered and rewarding reactions to their economic environment.

Some things, however, can be said on these topics at a high level of generality; and these questions will be touched on in the concluding chapters of this book.

1.9. The Use of Random Variates in Economic Models

There remain one or two topics of rather less importance, for whose discussion the accepted theories of probability and the recognised tools of the theoretical statistician are better adapted.

For example, the fact that the influences which may affect a particular economic variable (such as the output of an industry) are so numerous means that no simple determinist model of the development of any economic situation can avoid being glaringly incomplete. By introducing trends and random variation into a model, some representation can be restored to all those changing economic influences which are denied specific inclusion in the model.

Such devices are particularly useful in studies of consumers' behaviour, and models representing such behaviour usually include provision for random disturbances, and standard techniques of statistical estimation have been developed to enable estimates to be made of the 'parameters' such as elasticities of demand, which are built into these models.

1.10. Probability and Sample Surveys

The field of economic enquiry wherein the orthodox statistician is most at home is that which uses the tool of the sample survey. Civil service departments, marketing organisations and academic research centres, all need on occasion to estimate the statistical properties of some wide range of economic phenomena. This may be done by means of an exhaustive enquiry, such as a population census or a production census, or by getting people to fill in income tax forms; but very often better and cheaper estimates may be obtained by obtaining information about only a sample of the people or material in question.

In a carefully conducted sample enquiry the systematic sources of error in estimation can usually be made smaller than would be possible in a complete census. Such systematic errors might arise, for example, from inefficient or dishonest filling in of the forms on which information is to be provided; but in a sample there will be a new source of error, called *random sampling* error, and it is the task of the statistician to design the enquiry in such a way that this random sampling error should be moderate in size and its probable extent should be capable of estimation.

The design and analysis of such sampling enquiries has been highly developed, and it is in this rather minor branch of applied economics that the theory of probability is in fact most directly useful.

1.11. The Plan of the Book

The more important economic problems involving probability arise not directly from the 'stochastic' or 'chance' element in economic change, but from the ignorance and uncertainty influencing the economic decisions of individuals.

For this reason the 'objective' theories of probability, which are so useful and convenient in purely scientific applications, are quite ill suited to the solution of the important questions in economics. It is essential to use one of the 'personal' theories of probability, which deal with the effects of ignorance and uncertainty on the statements, choices or other behaviour of an individual.

Much of the apparatus of the theory of probability remains applicable, whether in the framework of an objective theory, or of a 'personal' theory. It is useful therefore to develop a probability model sufficiently versatile to suit a range of alternative interpretations, objective or personal, according to the field of application intended.

This is what will be done in the first part of this book. Because of the need to apply the theory to problems of individual decision, as well as to the philosophically simpler sampling and estimation problems, particular care will be necessary not to slur over the difficulties inherent in defining the basic concepts used in the theory. For this reason Chapters 2 and 3 and the Appendix to Chapter 2 will discuss the fundamentals of the subject such as the properties of orderings and the meanings of probability and irrelevance. A full discussion will be needed of the question whether it is sensible to represent probabilities by numbers which satisfy the 'addition convention' that the probability of the union of two disjoint events must equal the sum of their probabilities. The need for this discussion arises from the need to interpret the probability concepts in the context of economic decision-making in the face of uncertainty. Chapter 3 and the Appendix to Chapter 3 will examine this question.

Following this discussion of fundamentals, Chapters 4 and 5 will develop a model of probabilities for events (Chapter 4) and discrete variates (Chapter 5). In these chapters, events and discrete variates will be represented by vectors; the probabilities of events and the expected values of variates will be obtained by premultiplying the events or variates by a probability vector. This form of presentation has a superficial novelty. It is, however, a form closely analogous to the orthodox one of regarding events as sets of points in space, and probability as a measure over that space. The analogy is of the following nature. The points in space correspond to the various elements in the vectors: the membership or otherwise of a point in an event-set corresponds to whether an element in an event-vector is a one or a zero. The value of an element in the probability vector corresponds to the probability measure of a point in the space.

In Chapter 6, a model for the discussion of attributes and measurements will be developed on the same lines as those for events and variates, and the duality between the two pairs of models will be explained. A general discussion of the nature of random samples and of their place in the probability models follows in Chapter 7, although detailed discussion of the problems of estimation based on sample evidence is deferred until Volume 2. Then, in Chapter 8 an alternative and far more orthodox probability model is provided, regarding variates and their distribution functions as fundamental concepts obeying various axioms, and extending the concept of variates to include continuous variates, which are here regarded as the limiting form of sequences of discrete variates. Continuous variates are useful fictions, since their distribution functions provide simple approximations to the distribution functions of certain discrete variates.

Having provided the apparatus of probability models, it will now be possible to tackle the general problem of estimation on the basis of statistical observations. The final chapter of the first volume (Chapter 9) gives an account of Bayes' procedures for estimation: these are the most general and complete procedures available for revising probability models in the light of statistical evidence. A thorough understanding of the theory underlying these procedures is a valuable introduction to the less ambitious special-purpose statistical techniques which are the stock-in-trade of the practising statistician. Bayes' procedures are extremely powerful, provided sufficient initial information about probabilities is available; unfortunately, this proviso is not very often satisfied, and it is mainly for this reason that cruder techniques have often to be used, which avoid the need for so many initial assumptions about probability functions: another reason for using these cruder techniques is that they often involve less heavy computation and are easier to explain to beginners. The discussion of these particular procedures is left over to Volume 2.

In the second volume the whole approach is less general than in the first volume: a number of particular methods and particular areas of study are considered, now that the general model of probability and of variates has been established. Volume 2 is largely concerned with statistical techniques, whereas Volume 3 discusses more general topics concerning the influence of uncertainty on economic choices.

In Chapter 10 is given an introductory account of some of the statistical techniques—other than Bayes' procedures—which are available to the economic statistician who wishes to test his economic models in the light of statistical evidence and to fit parameter values to them. The next few chapters deal with various special areas of application of statistical analysis within the economic and social fields. Chapter 11 deals with many kinds of procedure for testing whether a proposed theory or model is discredited or not by fresh statistical evidence. This kind of test might be considered to have logical precedence over the estimation procedures learnt in Volume 1, but those (Bayes') procedures can be made so comprehensive as to include their own tests of alternative models and so to dispense with the need for any individual tests. But when there is insufficient initial information about probability functions to allow the use of Bayes' procedures, rough tests, such as those described in Chapter 11 are a useful substitute.

In Chapters 12 and 13, two kinds of estimation and decision procedure are developed for use once the general class of model has been decided upon. In Chapter 12, point estimators are considered: these provide single-value estimates for a parameter or the choice of a single decision out of an available set.

In Chapter 13 procedures are provided for obtaining interval estimates. These estimates provide a range of values within which it may be supposed, with some confidence, that the true value of a parameter (or set of parameters) is to be found. Considerable attention is paid to the provision of procedures which are robust, in the sense that they do not critically depend on the precise functional form assumed in the model for the probability distribution of 'disturbances'.

Chapter 14 discusses the design and analysis of social surveys and it takes up again, at a more practical level, some of the topics considered theoretically in Chapter 7 on random samples.

Chapter 15 deals with economic models based on regression equations. Such equations have been touched on already in Chapter 6, but here they are considered far more exhaustively. Chapter 16 is entirely devoted to the methods of interpreting information given in the form of time series. This chapter and the preceding one (on regression models) are largely concerned with the subject matter of what is often described as 'Econometrics'. Chapter 17 deals with sequential sampling schemes, and Chapter 18 discusses economic models involving queueing situations and the development of frequency distributions through time.

Volume 3 is less concerned with mathematical statistics and estimation techniques and are more concerned with the important economic effects of the prevalence of uncertainty and ignorance concerning the future: Chapter 19 provides an introductory survey of these effects: Chapter 20 discusses various situations where economic choices have to pay particular attention to risk and Chapter 21 considers how economic planning can make due allowance for risk. Finally, Chapter 22 considers some of the wider aspects of the economic effects of uncertainty, for example their effects on financial decisions and on policy concerning the management of the economy.

Chapter 2

PREFERENCE ORDERINGS AND PROBABILITY ORDERINGS

2.1. Ordinal and Cardinal Utility and Probability

Much economic theory is concerned with decisions. These decisions often involve judgements between two alternatives, with regard either to preference or to probability. In many cases, it is convenient to suppose that a set of alternatives can be arranged in some order of preference or of probability. In these cases, preferredness (usually termed 'utility') and probability are called *ordinal concepts*. In other cases, we may wish to go further and measure the utility or probability along a numerical scale; in such cases one would regard the utility or probability as a *cardinal concept*.

In this chapter we discuss utility and probability mainly as ordinal concepts, although in the final sections we consider a possible method of 'calibrating' a probability ordering. In later chapters we shall be considering probability and utility as cardinal concepts.

2.2. Orderings

An *ordering* of a set of N elements $A, B, \ldots N$ will mean a set of $\frac{1}{2}N(N-1)$ relations, one between each of the distinct pairs of elements (A, B) etc. The relation between any pair (A, B) may be any of the following:

$$A>B,\ A=B,\ A\geqslant B,\ A \parallel B,\ B>A,\ B=A,\ B\geqslant A,\ B \parallel A. \quad (2.2.1)$$

The meanings of the symbols $>$, $=$, $\parallel$ etc. will depend upon the context in which the ordering is set.

It will always be assumed that

1. $A \geqslant B$ denotes '$A > B$ or $A = B$'.
2. $A = B$ means the same as $B = A$.
3. $A \parallel B$ means the same as $B \parallel A$.
4. $A > B$ and $B > A$ are contradictory.
5. $A > B$ and $A = B$ are contradictory.

Further such assumptions about the meanings of the symbols will be introduced in particular contexts. Often the symbol $\parallel$ will be excluded and sometimes the symbols $\geqslant$ and $=$ also will be excluded, so that only two possible relations $A>B$ and $B>A$ remain for any pair of elements (A, B).

In order to shorten the exposition of certain ideas, it will sometimes be found convenient to introduce into an ordering relations of type $A=A'$ where A and A' in fact refer to the same element of the ordering. When relations of this type are included, the number of relations among N elements is thereby increased from $\frac{1}{2}N(N-1)$ to $\frac{1}{2}N(N+1)$.

If we wish to indicate that the expressions A and A' refer to the same element in an ordering, we shall use the identity symbol for this purpose and write

$$A \equiv A'. \tag{2.2.2}$$

The Appendix to Chapter 2 gives a fairly full discussion of orderings of many kinds in a variety of contexts. In the main text, we concentrate on preference orderings and probability orderings.

2.3. The Relation of Indecision

The introduction of the symbol $\|$ involves us in a departure from orthodox procedure. This symbol is needed to denote a state of indecision or uncertainty with regard to the ordering of a pair of elements in respect of preference or of probability.

It is a central theme of this book that the desire to simplify decisions by reducing uncertainty is a major factor to be allowed for, both in the study of business decisions and in the choice of technical methods for making policy recommendations and for drawing inferences from economic statistical data.

It is because of the importance attached to reducing or side-tracking uncertainty, that it is convenient to introduce into our formal discussions this symbol $\|$ to relate two alternatives A and B, when the decision between A and B, whether in preference or in probability, is apt to be made on an arbitrary basis, reflecting the attitude 'couldn't care less' or the attitude 'haven't a clue'. Unfortunately, very little can be said about orderings which contain many such relations of indecision; one can then only consider various 'illustrations' of such orderings, in which equalities or inequalities are arbitrarily substituted for the relations of indecision. These 'illustrations' may give some indication of the range of possibilities left open as a result of the existence of so much indecision.

The important distinction between the relation $A=B$ and the relation $A \| B$ is that in most contexts $A=B$ reflects a decision to place A and B on a footing of exact equality in an ordering, whereas $A \| B$

reflects no such decision, but rather indicates an inability to compare A and B at all.

2.4. Preference Orderings

The symbols, $>$, $=$, $\geqslant$ and $\|$ in any preference ordering will be interpreted as follows:

$A > B$ A would be accepted in exchange for B, but B would not be accepted in exchange for A.

$A = B$ A would (with the slightest inducement) be accepted in exchange for B; moreover, B would (with the slightest inducement) be accepted in exchange for A.

$A \geqslant B$ A would (with the slightest inducement) be accepted in exchange for B.

The words in parenthesis, 'with the slightest inducement', are inserted because, if a man regarded two objects as of absolutely identical utility to him, he presumably would *not* accept the one in exchange for the other without just that 'slightest inducement' involved in getting him to listen and utter his agreement.

$A \| B$ A would not be accepted in exchange for B, nor would B be accepted in exchange for A.

2.5. Inconsistent Preference Orderings

Even when only the symbol $>$ appears in the relations of a preference ordering, this does not ensure that the elements concerned can be placed in a definite order of preference. For there is no rule that forbids an ordering to contain a set of relations such as $A > B$, $B > C$ and $C > A$, which must at some point contradict any suggested order of preference for the three elements A, B and C.

There is, however, an appearance of inconsistency about any preference ordering which does include such a set of relations; for, with such preferences, one must be willing to accept B in exchange for C, A in exchange for B and then C in exchange for A.

There is even a slight flavour of inconsistency about the inclusion of such a set of preferences as

$$A > B,\ B \geqslant C,\ C \geqslant A \quad \text{or} \quad A \| B,\ B \geqslant C,\ C \geqslant A,$$

since, with either of these sets of preferences, one will with the very slightest inducement accept C for A and then B for C, although definitely unwilling to accept B for A directly.

For such reasons it is convenient to describe as *inconsistent* any cycle of preferences between three elements A, B, C, which is of type

$$A > B \geqslant C \geqslant A \quad \text{or of type} \quad A \parallel B \geqslant C \geqslant A; \qquad (2.5.1)$$

a fortiori, it is convenient to describe as inconsistent any relation of type

$$A > B = C = A, \ A > B > C \geqslant A, \ A \parallel B = C > A, \text{ etc.} \qquad (2.5.2)$$

or of any other type obtainable by replacing one or both of the signs $\geqslant$, in (2.5.1) by an $=$ or by a $>$.

We extend the use of the adjective 'inconsistent' to apply to orderings, by describing any ordering which contains an inconsistent cycle of preferences as an *inconsistent preference ordering*.

EXAMPLE 2.5.1. Show that an ordering containing the preference cycle

$$A \geqslant B \geqslant C = D > E = A \quad \text{is inconsistent.}$$

Solution. Consider the possible preference relations which may exist between C and A. The cycles $C > A \geqslant B \geqslant C$ and $C \parallel A \geqslant B \geqslant C$ are both inconsistent: hence, unless $A \geqslant C$, the whole ordering is inconsistent by virtue of containing an inconsistent cycle relating to the three elements C, A, B.

But even suppose that $A \geqslant C$; then, considering the cycle for C, D, A, we similarly conclude that it is inconsistent unless $A \geqslant D$. Hence, unless $A \geqslant D$, the whole ordering is inconsistent. But even if $A \geqslant D$, we are landed with the inconsistent cycle $D > E = A \geqslant D$.

It follows that, under all possible circumstances, the ordering which contains the preference cycle $A \geqslant B \geqslant C = D > E = A$ must be inconsistent.

The reader is invited to consolidate his familiarity with the definitions by solving the following similar example.

EXAMPLE 2.5.2. Show that an ordering is inconsistent if it contains any cycle of preferences $A{*}B{*}C{*}, \ldots, Y{*}Z$ where some particular * represents $>$ or $\parallel$ and where all the other * represent $>$, or $\geqslant$ or $=$.

It is because of the possibility of this extension—from the definition of inconsistency in terms of cycles of three elements to a conclusion relating to cycles of many elements—that it is sufficient to define inconsistent cycles in terms of three elements, and unnecessary to define inconsistency in terms of cycles of size three *or more*.

The consideration of inconsistency brings out an important distinction between the relations $A = B$ and $A \parallel B$. Whereas the relations $A = B$ and $B = C$ forbid the relation $C > A$ on pain of inconsistency, there need be nothing inconsistent about the cycle of relations $A \parallel B \parallel C > A$. For example, Pongo may well be prepared to accept 109 matches (C) in exchange for 108 matches (A), without being prepared either to accept an indiarubber (B) in exchange for 108 matches (A) or to accept 109 matches (C) in exchange for the indiarubber (B).

2.6. Probability Orderings

There is a close correspondence between the interpretation of the relations expressed by the symbols $>$, $=$, $\geqslant$ and $\parallel$ in a probability ordering and that of the corresponding relations in a preference ordering. Thus, we shall interpret

$A > B$ as 'A is more probable than B';
$A = B$ as 'A and B are equally probable';
$A \geqslant B$ as 'Either $A > B$ or $A = B$';
$A \parallel B$ as 'There is complete uncertainty as to the relative probability of A and B'.

As in the case of preference orderings, we may regard as *inconsistent* any cycle of probability relations among three elements, which is either of the type

$$A > B \geqslant C \geqslant A \quad \text{or of type} \quad A \parallel B \geqslant C \geqslant A. \tag{2.6.1}$$

As in the case of preference orderings, we also regard as inconsistent any of the more definite cycles, such as

$$A > B = C = A, \; A > B > C \geqslant A, \; A \parallel B = C > A, \text{ etc.} \tag{2.6.2}$$

obtainable from (2.6.1) by substituting $>$ or $=$ for one or both the symbols $\geqslant$, in one of the cycles.

The reader will be able to verify that any of these inconsistent cycles among three elements do have interpretations that imply a somewhat confused state of mind in the person asserting the probability relations involved. For example the cycle $A \geqslant B \geqslant C \geqslant A$ bears the interpretation

> 'Although B is no less probable than C, and C is no less probable than A, one cannot assert that B is no less probable than A.'

Again, the relation $A \parallel B \geqslant C \geqslant A$ would imply the same confusing conclusion.

We may extend the use of the term 'inconsistent' to apply to orderings by the same device as in the case of preference orderings. We shall in fact regard as inconsistent any probability ordering which contains an inconsistent cycle of relations among three elements.

In the case of probability orderings, however, we shall not confine the application of the term 'inconsistent' to such probability orderings; we shall shortly introduce other rules the violation of which will automatically lead to the application of the term 'inconsistent' to the offending probability ordering.

The above introductory discussion of probability orderings has contained virtually no indication of what is to be meant by such assertions as, 'A is more probable than B', or 'A and B are equally probable'. Several different meanings have been suggested for such phrases, and the discussion of these will occupy much of Chapter 3.

2.7. Intersections and Unions: Disjoint Events and Exhaustive Sets of Events

Suppose that $A, B, C, \ldots$ are events which form the subject-matter of a probability-ordering. We shall denote the '*intersection*' of the events A and B by $A \cap B$; this intersection of the events means the event that both A and B occur. Thus, if A is the event 'The clock struck one' and B is the event 'The mouse ran down', then $A \cap B$ is the event 'The clock struck one and the mouse ran down'.

Two events are said to be *disjoint* if their intersection is logically or physically impossible. Thus, if B is the event that the mouse ran down and C is the event that the mouse ran up, and if B and C refer to the same mouse and moment of time, we may regard B and C as disjoint events, since one cannot easily regard their intersection as physically possible.

We shall denote the *union* of two events B and D by $B \cup D$. The union of two events B and D means the event that either or both of B and D occur. Thus, if B denotes 'The mouse ran down' whereas D denotes 'The clock ran down' then the union $B \cup D$ denotes 'The mouse or the clock or both ran down'.

The notation for intersections and unions may be extended to apply to more than two events. $A \cap B \cap C \cap \ldots \cap Z$ denotes the event that all the events $A, B, C, \ldots, Z$ occur, whereas $A \cup B \cup C \cup \ldots \cup Z$ denotes the event that at least one of the events $A, B, C, \ldots, Z$ occurs. The set $A, B, C, \ldots, Z$ is said to be *exhaustive* if the union $A \cup B \cup C \cup \ldots \cup Z$ is certain to occur, i.e. if at least one of the events $A, B, C, \ldots, Z$ must inevitably occur. The set $A, B, C, \ldots, Z$ is said to be disjoint if every distinct pair A, B; A, C; $\ldots$, A, Z; B, C; $\ldots$, B, Z; $\ldots$ etc. is disjoint, i.e. if not more than one of the set can occur together. We may accordingly speak of a *disjoint exhaustive set* of events whenever *one and only one* of the events must occur. Thus we may regard the six possible results of throwing a (possibly loaded) die as a disjoint exhaustive set of six events, provided we know that the die is being once thrown.

2.8. A Second Form of Inconsistency in a Probability Ordering

We shall wish at times to refer to the following further requirements to be satisfied by the relations in a probability ordering containing the events $A, B, C, \ldots$

I. If A is certain and if B is impossible,

(*i*) $A > B$; (*ii*) $A \cap C = C = B \cup C$;

(*iii*) C is certain, if, and only if, $A = C$; this includes the case where $C \equiv A$;

(*iv*) C is impossible if, and only if, $B = C$; this includes the case where $C \equiv B$;

(*v*) $(A \cup C) = A \geqslant C \geqslant B = (B \cap C)$. (2.8.1)

In each of these four requirements C denotes any event in the ordering.

The last requirement, I (*v*), rules out any event from exceeding in probability any other event that is certain, or from being less probable than an impossible event; it also ensures that the union of an event which is certain with another event, will itself be certain, whereas the intersection of an impossible event with any event, is itself impossible; so that any pair of events of which one is impossible is disjoint.

II. If the pairs (A, B) and (A, C) are both disjoint, then

$$(A \cup B) \geqslant (A \cup C) \text{ if, and only if, } B \geqslant C. \qquad (2.8.2)$$

This requirement states that if A cannot occur when either B or C (or both) occurs, then the probability relation between 'either A or B' and 'either A or C' has to be the same as the probability relation between B and C. For instance the probability relation between 'either (A) there is no God or (B) God's in His heaven' and 'either (A) there is no God or (B) God is on the side of the big battalions' has to be the same as that between '(B) God's in His heaven' and '(C) God is on the side of the big battalions'.

An important Corollary of the satisfaction of Consistency Requirement II is the following:

If $A \geqslant C$ and $B \geqslant D$, and $A \cap B \leqslant C \cap D$, then $A \cup B \geqslant C \cup D$; in particular,

if $A \geqslant C$ and $B \geqslant D$ and if A, B are disjoint, then $A \cup B \geqslant C \cup D$. (2.8.3)

The method of proof of these results is indicated in EXERCISE 2.15 at the end of this Chapter.

If the reader cares later to interpret the requirements I and II in the light of any of the meanings of probability relations which will have been discussed in Chapter 3, he is likely to find that these requirements conform with common sense. Accordingly, we enlarge the scope of the term consistency in relation to a probability ordering, by saying that unless the two requirements I and II of this section are satisfied, the ordering is inconsistent.

2.9. Illustrations of a Probability Ordering

It will typically be found that in any realistic preference or probability ordering, a number of the relations of type $A \parallel B$ imply inconsistency in the ordering. This is because people are not in the habit of thinking out every one of the logical implications of their preference and probability judgements. In these typical cases, the inconsistencies can often be removed by replacing some of the relations of type $A \parallel B$ by others of types such as $A > B$. But even when all inconsistencies have in this manner been removed, the ordering may still be left containing many uncertainty relations of type $A \parallel B$.

Since $A \parallel B$ does reflect uncertainty and indecision, it is very difficult to draw useful or definite conclusions from any ordering which still contains many relations of this type. Therefore it is often instructive to replace all the relations of type $A \parallel B$ by others, such as $A > B$ or $A = B$, and to do so in such a way as to avoid inconsistency. Any new ordering derived in this way from a consistent ordering will be called an *illustration* of that ordering.

It seems plausible that every consistent ordering must have at least one (consistent) illustration: however, since no straightforward proof of this conjecture is available, it is convenient to adopt the further consistency requirement (III) that to be consistent, any ordering containing a relation of type $A \parallel B$ must have at least one consistent illustration. On the other hand, it is of course quite likely that a consistent ordering will allow a variety of different illustrations. However, provided that the number of relations of type $A \parallel B$ is finite, then the number of illustrations possible for the consistent ordering must clearly also be finite. This finite set of illustrations may be called the *illustration set* for the ordering. A consideration of the extreme cases within the illustration set for an ordering may be a useful substitute for any attempt to draw conclusions directly from the ordering itself.

2.10. The Fair Roulette-Wheel of Scope *n*

Suppose that a roulette-wheel, when spun, will produce any one of a set of n single disjoint results, R_j $(j = 1, 2, \ldots, n)$, and that each pair of the R_j are equal in probability. We shall call such a roulette-wheel a fair roulette-wheel of scope n, or, more briefly a fair n-roulette-wheel. Whereas we shall refer to the n disjoint results, R_j, as the *single* results of the spin, we shall speak simply of 'the results' of the spin when we mean all the 2^n events consisting of the various possible unions of the single results, together with the null-result: for example, supposing $n \geqslant 8$, one of the results would be $R_2 \cup R_8 \cup R_3$; another result would be the impossible event of no R_j at all. Between each pair of these 2^n results there will be a probability relation, and we shall refer to all these probability relations as 'the probability ordering' of the results of the spin.

In this case of a fair roulette-wheel of scope n, we can see exactly what this probability-ordering must be. Suppose that G' and G, any two results of the spin, are unions of a' and of a single results, respectively: then the probability relation between G' and G must be

$$G' >, =, \text{ or } < G \tag{2.10.1}$$

according as

$$a' >, =, \text{ or } < a. \tag{2.10.2}$$

2.11. The Probability-Calibration of an Event or Set of Events by means of a Fair Roulette-Wheel

It is often instructive to use the set of results of the spin of a fair roulette-wheel as a set of standards with which to compare an event or set of events in probability.

Let $G(j)$ $(j = 0, 1, \ldots, n)$ denote unions of j disjoint single results of the spin of a fair n-roulette-wheel. Then we may form the probability-ordering of the $(n+2)$ events $G(j)$ and E, where E is any other event. It will be either of the form

$$G(0) < G(1) < \ldots < G(p) < E < G(p+1) < \ldots < G(n) \tag{2.11.1}$$

or of the form

$$G(0) < G(1) < \ldots < G(p) = E < G(p+1) < \ldots < G(n). \tag{2.11.2}$$

In either case we may refer to the ordering as the probability-calibration of the event E; if the ordering is of the form (2.11.2) we shall call the calibration *exact.*

In the same way, if $E(r)$ $(r = 1, 2, \ldots, m)$ are m disjoint events, we may form the probability-ordering of the $(m+n+1)$ events $G(j)$ and $E(r)$, thereby calibrating in probability each one of the events $E(r)$. If it so happens that the calibration of each $E(r)$ is exact, we may call the calibration of the set of events $E(r)$ exact.

When the calibration of a set of m disjoint events is, in this manner, exact, much more can be said. For it is fairly clear that if the probability ordering of the (2^m-1) unions of these events and the $n+1$ events $G(j)$ is to be consistent, then the entire set of the (2^m-1) unions of the events can be exactly calibrated in probability by the fair n-roulette wheel. The proof of this depends on the fact that the union of any two or more results of the spin of the wheel is itself a result of the spin. As the proof requires a little care, it is not given in the text but is made available in the appendix to Chapter 3.

2.12. Irrelevance of an Event to the Spin of a Fair Roulette-Wheel

Much of certain theories of statistics, and quite a lot of the technique of statistical experimental design, depend on the idea that there are available bits of apparatus, such as fair-roulette-wheels, dice, coins which can be spun, packs which can be shuffled, etc., which have the property that pretty well everything apart from the apparatus is irrelevant to the results which will be obtained by using the apparatus. For example, what President X says or does about North Vietnam is quite irrelevant to the result I may get from spinning a fair roulette-wheel of scope 37; whether Christ ascended into heaven is irrelevant to whether this fair coin will come down heads or tails.

In the next chapter we shall see that, although irrelevance is about the most fundamental concept in the whole theory of probability, it is extremely difficult to decide what it means, without arguing in a circle. But in the case of apparatus like fair roulette-wheels, it is fortunately quite easy to explain what is meant by the irrelevance of an event E to the result of using the apparatus. We can sum the explanation up, by saying that an event E' is irrelevant to the result of spinning a fair roulette-wheel, if for every pair $R(j)$, $R(k)$, of single results of spinning the wheel,

$$E' \cap R(j) = E' \cap R(k). \tag{2.12.1}$$

This condition may be expressed in everyday language by saying that the fairness of the roulette-wheel is in no way affected by the event E'.

But we can also explain, almost as clearly, what is meant by the irrelevance of one event E' to another event E, whose probability has been exactly calibrated by a fair roulette-wheel. In this case, we can assert that

$$E = G, \tag{2.12.2}$$

where G is some specified result of spinning the wheel: the event E' is then irrelevant to E, if it is also true that

$$E' \cap E = E' \cap G. \tag{2.12.3}$$

This condition can be expressed in ordinary language, by saying that the probability of E is in no way affected by the occurrence of E'; if we anticipate the idea that an unaltered probability-calibration corresponds to what is ordinarily understood by unaltered probability.

Two points should be noted about this definition. The first is that it only enables us to define the irrelevance of an event E' to an event (or events) which is *exactly* calibrated in probability by some fair roulette-wheel. The second point is that the definition is entirely concerned with probability-relations and probability-orderings, and that no mention has yet been made of any numerical measure of probability. The nearest we have been to mentioning numerical probability is in alluding to probability-calibration by comparison, in probability, with the results of spinning a fair roulette-wheel. This does not bring in numerical probability, although it may be difficult to overlook the fact that one particular conventional method of assigning numerical measures to the probabilities of such results is particularly popular.

Appendix to Chapter 2

ORDERINGS IN A MORE GENERAL CONTEXT

A2.1. The Relation >

Where A and B are any two elements which are to be compared in an ordering, we may use the symbol > to connect them either as $A > B$, or as $B > A$.

For certain orderings, > may be used between every pair of elements. For example, if $A, B, \ldots, N$ are the contestants in an American-type tennis tournament in which each contestant plays every other only once, we could write for each pair (A, B), either $A > B$ denoting the defeat of B by A, or $B > A$ denoting the defeat of A by B; but we could not write both $A > B$ and $B > A$. For this ordering, provided all pairs played their matches to a finish, the symbol $>$ would suffice. But for most orderings a richer armoury of conjunctive symbols is needed, or is at least a great convenience.

A2.2. The Relations $<$, $=$, $\geqslant$, $\leqslant$

The symbol $<$ will be used to mean the opposite of $>$. Thus $A < B$ will mean $B > A$ for all pairs of elements (A, B).

The symbol $=$ will be read as 'equals' and its interpretation will depend on the field of application. $A = B$ will always be inconsistent with $A > B$, and the interpretation will usually be in correspondence with one's intuitive notion of the relation of equality in the context.

The symbol $\geqslant$ will mean '$>$ or $=$', and the symbol $\leqslant$ will mean '$<$ or $=$'.

A2.3. The Relation $\|$

The interpretation of the symbol $\|$ will vary considerably from one context to another. In general, it will denote an inconclusive relationship or perhaps the absence of any definite relationship. It will always imply the non-assertion of any of the relations $>$, $<$, $=$, $\geqslant$ or $\leqslant$ between the two elements concerned.

The relation $A \| B$ is quite distinct from the relation $A = B$. The manners in which the relation $A \| B$ may arise are best explained by a series of examples.

A2.4. The Ordering of a Set of Points in a Plane

Let $A > B$ denote that the direction of A from B lies between north-east and north-west. Let $A = B$ denote that A and B coincide. Then we could use $A \| B$ to denote the denial of $A > B$, $B > A$ and $A = B$. As always, $A < B$ will mean $B > A$. Thus the possible relations between A and B are

(*i*) $A = B$ (A and B coincide);
(*ii*) $A > B$ (A lies between N.E. and N.W. of B);
(*iii*) $A < B$ (A lies between S.W. and S.E. of B);
(*iv*) $A \| B$ (None of *i*, *ii* or *iii* hold, so that A must lie somewhere in the range of directions N.E. and S.E. inclusive, or in the range N.W. and S.W. inclusive, from B).

For example, the ordering for the five points A, B, C, D, E in the diagram below is

$A \parallel B$, $A > C$, $A > D$, $A > E$, $B \parallel C$, $B > D$, $B \parallel E$, $C \parallel D$, $C > E$ and $D \parallel E$.

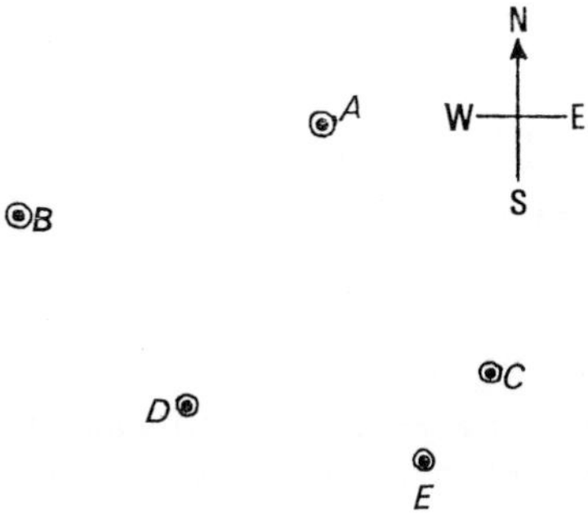

Fig. 1

A2.5. The Ordering of a Set of Events in Space-Time

Let $A, B, \ldots, N$ denote a set of events in space-time. Let $A > B$ denote that A occurred after B in the sense that a signal sent from B would have had time to arrive at A, it being understood that signals can travel at any velocity up to, but not exceeding, that of light.

Four relations may exist between a pair of events A and B.

(*i*) $A = B$ (A and B coincide at the same time and place),

(*ii*) $A > B$ (A occurs after B),

(*iii*) $A < B$ (A occurs before B),

(*iv*) $A \parallel B$ (A does not coincide with B but occurs neither before nor after B).

For example if A were an event on earth, and if B and C were two explosions on the sun taking place with C one second after B, it could quite well be that the ordering of the three events was $A \parallel B$, $A \parallel C$, $B < C$. This example illustrates an important contrast between the relations $=$ and $\parallel$: whereas it is always convenient to postulate that $[A = B]$ and $[A = C]$ together imply $[B = C]$, there are many types of orderings in which it would be disastrous to postulate that $[A \parallel B]$ and $[A \parallel C]$ together imply $[B \parallel C]$.

A2.6. The Ordering of Sets of Elements

Consider the ordering of the 2^n possible sets of elements which can be composed of none, of some, or of all of the n distinct elements $A, B, \ldots, N$. Such an ordering may be called the 2^n-ordering based on the disjoint set $A, B, \ldots, N$. Let us define the possible relations between any two sets P and Q in the ordering as follows.

(*i*) $P \geqslant Q$. P includes every element of Q;
(*ii*) $P = Q$. $[P \geqslant Q]$ and $[Q \geqslant P]$;
(*iii*) $P > Q$. $[P \geqslant Q]$ but not $[Q \geqslant P]$;

then the proper interpretation of $P \parallel Q$, namely

(*iv*) $P \parallel Q$. neither $[P \geqslant Q]$ nor $[Q \geqslant P]$,

is that there are some elements of Q not included in P and also some elements of P not included in Q.

This example again illustrates the point that although $[P = Q]$ and $[P = R]$ imply $[Q = R]$, yet $[P \parallel Q]$ and $[P \parallel R]$ need not imply $[Q \parallel R]$. For example if P = whisky and soda, Q = whisky and milk and R = whisky and milk and ink, then $P \parallel Q$ and $P \parallel R$ but $Q < R$.

A2.7. Postulates of Consistency in an Ordering

In any particular field of application, one is likely to lay down various requirements which an ordering must satisfy if it is to be regarded as consistent. To begin with, one is likely to require that every term in the ordering should satisfy the relation of equality with itself. In some applications one may wish to regard cyclic sets of relations such as $A > B$, $B > C$ and $C > A$ as being inconsistent. Other examples will be found in §§ **2.5**, **6** and **8**.

Whatever may be the particular set of requirements for consistency which we choose to adopt, they may serve to enable us to convert some of the unspecified relations of type $A \parallel B$, to one of the more definite types $A < B$, $A \leqslant B$, $A = B$, $A > B$, or $A \geqslant B$. They may also reveal inconsistencies in an ordering by showing that a relation of type $A > A$ is implied.

In our discussion of consistency postulates we shall use the symbol $\rightarrow$ to denote 'implies' and the symbol $\equiv$ to denote 'implies and is implied by'. The symbol $\sim$ will denote 'not' and the pair of symbols $\rightarrow\sim$ will denote 'rules out'.

A2.8. Consistency Postulates for Ordering of Events in Space-Time

Consider again the example of section **A2.5**. $A = B$ denoted that A and B occurred at the same time and place, whereas $A > B$ denoted that A occurred after B and $A < B$ that A occurred before B and $A \parallel B$ that A occurred neither before nor after B but was not at the same time and place as B.

The following three sets of postulates would be natural requirements of an ordering for it to be consistent.

For every three events A, B and C in the ordering

P. 1 $[A \equiv B] \rightarrow [A = B]$
P. 2 $[A > B]$ and $[B \geqslant C] \rightarrow [A > C]$
$[A \geqslant B]$ and $[B > C] \rightarrow [A > C]$
$[A \geqslant B]$ and $[B \geqslant C] \rightarrow [A \geqslant C]$
P. 3 $[A = B] \rightarrow \sim [A > B]$
$[A = B] \rightarrow [B = A]$
$[A = B] \rightarrow \sim [A \parallel B]$
$[A > B] \rightarrow \sim [A \parallel B]$

The reader is invited to express these requirements in verbal form.

Inconsistencies in an ordering may arise solely out of the inclusion of one or more relations of type $A \parallel B$, which conflict with other relations and the consistency postulates. In this case, the ordering may be capable of being restored to consistency by the replacement of these conflicting relations of type $A \parallel B$ by other relations of type $A > B$ or $A < B$ etc.

Suppose, for example, that we were faced with the ordering

$$A > B, A > C, A < D, A \parallel E, B < C, B \parallel D, B < E, C \parallel D, C > E, D \parallel E.$$

We could detect the following inconsistencies:
In order to satisfy P. 2 and P. 3 we should need $A > E$, since $A > C$ and $C > E$; but $A > E$ is inconsistent with $A \parallel E$. We should further require $B < D$, since $A > B$ and $A < D$, but this is inconsistent with $B \parallel D$. Again we should require $C < D$, since $A > C$ and $A < D$, but this is inconsistent with $C \parallel D$. Finally, we should require $D > E$, since $A < D$ and we have required $A > E$.

If we replace $A \parallel E$, by $A > E$, $B \parallel D$ by $B < D$, $C \parallel D$ by $C < D$ and $D \parallel E$ by $D > E$, we should obtain the consistent ordering

$$A>B,\ A>C,\ A<D,\ A>E,\ B<C,\ B<D,\ B<E,\ C<D,\ C>E,\ D>E.$$

One can summarise this ordering by the statement

$$D > A > C > E > B.$$

In this example there was only one possible consistent illustrative ordering obtainable by replacing the relations of type $P \parallel Q$ by strict inequalities; but in many cases there will be a choice of several such illustrative orderings. Consider again the ordering given at the end of section **A2.4**

$$A \parallel B, A > C, A > D, A > E, B \parallel C, B > D, B \parallel E, C \parallel D, C > E \text{ and } D \parallel E.$$

The following illustrative orderings are possible.

$A>B, A>C, A>D, A>E, B>C, B>D, B>E, C>D, C>E$ and $D>E$;
or
$A<B, A>C, A>D, A>E, B>C, B>D, B>E, C<D, C>E$ and $D<E$.

These two illustrative orderings may be summarised as $A>B>C>D>E$ and $B>A>D>C>E$. They are by no means the only possible ones, since we could also take those which may be summarised by

$A>B>C>E>D$, $A>B>D>C>E$, $A>C>B>D>E$,
$A>C>B>E>D$, $A>C>E>B>D$, $B>A>C>D>E$
and $B>A>C>E>D$.

One may think of these illustrative orderings as attempts by different people to place the five points in order, where all agree that to be between N.E. and N.W. of another point certainly entitles one to priority over it in the ordering, but where opinions differ about what other criteria should be used to decide priority.

A2.9. Preference Orderings

We have looked very briefly at preference orderings in § **2.4**. Here we will follow up one or two further points.

Let $A, B, \ldots, N$ denote the 2^n sets of objects which may be formed of none, some, or all, of n specified objects. Let $A \supset B$ denote [A includes B], and let $A \subset B$ denote [$B \supset A$].

Let $A \cup B$, the union of A and B, denote the set of those objects which are included in at least one of A and B, and let $A \cap B$, the intersection of A and B, denote the set of those objects which are included both in A and in B.

Let 0 denote the set containing nothing. Then A and B will be termed disjoint if $A \cap B \equiv 0$, i.e. if A and B contain no object in common.

A preference ordering among $A, B, \ldots, N$ will mean an ordering of them in which

$A > B$ denotes that A is preferred to B,
$A = B$ denotes that A and B are regarded as exchangeable,
$A \parallel B$ denotes an inability to state a preference either way or to regard A and B as exchangeable.

This still leaves preference and exchangeability as undefined concepts. For an operational interpretation of these terms one may use the definitions of § **2.4**.

As usual, $A \geqslant B$ denotes [$A > B$ or $A = B$] and $A < B$ and $A \leqslant B$ denote $B > A$ and $B \geqslant A$.

It is natural to adopt the identical consistency requirements for such a preference ordering, as we adopted them in **A2.8** for the ordering of events, namely the requirements P. 1, P. 2 and P. 3 of that section.

We might add the fourth requirement

P. 4 If A and B are disjoint and B is not empty, $A \cup B > A$.

This places a real restriction on the preference ordering, since it requires that one always prefers to include a further object.

Suppose now that a certain amount is known about the ordering, but that between many pairs of sets of objects the preference is not known. Then the four requirements, P. 1 to P. 4, may enable one to derive preference relations for some or all of these pairs from the known preferences with regard to other pairs.

Suppose, for example, that one was given

$$A = B \cup C \quad \text{and} \quad B > C \cup D \quad \text{and} \quad C > D.$$

Then we can derive from P. 4

$$D > 0 \text{ (the empty set)},$$

since $D = D \cup 0$.

By similar application of P. 4, information can rapidly be obtained about preferences relating to sets excluding A. For example,

$$B \cup C \cup D > A = B \cup C \qquad A \parallel B \cup D;$$
$$B \cup C \parallel B \cup D \qquad B \cup C > B;$$
$$B \cup D > B \geqslant C \cup D > C > D > 0.$$

However, very little can be deduced about preferences concerning sets which include A. It is left to the reader as an exercise to determine what little can be deduced concerning these preferences.

Owing to the paucity of information, a great many consistent illustrations of the above preference ordering are possible. One such illustration is as follows:

$$A \cup B \cup C \cup D > A \cup B \cup C = A \cup B \cup D = A \cup C \cup D = B \cup C \cup D >$$
$$A \cup B = A \cup C = A \cup D > A = B \cup C = B \cup D > B = C \cup D > C > D > 0$$

A2.10. Preferences between Baskets of Goods

A particular type of preference ordering among objects is the following: A_i $(i = 1, 2, \ldots, n)$ denote distinct objects, and we consider sets of the form $\sum_{i=1}^{n} r_i A_i$, where each r_i $(i = 1, 2, \ldots, n)$ is either zero or a positive integer, and where, for example, $3A_1 + 0.A_2 + 17A_3 + 142\,857A_4$

denotes 3 objects of type A_1, 17 objects of type A_3 and 142 857 objects of type A_4.

In order to keep the number of sets considered within bounds, one may impose some upper limit on $\sum_{i=1}^{n} r_i$, for example $\sum_{i=1}^{n} r_i < 10^{[10^{10}]}$.

We may retain postulates P. 1 to P. 3 unmodified and express P. 4 as

$$\text{P. 4} \quad \left[r_i \geqslant r_i' \ (i = 1, 2, \ldots, n) \text{ and } \sum_{i=1}^{n} r_i > \sum_{i=1}^{n} r_i' \right] \rightarrow \left[\sum_{i=1}^{n} r_i A_i > \sum_{i=1}^{n} r_i' A_i \right].$$

Economists have devoted much time to pondering on what might be learned about such orderings from market information of the following type:

'When p_i $(i = 1, 2, \ldots, n)$ are the unit-prices of the A_i, and when C is the highest cost-value allowed for the chosen set, the actual choice was for the set $\sum_{i=1}^{n} R_i A_i$.'

The conclusion to be drawn from any one such item of information is of course that

$$\sum_{i=1}^{n} R_i A_i > \sum_{i=1}^{n} r_i A_i$$

for all r_i such that

$$\sum_{i=1}^{n} r_i p_i \leqslant C \quad (r_i \geqslant 0, \ i = 1, 2, \ldots, n).$$

The consistency requirements P. 1 to P. 4 will then add something to this information, and, in particular, they might reveal that the preference ordering was inconsistent. In cases where a great many such items of information are provided, it may be possible, with the aid of P. 1 to P. 4 to learn a great deal about the preference ordering.

EXAMPLE A2.10.1. Goods of two kinds X and Y are available for sale. The following information is provided about the purchases observed at various levels of the price P_Y of the good Y (expressed in units of the other good X). The pairs of amounts purchased of X and Y at various levels of P_Y and of 'purchasing-power in terms of X' were as shown in the following table.

Level of purchasing-power	2	3	4
Price-ratio P_Y	*Pairs of amounts purchased*		
1	0,2	0,3	—
2	0,1	1,1	2,1
3	—	3,0	—

If we assume that the postulates 1, 2, 3 and 4 are satisfied, the information in the table enables us to learn a great deal about the preference ordering of the 10 bundles of goods containing less than 4 units each, namely, (0,0), (0,1), (0,2), (0,3), (1,0), (1,1), (1,2), (2,0), (2,1), (3,0), where (x,y) denotes the bundle containing x units of X and y units of Y.

We reason as follows: where Z denotes purchasing-power,

(0,3) > (1,2) since (0,3) is bought when $P_Y = 1$ and $Z = 3$
(0,3) > (2,1) since (0,3) is bought when $P_Y = 1$ and $Z = 3$
(2,1) > (0,2) since (2,1) is bought when $P_Y = 2$ and $Z = 4$
(0,2) > (1,1) since (0,2) is bought when $P_Y = 1$ and $Z = 2$
(1,1) > (3,0) since (1,1) is bought when $P_Y = 2$ and $Z = 3$
(3,0) > (0,1) since (3,0) is bought when $P_Y = 3$ and $Z = 3$
(0,1) > (2,0) since (0,1) is bought when $P_Y = 2$ and $Z = 2$
(2,0) > (1,0) due to postulate 4; (1,2) > (1,1) due to postulate 4;
(1,0) > (0,0) due to postulate 4; (2,1) > (1,1) due to postulate 4.

These conclusions may be summed up in the form:

(0,3) > (2,1) > (0,2) > (1,1) > (3,0) > (0,1) > (2,0) > (1,0) > (0,0)
(0,3) > (1,2) > (0,2)
(2,1) || (1,2).

Thus, apart from the preference-relation between (2,1) and (1,2), we have obtained the complete preference-ordering of the ten bundles of goods. Later in this Appendix we shall see that if we adopt a fifth postulate, the convexity postulate, we can also derive a definite preference relation between (2,1) and (1,2).

Economists often regard the amounts purchased as continuous real variables instead of as integers. In this case, the ordering is no longer between a finite number of bundles, and more care is needed in handling and interpreting the preference relations. The following example relates to this case.

EXAMPLE A2.10.2. There are 2 goods, X and Y, and P_Y denotes the price of Y in terms of units of X. It is known that at every positive price, P_Y, and at all levels of purchasing-power Z, in terms of units of X, the amounts (x,y) bought of the two goods satisfy

$$x = P_Y y = Z/2.$$

What can there be learnt about the preference-ordering?

Solution. We show first that if (x,y) and (x',y') are any two bundles, and if $2x \geqslant x' + \dfrac{y'x}{y}$, then $(x,y) > (x',y')$. (A2.10.3)

This follows immediately from the fact that, when $P_Y = \frac{x}{y}$ and $Z = 2x$, the bundle (x,y) is chosen, although Z could have purchased the bundle (x',y'). More than this can be learned about the ordering by making use of the following lemma:

LEMMA. If a and b are any real numbers satisfying $b > a > 0$, and n is any integer satisfying $n > \dfrac{2ab}{b-a}$, then $e^{a/n} + e^{-b/n} < 2$. (A2.10.4)

Proof of Lemma:

Since
$$n > \frac{2ab}{b-a}, \quad \frac{a}{n} > 1$$

$$e^{a/n} = 1 + \frac{a}{n} + \frac{a^2}{2n^2} + \frac{a^3}{6n^3} + \ldots < 1 + \frac{a}{n} + \frac{a^2}{n^2} + \ldots = \frac{n}{n-a}$$

and
$$e^{b/n} > 1 + \frac{b}{n} \text{ so that } e^{-b/n} < \frac{n}{n+b}$$

and $e^{a/n} + e^{-b/n} < \dfrac{n}{n-a} + \dfrac{n}{n+b} = 2 - \dfrac{(b-a)n - 2ab}{(n-a)(n+b)} < 2$ since $n > \dfrac{2ab}{b-a}$.

(A2.10.5)

It can now be seen that

$$(x,y) > (xe^{a/n}, ye^{-b/n}) \text{ and that } (x,y) > (xe^{-b/n}, ye^{a/n}) \quad \text{(A2.10.6)}$$

(A2.10.6) follows at once from (A2.10.4) and (A2.10.5), by substituting for (x',y') in turn the two R.H.S. of the inequalities in (A.2.10.6).

We are now in a position to prove that

$$(x,y) > (x'',y'') \text{ whenever } xy > x''y''. \quad \text{(A2.10.7)}$$

Three cases may be distinguished:

(*i*) $x \geqslant x''$ and $y \geqslant y''$ and $xy > x''y''$; in this case $(x,y) > (x'',y'')$ because of P. 4.

(*ii*) $x \leqslant x''$ and $y > y''$; in this case, we can choose a and b so that $x'' = xe^{a}$ and $y'' = ye^{-b}$ and $b > a \geqslant 0$: we can then define the bundles (x_j, y_j); $j = 0, 1, 2, \ldots, n$; where n is some integer greater than $\dfrac{2ab}{b-a}$ by

$$x_j = xe^{ja/n},\ y_j < ye^{-jb/n};\ j = 1, 2, \ldots, n.$$

It then follows from (A2.10.6) that

$(x,y) \equiv (x_0,y_0) > (x_1,y_1) > (x_2,y_2) \ldots > (x_n,y_n) \equiv (x'',y'')$ so that $(x,y) > (x'',y'')$.

(*iii*) $x > x''$ and $y \leqslant y''$: in this case we can choose a and b so that $x'' = xe^{-b}$ and $y'' = ye^{a}$ and $b > a \geqslant 0$:
we may then prove, by the same method as in (*ii*), that $(x,y) > (x'',y'')$.
We may summarise the preference ordering as follows:
$(x,y) > (x',y')$ whenever $xy > x'y'$, and $(x,y) \parallel (x',y')$ whenever $xy = x'y'$.

Some economists would express this by saying that the indifference map is completely determined by the market data as consisting of the family of indifference curves, $xy = \lambda$; but the use of the word 'indifference' in this context is better avoided.

EXAMPLE A2.10.3. Suppose that, as in the last example, the amounts (x,y) bought of the two goods satisfy $x = P_Y y = \frac{1}{2}Z$ at all levels Z of purchasing-power and at every price $P_Y \geqslant 1$, but that when $P_Y < 1$, $x = 0$ and $y = Z/P_Y$.

What can we now learn about the preference-relations between bundles containing the Y-good only and other bundles?

Solution. First, we note that the methods used in the last example suffice to show that whenever $x \geqslant y$ and $x' \geqslant y'$ and $xy > x'y'$ then $(x,y) > (x',y')$; also that whenever $x \geqslant y$ and $x' \geqslant y'$ and $xy = x'y'$, then $(x,y) \parallel (x',y')$.

Next, we can find the preference relations concerning all bundles of type $(0,y)$, consisting only of the Y-good.

First consider the relations of type $(0,y) > (x',y')$. Since $(0,y)$ is certainly bought if $P_Y < 1$ and $Z = P_Y y$, we can show that $(0,y) > (x',y)'$ whenever $(x'+y') < y$. For let $x'+y' = (1-c)y$, $(c > 0)$; then let $P_Y = (1-c)$ and $Z = (1-c)y$. The cost of (x',y') is

$$x'+P_Y y' = Z-cy' < Z.$$

The fact that $(0,y)$ was bought although (x',y') cost less, establishes the inequality $(0,y) > (x',y')$.

In particular this shows that $(0,y) > (x,x)$ whenever $2x < y$.

But $(x,x) > (x',y')$ whenever $x' \geqslant y'$ and $x^2 > x'y'$; so, combining these two results, $(0,y) > (x',y')$ whenever $x' \geqslant y'$ and $y^2 > 4x'y'$.

Again, by P. 4, $(x,x) > (x',y')$ whenever $x \geqslant y' > x'$, so that

$$(0,y) > (x',y') \text{ whenever } y > 2y' > 2x'.$$

In turn, we now consider relations of type $(0,y) < (x',y')$.

First we note that $(0,y) < (x,x)$ when $2x > y$, since $P_Y = 1$, and $Z = 2x$, (x,x) will be bought, although $(0,y)$ could be bought for Z or less.

But we also know that $(x,x) < (x',y')$ whenever $x' \geqslant y'$ and $x'y' > x^2$; hence $(0,y) < (x',y')$ whenever $4x'y' > y^2$ and $x' \geqslant y'$. It also follows from P. 4 that $(0,y) < (x',y')$ whenever $y' \geqslant y$ (except when $x' = 0$ and $y' = y$).

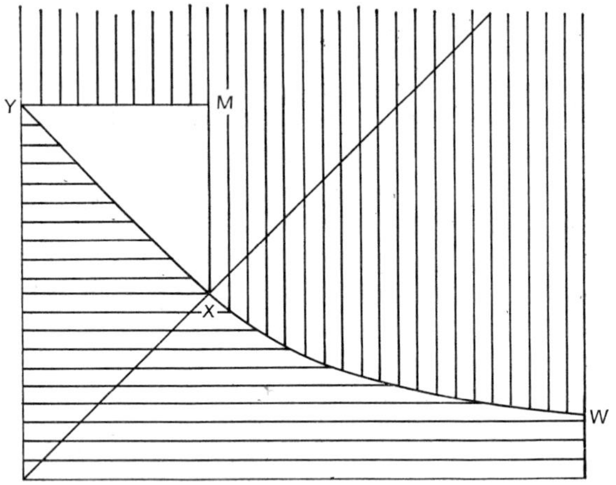

Fig. 2

We can sum up what we have learnt about the preference relations with $(0,y)$ in Fig. 2. When $(x'y')$ is in the vertically shaded area or on MX or MY, then $(0,y) < (x',y')$. When (x',y') is in the horizontally shaded area, then $(0,y) > (x',y')$. But when (x',y') is on the part XW of the curve $4x'y' = y^2$, or on YX, or inside the triangle YMX, then $(0,y) \parallel (x',y')$.

This example illustrates the point that Postulates 1, 2, 3, 4 together with complete market information about what will be bought at every combination of relative price P_Y and Z income, in terms of X, may still leave in the preference-ordering a great many uncertainty relations type $(x,y) \parallel (x',y')$. This could be all the more true in cases involving bundles of many goods instead of bundles of only two goods.

Having discovered in this example the preference relations involving a bundle containing only the good Y, we do not require much further work to discover the whole preference ordering (see *Exercises* 2.12, 2.13, 2.14).

A2.11. The Postulate of Convexity in a Preference Ordering

The following further consistency requirement is occasionally considered in theories of consumers' expenditure-decisions.

P. 5 If for some w $(0 < w < 1)$, for all i $(i = 1, 2, \ldots, n)$ $s_i \geqslant wr_i + (1-w)r_i'$, and for some i $(i = 1, 2, \ldots, n)$ $r_i \neq r_i'$, then either $(s_1, s_2, \ldots, s_n) > (r_1, r_2, \ldots, r_n)$ or $(s_1, s_2, \ldots, s_n) > (r_1', r_2', \ldots, r_n')$ where $(x_1, x_2, \ldots, x_n)$ denotes a bundle containing amounts x_j of the n goods $(j = 1, 2, \ldots, n)$.

This postulate P. 5 requires, in effect, that a weighted average of two distinct bundles of goods will always be preferred either to the one bundle or to the other (or to each). In real life, this obviously does not always hold, but in so far as it does hold, it corresponds to the tendency for one to become willing to pay less in terms of other goods for more of one particular good A, the more one already has of A.

When one is prepared to impose the requirement P. 5, one is often enabled to add considerably to what can be learnt about the preference ordering from a given set of market information.

EXAMPLE A2.11.1. *Reconsideration of Example* A.2.10.1 *when P.* 5 *is assumed.* In Example A2.10.1, postulates P. 1, P. 2, P. 3 and P. 4 were insufficient to determine any definite preference between the bundles (1,2) and (2,1).

But P. 5 enables one to determine such a preference. Since the bundle (1,2) is the unweighted average of the bundles (0,3) and (2,1) (i.e., $(1,2)+(1,2) \equiv (0,3)+(2,1)$), P. 5 requires that either $(1,2) > (0,3)$ or $(1,2) > (2,1)$. But in the solution of example A2.10.1 we showed that $(0,3) > (1,2)$, and it follows that P. 5 ensures $(1,2) > (2,1)$.

Hence P. 5 enables us to amend our earlier conclusions and to obtain the completely definite ordering

$$(0,3) > (1,2) > (2,1) > (0,2) > (1,1) > (3,0) > (0,1) > (2,0) > (1,0) > (0,0).$$

EXAMPLE A.2.11.2. *Reconsideration of example* A2.10.3 *assuming P.* 5 When P. 5 was not assumed, it was possible to establish only the preference ordering summarised in Fig. 2 and in the accompanying text. Now we will reconsider the relations of type $(0,y) < (x',y')$ when P. 5 can be assumed.

We know already that

$$(0,y) < (x,x) \text{ when } 2x > y, \text{ and } (0,y) < (0,y'') \text{ when } y'' > y.$$

It follows from P. 5 that, if $0 < w < 1$, and $x' = wx$ and $y' = wx + (1-w)y''$, then either $(x',y') > (0,y'')$ or $(x',y') > (x,x)$.

Combining these results, we learn that, when P. 5 is assumed, if $x' > \frac{1}{2}wy$ and $y' > y - \frac{1}{2}wy$, then $(0,y) < (x',y')$.

Hence, whenever $y < (x'+y')$ and $y < 2y'$ then $(0,y) < (x',y')$.

This result may be expressed as follows in terms of Fig. 2. When P. 5 is assumed in addition to P. 1, P. 2, P. 3 and P. 4, then, when (x',y') is above the curve YXW or at the point X, $(0,y) < (x',y')$.

We know already that when (x',y') is below the curve YXW, then $(0,y) > (x',y')$; however, when (x',y') is on the curve YXW (but not at Y or X), no definite preference can be established and $(0,y) \parallel (x',y')$.

On the basis of these results it is easy (see *Exercise* 2.11) to write down the complete preference ordering for this situation.

A2.12. Pareto-Optimal Situations

Consider a situation where several individuals may enjoy the fruits of economic enterprise. Let a_{ij} $(j = 1, 2, \ldots, k)$ denote the collection of goods received by individual j in situation i $(i = 1, 2, \ldots, n)$.

Let $a_i > a_h$ denote 'a_{ij} is preferred to a_{hj} by at least one individual j, whereas $a_{hj'}$ is not preferred to $a_{ij'}$ by any individual j'.' Let $a_i \geqslant a_h$ denote 'a_{hj} is not preferred to a_{ij} by any individual j.' Let $a_i \parallel a_h$ denote 'at least one individual j prefers a_{ij} to a_{hj} and at least one individual j' prefers $a_{hj'}$ to $a_{ij'}$.'

Then, provided k and n are finite, it is easy to prove that there must be at least one a_i such that for no h does $a_h > a_i$. Such a situation, i, is called *Pareto-optimal* among the n situations. Unfortunately, the usual relation between two possible situations about which policy decisions have to be made is $a_h \parallel a_i$, and it is found that there are very many different Pareto-optimal situations from which to choose. Moreover, a great many situations which are not Pareto-optimal may well be preferred by all but a few to several of the Pareto-optimal situations. Such facts as these prevent Pareto-optimality from being a very helpful concept.

Exercises 2

2.1. Explain the distinction between ordinal and cardinal concepts.

2.2. Explain the distinction between the relations $\parallel$ and $=$ in preference orderings and probability orderings.

2.3. What conditions must (*i*) a preference ordering, (*ii*) a probability ordering, satisfy if it is to be consistent?

2.4. Two dice, respectively marked A to F and 1 to 6 are each thrown once only. $[A]$, $[B]$, etc. $[1]$, $[2]$, etc. denote the events that A, B, etc. 1, 2, etc. appear uppermost. Explain the meaning of
(*i*) $[C] \cap [3]$ (*ii*) $[C] \cup [3]$ (*iii*) $([A] \cup [B]) \cap [3]$
(*iv*) $([A] \cup [B]) \cap ([B] \cup [C])$.

2.5. Under the conditions of question 4, which of the following sets are (*a*) exhaustive, (*b*) disjoint:
(*i*) $[A], [B]$ (*ii*) $[A], [1] \cap [2]$
(*iii*) $[A] \cup [B] \cup [C] \cup [D], [E] \cup [1] \cup [F]$
(*iv*) $[A] \cup [B], [C] \cup [D], [E] \cup [F], [1] \cap [2]$?

2.6. Provide 2 illustrations of the following ordering based on the disjoint exhaustive set of 4 events A, B, C and D.
$A > B, B > C, D > C, [A \cup C] > [B \cup D], [B \cup D] > A, [B \cup C] \geqslant D$

2.7. Describe the probability ordering for the result of one spin of a fair roulette-wheel which has 37 compartments.

2.8. What is meant by the statement that a set of events is irrelevant to the result of spinning a fair roulette-wheel?

2.9. How may the concept of a fair roulette-wheel be used in calibrating the events in a probability ordering?

2.10. What is meant by the irrelevance of a set of outside events to a calibrated probability ordering of events?

2.11. Given the market information assumed in Examples A2.10.2 and A2.11.2 (in Appendix to Chapter 2), and assuming Postulates 1 to 5 of sections **A2.10** and **A2.11**, obtain the complete preference ordering in the following form: 'Let $\phi(x,y) = xy$ if $x > y$ and $\phi(x,y) = (x+y)^2/4$ if $x < y$. Then

$(x',y') > (x'',y'')$ when $\phi(x',y') > \phi(x'',y'')$;
$(x',x') > (x'',2x'-x'')$ when $x' > x''$;
and $(x'',x'') > (x',2x''-x')$ when $x'' > x'$;
and otherwise $(x',y') \parallel (x'',y'')$ when $\phi(x',y') = \phi(x'',y'')$.'

Hint: Use the solution of Example A2.11.2 and the fact that $(x',y') > (x'',y'')$ whenever there is a y such that

$$(x',y') > (0,y) > (x'',y'').$$

2.12. Under the conditions assumed and with the market information provided in Example A2.10.3 and using Fig. 2, copy out the diagram and then mark out and describe the 3 sets of points (x_i, y_i) $(i = 1, 2, 3)$ such that

$(x_1,y_1) > (x',y')$, $(x_2,y_2) < (x',y')$ and $(x_3,y_3) \parallel (x',y')$, where (x',y') is a point P′ on the line YX in the diagram.

2.13. Under the conditions assumed and with the market information provided in Example A2.10.3, let (x',y') be any point P′ on XW. Show that if (x'',y'') is any other point P″, then $(x'',y'') > (x',y')$ if P″ is in the vertically shaded area, $(x'',y'') < (x',y')$ if P″ is in the horizontally shaded area and $(x'',y'')\ (x',y')$ if P″ is inside, or on any side of, triangle YMX.

2.14. Use the solutions of Exercises 12 and 13 to obtain the complete preference ordering for Example A2.10.3.

2.15. If A, C are disjoint and B, D are disjoint, and if $A \geqslant B$ and $C \geqslant D$, show that Consistency Postulate 2 of § **2.8.2** ensures that $A \cup C > B \cup D$.

[*Method of Solution.* Let a, b, c, d denote not-A, not-B, not-C, not-D and let Ab etc. denote $A \cap b$ etc. Then

$$\begin{aligned} & Ab \geqslant aB \text{ (Why?)}; \\ & Ab \cup bC \geqslant aB \cup bC = aBc \cup aBC \cup bC \\ & \qquad = aBc \cup BC \cup bC \text{ (Why?)} \\ & \qquad = aBc \cup C \geqslant aBc \cup D \text{ (Justify each step)} \\ \therefore\ & AB \cup BC \cup Ab \cup bC \geqslant AB \cup BC \cup aBc \cup D \\ \therefore\ & (B \cup b)(A \cup C) \geqslant \{(A \cup C \cup ac)B\} \cup D \\ \therefore\ & (A \cup C) \geqslant (B \cup D). \end{aligned}$$

Explain each stage which depends on Consistency Postulate 2.]

2.16. Use the result of Exercise 15 to show, that assuming Consistency postulate 2 in § **A2.8.2**, then if $A \geqslant B$, $C \geqslant D$ and $A \cap B \leqslant C \cap D$, it follows that $A \cup C \geqslant B \cup D$.

2.17. Extend the result of Exercise 16 to the case of 3 pairs of events such that $A \geqslant B$, $C \geqslant D$ and $E \geqslant F$.

Chapter 3

PROBABILITY AND IRRELEVANCE

3.1. The Meaning of an Event

Any probability ordering consists of a set of probability comparisons between pairs of events. An event may lie in the future; it could be that the result of a certain experiment will be such-and-such; for example, one could speak of the event that the three of a certain die will lie uppermost after it has been thrown, or that the spin of a certain halfpenny will result in heads coming uppermost. But an event may also refer indirectly to the present. For example one might make a probability comparison between the event that it is raining in Manchester and the event that it is not raining in Manchester: in explanation one might add that if one made efficient enquiries about the present weather in Manchester this would lead to the (future) event of discovering that it had indeed now been raining in Manchester or else that it had not now been raining there. Thus an event can relate to hypotheses about the present or about the past as well as about the future. There is no very important logical distinction between an event concerning an unknown future and an event concerning an unknown present or an unknown past.

3.2. Disagreement about the Meaning of Probability

There is much disagreement among statisticians as to what is the correct meaning to attribute to probability comparisons. This does not affect the theoretical discussion of a given probability model, but it does affect the decision how to relate the real situation and the probability model. Furthermore, it is difficult to digest a lengthy discussion of a probability model without having some idea of the real world situations to which it might apply.

In this chapter we describe some of the meanings which various statisticians have applied to a comparison of probabilities.

3.3. A Crude Interpretation of Probability as Relative Frequency

The most down-to-earth approach is the following. Suppose that an experiment can be repeated several times, like that of firing shots at a target, and that it can have various results, such as 1, 2, 3, 4, 5, and there is no reason to expect every performance to yield the same result. By saying that 4 is more probable than 5 we should be predicting a tendency for 4 to occur more often than 5. This could be roughly tested by firing several shots: if 4's heavily outnumbered 5's, the statement would be strongly confirmed; if 5's heavily outnumbered 4's the statement would be seriously discredited. It is in the nature of probability statements that many can never be finally proved or disproved; but the longer one goes on experimenting, the stronger the confirmation or discrediting tends to become. The probability comparison, according to this approach, relates to a property of the experiment and predicts something about a whole sequence of performances of the experiment.

Our particular example illustrates a difficulty: with practice the performance might improve, so that, whereas at first 4's outnumbered 5's, eventually 5's would outnumber 4's. Would one in this case say that at a certain stage in the sequence of experiments the probability relation would be going to change from $P_4 > P_5$ to $P_5 > P_4$?

The answer is that probability comparisons of this particular kind relate only to experiments where the conditions are *not* allowed to change systematically. Thus the probability comparison could be confirmed or discredited only if one could prevent any change or assume no change in the skill of the marksman.

This type of probability comparison is therefore usually confined to experiments which can be repeated several times without any systematic changes in conditions.

Statements of the kind 'heads and tails are equally probable when this coin is spun' can be similarly interpreted as forecasting a tendency for heads and tails to come down about equally often in a sequence of spins. Clearly, such a statement cannot be confirmed; but it can be seriously discredited if we observe that heads for example come down much more often than tails in several spins.

According to this interpretation of probability comparisons, one can always attach a probability number to any particular result of an experiment: this number can be roughly estimated as the observed ratio of performances of the experiment which give rise to that particular result to the total number of performances of the observed experiment. Then the assertion of any probability comparison between two results, P and Q, is equivalent to the assertion of the corresponding inequality or equality between their probability numbers. To say that 6 is more probable than 5 with a die, is to assert that the probability number for 6 exceeds that for 5; and this in turn is to assert a bias in the die in the sense of a tendency for 6 to turn up more often than 5.

This down-to-earth explanation of probability comparisons is superficial, in that it leaves unexplained what is meant by a tendency for something to turn up more often than something else: nor can it ever explain the meaning of one sentence involving the word 'probability', without introducing another sentence involving the same word or some synonym for it. Despite this, it provides the most convenient practical interpretation of probability comparisons in the context of scientific experiment and the design of sample surveys. Its obvious limitation is that it applies only to experiments which can be performed several times without systematic change in the underlying conditions.

3.4. Probability as a Measure of Evidence

Another possible interpretation of 'the event A is more probable than the event B' is that there is more evidence for, or more reason to expect, A than B. This would clearly be the case, for example, if the occurrence of B necessarily involved the occurrence of A also, and if there was reason to believe that A might possibly occur even if B did not. Again, this would clearly be the case if every argument that would be advanced for expecting B could be matched by a correspondingly equally good argument for expecting A, and if there was an additional reason for expecting A which did not apply to B.

Again, one would have more reason to expect A than B, if it was clear that there was more reason to expect C than B, and if every argument and item of evidence for believing C applied with equal force to A.

On this interpretation, to say that events A and C were equally probable, would mean that the evidence for A and the evidence for C were perfectly balanced or symmetrical. For example, a coin might be found by various tests to be perfectly symmetrical, apart from the fact that it was possible to identify the two sides; furthermore, in four spins it had come down twice on each side. In this case, one might say that the probability that one side (H), was the same as the probability that the other side (T), would be uppermost after the next spin.

Two points should be noticed about this example. First, the statement concerns the next spin and thus differs from the statement that would be made according to the frequency theory. Second, the statement rests on appeal to evidence, and not on an absence of evidence. No appeal is made to what is known as the principle of *insufficient* reason: indeed, the *more* evidence there is, provided it is symmetrical, the more weight will be carried by the statement that the two sides are equally probable.

Here is another example, which is very similar. I am told that a bag contains nineteen balls, of which some are red and the rest green. The contents are then stirred with a suitable tool, and six balls are then taken out. If three of these are red and three are green, I might well say that the probabilities that the next ball will be red or green are equal. This points the contrast to the frequency school terminology; for in their sense the two probabilities depend on how many balls of each colour remain in the bag and they cannot in any case be equal since the number of remaining balls is odd.

3.5. Probability as a Measure of Frequency of Past Occurrence

Somewhere between the above two interpretations of a probability comparison is that which relies only on the evidence of past performances of an experiment, and which *defines* the probability number of any particular result as the proportion of performances of the experiment which has led to this result. This interpretation treats a probability comparison as being a statement about evidence, but in this case the evidence is *confined* to that yielded by past performances of the same experiment.

All the interpretations so far discussed attempt to make probability comparisons objective and not dependent on personal judgement.

3.6. Probability as a Measure of Belief

Opposed to such interpretations are those which take '*A* is more probable that *B*' to mean that 'I believe' (or that some specified individual believes) 'that *A* will occur rather than *B*'. On the borderline is that interpretation which expresses it as a recommendation to believe *A* rather than *B*. Such an interpretation would attribute a higher *rational* degree of belief to *A* than *B*.

For example, it is somehow known that the weight of an object lies between 0 units and 10 units; it is also known that the margin of error of a weighing machine is 1 unit either way, and that the probabilities of overweighing and underweighing are equal. Finally, the object is weighed and its apparent weight is 3·4 units. One might then say that it was equally as probable that the true weight was less than 3·4 units as that it was more than 3·4 units: this would amount to a recommendation to (act as though one) put equal belief in the two possibilities. *Why* one should be led to this statement comparing the probabilities is another matter.

3.7. Tests of Belief

The completely personal interpretation of '*A* is more probable than *B*', as referring to someone's stronger belief in event *A* than in event *B*, is open to the objection that personal beliefs are unobservable. However, it is possible to ask people what they believe and it is also possible to invite them to make wagers. The most satisfactory attempt to explain the meaning of a personal probability comparison was that of F. P. Ramsey. He asserted that to discover whether an individual considers *A* or *B* more probable, only the following three steps are necessary.

(1) Discover whether, given the choice, he would choose enormous wealth or death by slow torture. Without loss of generality, we may suppose that he would choose wealth.

(2) Offer him the option of a wager, yielding wealth if *A* but death by slow torture if *B*, **or** a wager, yielding wealth if *B* but death by slow torture if *A*; convince him that you mean the offer seriously and that he must choose one or the other.

(3) Observe which choice he makes. The former choice will indicate that he considers *A* more probable than *B*, the latter choice that he considers *B* more probable than *A*.

There is a slight difficulty about interpreting the statement that *A* and *B* are *equally* probable; but we will not attempt to resolve this, except to the extent of saying that the case of equal probability represents the borderline between cases where *A* is more probable than *B* and cases where *B* is more probable than *A*.

3.8. Possible Objections to Ramsey's Method of Measuring Probability

It may be objected that Ramsey claimed to be able to do the impossible in finding a definite probability ordering for every pair of events. For people's beliefs about probability comparisons must very often be of the type $A \parallel B$, where they have no clue how to compare *A* and *B* in probability. If they were asked to stake their life either on *A* or on *B* in such circumstances, no doubt they would yield to *force majeure* and stake either on *A* or on *B*, but their choice would be entirely arbitrary and would not really indicate either that they regarded *A* as more probable than *B*, or, by their long hesitation, that they regarded *A* and *B* as equally probable. In fact, they would have no idea whether they should regard *A* as more probable, or less probable than *B*, or equally probable. Ramsey's procedure is thus in danger of giving a completely unrealistic impression of orderliness in probability comparisons, when in fact there is nothing of the sort in the mind of the chooser.

3.9. Relevance

The difficulty of economics arises largely from the fact that very many different matters are relevant to any one decision in economic affairs. Thus the economist, seeking to account for the unemployment in Wales in 1935, may have to consider such apparently unrelated phenomena as:

(*i*) the loss of markets for British coal during the 1914-18 war;
(*ii*) the working out of the best coal seams in Welsh pits;
(*iii*) the demand for anthracite as opposed to steam coal;
(*iv*) the general strike of 1926;
(*v*) the piling up of stocks of world raw materials in the late 1920s;
(*vi*) the Wall Street crash of 1929;
(*vii*) the German banking system;
(*viii*) the policies of Mr Winston Churchill and Mr Snowden as successive Chancellors of the Exchequer;
(*ix*) the New Deal in the U.S.A.;
(*x*) factors limiting emigration from Wales to England.

3.10. Postulates about Irrelevance

It is a great advantage to be able to limit the number of factors relevant to a subject of study, and then to control the variation of those relevant factors as far as possible, so as to render comparisons clear-cut and calculation simple.

But, 'How,' it may well be asked, 'does one ever know that any factor is irrelevant to the question at issue?'

The honest answer must be that one cannot know this for certain, but that sensible men can often agree that all except a fairly small number of factors are wholly, or at any rate mainly, irrelevant to a particular question; and with luck they will agree on quite a short list of factors as being the mainly relevant ones. In other cases, many will agree that certain factors are probably irrelevant, and will be willing to admit that they are really irrelevant, if some fairly simple test or experiment fails to suggest that they are, after all, highly relevant.

3.11. Irrelevance in the Natural Sciences

In science, a general hypothesis may sometimes be put forward that X is irrelevant to the result of some class of experiment; this may be a new and startling idea, and will lead to a series of experiments seeking to test and disprove the hypothesis. If the hypothesis survives all these attempts, then the theory that X is indeed irrelevant is likely to become generally accepted, and may give rise to yet other such hypotheses and theories.

For example, in many branches of science, it is a common assumption that *mere* position in time and space are irrelevant to the operation of general scientific laws, such as those of gravitation and the conservation of energy.

Again, the basis of the theory of relativity is that the movement of an observer is irrelevant to his measurement of the velocity of light relative to himself.

Advance in knowledge would be impossible, if scientists did not constantly assume all kinds of factors to be irrelevant, on the grounds that it was obvious to common sense that they must be, and therefore did not bother to test every such assumption, and tested only those where something happened to suggest reason for doubting them.

The assumption that the passage of time is irrelevant to the operation of the basic laws of science, is an assumption fundamental to all arguments based on experience. For what guidance for the future can mere experience give, unless we know in what respect the behaviour of matter in the future will resemble its behaviour in the past? The correct answer to this rhetorical question is that it can be of no guidance. However, one assumes that the fact that the future is distinct from the past is irrelevant to the operation of scientific laws, and that no Creator will choose tomorrow to shuffle all the cards and substitute a new deal. However groundless this assumption may be, one may safely assume that most readers will be prepared to act on it.

3.12. Postulates of Irrelevance in the Study of Preferences

In most branches of science the postulates of irrelevance are so widely acceptable that they are usually made without notice or remark; but in economics, the interconnections are so intricate that postulates of irrelevance are likely to be far more controversial. However, it is almost essential to make free use of such postulates, even in some cases where they are not strictly realistic, if any progress in study is to be made through the jungle of interwoven phenomena.

Consider for example some of the postulates of irrelevance that one might adopt with regard to preference.

'The type of label stuck on the bottles is irrelevant to one's preference with regard to drinking two wines of known locality and vintage.'

'The days of the week are irrelevant to the question of whether an increase of real consumption from A to B on one day, would be accepted in return for a decrease from C to D on another day.'

The identity of the individuals concerned is irrelevant to the preference one would have for providing food and shelter to one who lacked them, rather than providing more wine to one who had already drunk enough.

One's expenditure on going to concerts is irrelevant to the enjoyment one gets from drinking three half-pints of beer a day instead of two.

None of these postulates is self-evident; indeed a good case could be made out against each of them, and the meaning of some of them is decidedly hazy. Nevertheless, as first approximations, such postulates may turn out to be very fruitful in enabling quite sweeping conclusions to be derived from a small amount of information. This is of course a danger, as well as an advantage; but if it is suspected that the conclusions are false, then an enquiry should be pursued into the manner in which the postulates break down, and into the direction and degree of the resulting disturbances to the conclusions.

3.13. Random Events

The frequency theory of probability and much of experimental design is based on the existence of a certain class of procedures; namely, those whose results would be admitted by practically everyone to be entirely irrelevant to the subject-matter of scientific enquiry. Such procedures are the spinning of a coin or a roulette-wheel, the throw of a die, and the deal of a hand from a thoroughly shuffled pack of cards. Furthermore, there would be practically universal agreement that, with such apparatus, a pair of experiments can be devised and carried out in which the result of one experiment is irrelevant to the result of the next experiment. Thus, suppose a roulette-wheel, numbered 0 to 36 inclusive, is spun twice and a 'success' is recorded if the second number exceeds the first, a 'failure' if the first exceeds the second, and, if the second number equals the first, the second pair of spins does not count, and further pairs of spins are performed until a success or failure is obtained. Call this the first experiment; then there would be practically universal agreement that the success or failure of this first experiment would be entirely irrelevant to the success or failure of a second experiment of the same kind even if it were performed with the same apparatus.

The possibility of using such apparatus enables 'randomisation' to be introduced into experimental design, so that observers, who may differ in their judgement of what results are most to be expected from a scientific statistical enquiry, may yet agree beforehand on a mutually acceptable procedure for interpreting and acting on the results when they are obtained.

3.14. The Meaning of Irrelevance

So far, the concept of irrelevance has been discussed as though its meaning was commonly understood and agreed. For most practical purposes this may be a safe assumption, but for the purposes of discussing the use of probability in economics, it will be necessary to consider a little more closely just what is to be meant by irrelevance.

After the concept of a probability ordering, that of irrelevance is about the most fundamental concept in the whole field of probability. Indeed the frequency theory of probability is built up from the notion that probability properly applies only to the results of experiments so special that they can be repeated to yield an indefinitely long sequence, in which each result is irrelevant to every other.

We shall be content to develop the concept of irrelevance in the context of a probability ordering. It may well be that a more general approach, defining irrelevance in the context of orderings of any kind would be possible; but since in this book we shall use the concept only in relation to probability orderings, and since the restriction of the concept to this context avoids certain complications, the restriction will be accepted.

3.15. Irrelevance in the Context of a Probability Ordering

The essential characteristic of the irrelevance of one event E' to other events $E(r)$ $(r = 1, 2, \ldots, N)$ is that the occurrence of E' should leave unchanged the probability ordering of the $E(r)$. But the matter is a little more complicated than this.

In order to approach the definition of the irrelevance of one event E' to other events $E(r)$, it is convenient first to define the irrelevance of the event E' to the probability ordering of the events $E(r)$: we shall do this so as to convey the meaning, that the occurrence of E' leaves the probability ordering of the $E(r)$ unaltered.

The event E' is said to be irrelevant or relevant to the probability ordering of the events $E(r)$, according as the probability ordering of the intersections $E' \cap E(r)$ does or does not exactly correspond to the probability ordering of the events $E(r)$.

That is to say that for every pair $E(t)$, $E(s)$, the probability ordering of $E' \cap E(t)$ and $E' \cap E(s)$ must be

$$E' \cap E(t) >, =, \text{ or } < E' \cap E(s)$$

according as that between $E(t)$ and $E(s)$ is

$$E(t) >, =, \text{ or } < E(s),$$

if E' is to be irrelevant to the probability ordering of the events $E(r)$.

To take an example of the irrelevance of one event E' to the probability ordering of just two events E and $\sim E$, consider the following situation. If E (she loves me) is less probable than $(\sim E)$ (she loves me not), and if E' is the event that the number of petals on this daisy is odd, then, of course, E' is irrelevant to the probability ordering of E and $(\sim E)$; for the petals of the daisy have no effect whatsoever on her attitude, so that $E' \cap E$ is less probable than $E' \cap (\sim E)$. But if one were superstitious, one might not admit this fact.

Unfortunately, the irrelevance of an event E' to the probability ordering of a set of events $E(r)$ is not sufficient evidence for the assertion that E' is irrelevant to each of the events $E(r)$, in any sense in which irrelevance is ordinarily understood.

Suppose, for example that E' is the event that 'I take up smoking again'. Let E be the event that 'I shall die of lung cancer', and $\sim E$ the event that 'I shall not die of lung cancer'. Suppose that E is less probable than $\sim E$, and also that $E' \cap E$ is less probable than $E' \cap (\sim E)$; this means, roughly, that 'I am less likely to die of lung cancer than not, and that this will still be so even if I take up smoking again'. In this case, the event E' is irrelevant to the probability ordering of E and $(\sim E)$. But this could quite well be so, even if my taking up smoking, in some every-day sense increased the risk of my dying of lung cancer. Thus, even though E' may be highly relevant to E, in the ordinarily understood sense of relevance, it may be irrelevant to the probability ordering of E and $(\sim E)$.

3.16. The Irrelevance of one Event to Other Events

Although the irrelevance of one event E' to the probability ordering of other events $E(r)$ is no guarantee of the irrelevance of E' to each of the events $E(r)$, there is no clash with the accepted meaning of relevance if we decree that whenever E' is relevant to the probability ordering of two events $E(t)$ and $E(s)$, then it is relevant to at least one of the events $E(t)$ and $E(s)$. In this sense the irrelevance of E' to the probability ordering of events $E(r)$ is a necessary, although not a sufficient, condition for the irrelevance of E' to all the events $E(r)$.

Fortunately, there are circumstances where the irrelevance of E' to the probability ordering of a set of events $E(r)$ does imply the irrelevance of E' to each one of the $E(r)$, but these are the very special circumstances that the $E(r)$ $(r = 1, 2, 3, \ldots, N)$ are a disjoint exhaustive set of equiprobable events. As will be argued below, the following definition is in fact consistent with the ordinary use of the word irrelevant:

If $E(r)$ $(r = 1, 2, \ldots, N)$ are a disjoint exhaustive set of equiprobable events, then the event E' is either irrelevant to the events $E(r)$ or relevant to at least one of them, according as E' is irrelevant or relevant to the probability ordering of the $E(r)$.

Since the probability ordering of the $E(r)$ is one of complete equality, this amounts to saying that E' is, or is not, irrelevant to all the $E(r)$, according as the intersections $E' \cap E(r)$ $(r = 1, 2, \ldots, N)$ are, or are not, equiprobable.

This all becomes much easier to understand, if the events $E(r)$ are visualised as the set of individual equiprobable results of spinning once a fair roulette-wheel. We are saying that E' is irrelevant to each of these results, if the occurrence of E' does not affect the fairness of the roulette-wheel.

So far, all we have done is to explain how relevance can sometimes be detected, and to define the irrelevance of an event E' to sets of events which are like the alternative results of spinning a fair roulette-wheel. But one wishes to extend the definition to cover the irrelevance of E' to other events. A large step in this direction can be taken by adopting the following definitions.

(*i*) If E' is irrelevant to each of a set of events, then it is irrelevant also to the union of any subset of these events.
(*ii*) If E' and E are both irrelevant to an event A, and if E and A are equiprobable, then E' is irrelevant to E if, and only if, it is irrelevant to the probability ordering of E and A.

The effect of the first definition is that, if E' is irrelevant to each disjoint result of spinning a roulette-wheel, it is also irrelevant to the event of getting one or other of any given set of these results: for example it is irrelevant to getting a result less than 23. This definition is consistent with the usual use of the word irrelevant.

The effect of the second definition is that, if E and E' are irrelevant to the results from spinning a roulette-wheel, and if E is equiprobable with some set A of the results of spinning the wheel, then E' is irrelevant to E if, and only if, the occurrence of E' leaves E equiprobable with A. The justification for this is, roughly speaking, that under these circumstances the probabilities of A, and also of E, are unaffected by the occurrence of E'.

Whenever, then, we can find a fair roulette-wheel, such that the event E equals in probability some set A of results of spinning the roulette-wheel, and if E and E' are both irrelevant to the spinning of the roulette-wheel, then for E' to be irrelevant to E it is necessary and sufficient that the events $E' \cap E$ and $E' \cap A$ should be equiprobable.

Unfortunately, there is no reason to suppose that for any event E it will be possible to find a fair roulette-wheel such that E is equiprobable with some set A of results of spinning it. We can, however, make sure that E is almost equiprobable with such an A, by taking a fair roulette-wheel with a sufficiently large number of disjoint possible results. This leads us to the notion of E' being practically irrelevant to E.

3.17. The Practical Irrelevance of One Event to Other Events

In order to explain practical irrelevance we introduce the idea of a practically impossible event. By an event being practically impossible we mean that it is so improbable that one would act as though it were impossible. For example, B might be the event that within the next five minutes a creature from outer space will make a safe landing in one's back garden and tap on the window to attract attention. Or the event B might be that, in ten spins of a fair roulette-wheel with N disjoint possible results, the number b occurred every time. Provided N were fairly large, this would be practically impossible. Similarly B might be the event that with one spin of a fair roulette-wheel with N^{10} disjoint possible results, the number 1984 occurred. This also would be practically impossible.

Suppose, then, that A and B are two disjoint possible results of spinning a fair roulette-wheel, that B is practically impossible, and that where E is some event irrelevant to the result of spinning the wheel, E is also such that

$$A < E < A \cup B.$$

E may then be said to be practically equiprobable with A; for, since we should act as though B were impossible, we should act as though $A \cup B$ were equiprobable with A, and hence it would be reasonable to act as though E were equiprobable with A.

Let E' be such an event that it is irrelevant to the spin of the wheel, and that

$$E' \cap A < E' \cap E < E' \cap (A \cup B).$$

Then $E' \cap E$ is practically equiprobable with $E' \cap (A \cup B)$, by the same kind of reasoning. In this case we may say that E' is practically irrelevant to E, since we should be prepared to act as though the conditions for its complete irrelevance were satisfied.

This definition of practical irrelevance may be put a little more formally as follows:

Where n is so large that a single result of spinning a fair n-roulette-wheel is practically impossible, and where the event E' is irrelevant to the probability ordering of an event E and the results of spinning the wheel, we describe E' as practically irrelevant to E.

3.18. Asymptotic Irrelevance

We may take the idea of practical irrelevance to its logical conclusion, by supposing the scope of the roulette-wheel to increase without limit, and E' to remain irrelevant to the probability ordering of E and of the results of the spin, however large that scope becomes. In this case, E' may be termed asymptotically irrelevant to E. Thus,

> E' is asymptotically irrelevant to E, if E' is irrelevant to the probability ordering of E and the results of spinning a fair n-roulette-wheel, for all finite n.

Asymptotic irrelevance has almost, but not quite, all the properties one would wish irrelevance to have. As an example of the shortcomings of this concept, it is shown in the Appendix to Chapter 3 that the knowledge that E' is asymptotically irrelevant to E, and that $E' > \sim E'$ and that $E > \sim E$, does *not* enable us to conclude that 'Both or neither E and E' is more probable than just one of them', i.e.

$$[(E' \cap E) \cup \{(\sim E') \cap (\sim E)\}] > [\{(\sim E') \cap E\} \cup \{E' \cap (\sim E)\}]. \tag{3.18.1}$$

Yet this is something which we should certainly expect to be true, if E' were really irrelevant to E. For example, it is very easy to prove, on the basis of the most simple models involving numerical probability, that if E' and E are any two *independent* events, and if the numerical probabilities of E' and E exceed the numerical probabilities of $(\sim E')$ and of $(\sim E)$, then the numerical probability of $[(E' \cap E) \cup \{(\sim E') \cap (\sim E)\}$ exceeds that of $\{(\sim E') \cap E\} \cup \{E' \cap (\sim E)\}$; and independence in such models corresponds to what is ordinarily meant by irrelevance.

It is difficult to say more about this at an elementary level, but some further discussion of this point is given in the Appendix to Chapter 3.

3.19. Numerical Probability

The device of calibrating events in probability by comparing them in probability with the results of a spin of a fair roulette-wheel of large scope, provides a means of obtaining a convenient numerical measure of their probability.

To attribute to $G(t)$, the union of t single disjoint results of spinning a fair n-roulette-wheel, the numerical probability

$$\text{Prob. } G(t) = t/n \tag{3.19.1}$$

is such a hackneyed procedure that it may seem obviously right. It is, of course, only a convention to attribute this probability number, t/n, to $G(t)$ and to other unions of the same number, t, of disjoint single results.

Like some, but not all, other widely accepted conventions, this one has a number of advantageous consequences. These will be asserted here in the text, and the justification for these assertions will be given in the Appendix to Chapter 3. The advantages are that the numerical probabilities calculated by these conventions satisfy the following relations.

(*i*) Where G' and G are two results of spinning one fair roulette-wheel, or two fair roulette-wheels of the same or different scope, Prob. $(G') >, =,$ or $<$ Prob. (G) according as $G' >, =,$ or $< G$.

(*ii*) Prob. $\{G(n)\} = 1$, where $G(n)$ is the (certain) event that the union of all the disjoint single results of the spin occurs; similarly $G(0)$ the impossible null-event that no result is obtained, has numerical probability zero.

(*iii*) Where G' and G are a pair of disjoint results which may follow the spinning of a fair roulette-wheel,

$$\text{Prob. } (G' \cup G) = \text{Prob. } (G') + \text{Prob. } (G).$$

(*iv*) If the result of spinning one fair roulette-wheel is irrelevant to the result of spinning another, and if G' and G are results of spinning the one wheel and of spinning the other, then

$$\text{Prob. } (G' \cap G) = \text{Prob. } (G') \,.\, \text{Prob. } (G).$$

Having established the convention for the numerical probabilities of the various results of spinning fair roulette-wheels, it is then a simple matter to obtain the numerical probabilities of other events almost exactly, by comparing them in probability with the results of spinning a fair roulette-wheel of large scope. Thus, if E satisfies the probability-relations $G(t) \leqslant E < G(t+1)$, where Prob. $\{G(t)\} = t/n$ and Prob. $\{G(t+1)\} = (t+1)/n$, we may assert that

$$t/n \leqslant \text{Prob. } (E) < (t+1)/n.$$

3.20. Exact Properties of Numerical Probability

If each of the events E satisfied probability relations of *equality* with a result of spinning a fair roulette-wheel, one would be able to allot exact probability numbers to each of the events E. Moreover, these probability numbers would have all the convenient properties of the probability numbers (of the results of spinning fair roulette-wheels) given in the last section. Thus, provided the probability ordering was consistent, writing P as short for Prob.,

(*i*) $P(E') >, =, \text{or} < P(E)$, according as $E' >, =, \text{or} < E$;
(*ii*) $P(E') = 1$, if E' is certain; $P(E') = 0$ if E' is impossible;
(*iii*) If E' and E are disjoint, $P(E' \cup E) = P(E') + P(E)$;
(*iv*) If E' is irrelevant to E, $P(E' \cap E) = P(E') \,.\, P(E)$.

It is also shown in the Appendix to Chapter 3 that, whether E' is irrelevant to E or not,

(*v*) $P(E' \cup E) = P(E') + P(E) - P(E' \cap E)$.

3.21. Approximate Properties of Numerical Probability

If the events satisfy only *inequalities* in their probability ordering with the results of spinning the roulette-wheel, then the conclusions of the last section no longer follow exactly. Nevertheless, if one considers a series of roulette-wheels of greater and greater scope, these conclusions become more and more nearly true. One can therefore say that they are true to any required degree of approximation, provided the events E can be compared in probability with the results of a spin of a roulette-wheel of sufficiently large scope.

3.22. A Case where Numerical Probabilities Cannot be Found

The probability relations between the events E and the results of spinning the roulette-wheel must of course all be definite relations of types $>$, $\geqslant$, $<$, $\leqslant$, or $=$, and must not be of type $\|$. If some of the relations are of type $\|$, it must be possible to replace them with ones of the more definite types and obtain an illustration of the ordering. If it is not possible to obtain a consistent illustration, then it will certainly not be possible to obtain numerical probabilities with the properties described in § **3.21**.

But, even if a consistent illustrative ordering can be found, it does not *necessarily* follow that *numerical* probabilities with the desired properties can be assigned. In the Appendix to this Chapter, an example is in fact provided of a perfectly consistent probability ordering of the various unions of 10 disjoint events, to which it is *impossible* to attribute numerical probabilities with the desired properties described in § **3.21**.

Appendix to Chapter 3

A3.1. A Consistent Probability Ordering for which Numerical Probabilities cannot be Found

Consider a probability ordering based on 10 elements $A, B, \ldots, J$. It will be termed consistent if it satisfies the requirements

P. 1 $[A > B] \equiv [B < A]$, $A = A$;
P. 2 $[A > B]$ and $[B > C] \rightarrow [A > C]$
P. 3 If A and $[B \cup C]$ are disjoint,
$$[B > C] \equiv [(A \cup B) > (A \cup C)]. \qquad (A3.1.1)$$

The probability ordering will be termed *potentially numerical*, if there exists a real function ϕ of $A, B, \ldots$ etc. satisfying

R. 1 $[\phi(A) > \phi(B)] \equiv [A > B]$;
R. 2 If A, B are disjoint,
$$\phi(A \cup B) = \phi(A) + \phi(B). \qquad (A3.1.2)$$

It might be hoped that every consistent probability ordering would be potentially numerical. This hope is dashed by the following example. Two probability orderings X and Y will be fully defined below: the first is merely a stepping-stone to the second and is both consistent and potentially numerical; but Y, the second, although it is consistent, is *not* potentially numerical.

Let $A, B, C, D, E, F, G, H, I$ and J be disjoint. Consider the 1024 sets obtainable by taking 0 or 1 of each of $A, B, C, D, E, F, G, H, I$ and J. Associate with A–J the 10 numbers

$$n(A) = 5,\ n(B) = 40,\ n(C) = 300,\ n(D) = 2000,\ n(E) = 10\,000,$$
$$n(F) = 14\,700,\ \ n(G) = 21\,296,\ \ n(H) = 24\,330,\ \ n(I) = 78\,628,$$
$$n(J) = 80\,000. \qquad (A3.1.3)$$

With each of the 1024 sets associate the number obtained by adding the numbers associated with its constituents. For example, associate with the set $[A \cup C \cup D \cup G]$ the number

$$n(A) + n(C) + n(D) + N(G) = 23\,601$$

Consider the 512×1023 pairs of these 1024 sets. Probability ordering X is to be obtained by putting for each pair $[Q, R]$, $Q < R$ if $n(Q) < n(R)$ and $Q > R$ if $n(Q) > n(R)$. (The case of equalities will be found not to arise.) It is fairly clear that the probability ordering X is both consistent and potentially numerical, since it can be directly associated with sets of results of a spin of a fair roulette-wheel of scope $\sum_{A}^{J} n(\ ) = 231\,299$.

Probability ordering Y is to be derived from probability ordering X as follows. We replace $<$ by $>$ in exactly 8 of the probability comparisons so as to obtain

$$\begin{aligned} &[A \cup E \cup F \cup I] > [B \cup D \cup G \cup J] \\ &[B \cup D \cup H \cup I] > [C \cup E \cup F \cup J] \\ &[A \cup C \cup E \cup F \cup I] > [B \cup C \cup D \cup G \cup J] \\ &[A \cup E \cup F \cup H \cup I] > [B \cup D \cup G \cup H \cup J] \end{aligned} \qquad \text{(A3.1.4)}$$

$$\begin{aligned} &[A \cup C \cup E \cup F \cup H \cup I] > [B \cup C \cup D \cup G \cup H \cup J] \\ &[A \cup B \cup D \cup H \cup I] > [A \cup C \cup E \cup F \cup J] \\ &[B \cup D \cup G \cup H \cup I] > [C \cup E \cup F \cup G \cup J] \\ &[A \cup B \cup D \cup G \cup H \cup I] > [A \cup C \cup E \cup F \cup G \cup J] \end{aligned} \qquad \text{(A3.1.5)}$$

The reader will be able to confirm that this set of modifications is complete in itself, in the sense that it includes all the consequences required for consistency by any subset of itself. The modifications do not introduce any inconsistency into the ordering.

But now consider the possibility that numbers can still be allotted to the 1024 sets of the modified ordering Y such as will still satisfy R. 1 and R. 2. If so, the 10 numbers $N(A)$ to $N(J)$ associated with A to J must certainly satisfy

$$\begin{aligned} &N(A)+N(E)+N(F)+N(I) > N(B)+N(D)+N(G)+N(J) \\ &N(B)+N(D)+N(H)+N(I) > N(C)+N(E)+N(F)+N(J) \end{aligned} \qquad \text{(A3.1.6)}$$

However they must also satisfy

$$\begin{aligned} &N(A)+N(D)+N(G)+N(J) > N(B)+N(C)+N(H)+N(I) \\ &N(C)+N(E)+N(H)+N(J) > N(A)+N(F)+N(G)+N(I) \\ &N(B)+N(C)+N(F)+N(G) > N(A)+N(D)+N(E)+N(H) \end{aligned} \qquad \text{(A3.1.7)}$$

(A3.1.6) follows from (A3.1.4) and (A3.1.7) follows from the fact that since the corresponding 3 inequalities with n replacing N throughout hold, due to (A3.1.3); and since (A3.1.4) and (A3.1.5) are the only alterations to ordering X in obtaining ordering Y, the latter ordering Y must satisfy

$$\begin{aligned} &A \cup D \cup G \cup J > B \cup C \cup H \cup I \\ &C \cup E \cup H \cup J > A \cup F \cup G \cup I \\ &B \cup C \cup F \cup G > A \cup D \cup E \cup H \end{aligned} \qquad \text{(A3.1.8)}$$

and P. 1 then implies (A3.1.7).

It is clear that by adding the 5 inequalities (A3.1.6) and (A3.1.7) we achieve a contradiction of the form

$$2\sum_{A}^{J} N(\ ) > 2\sum_{A}^{J} N(\ ). \tag{A3.1.9}$$

We may therefore conclude that although the modified ordering Y is consistent it is *not* potentially numerical.

Although the ordering Y is consistent according to our definitions, it may be noticed that there is nevertheless something rather peculiar about it. The existence of this ordering apparently rules out the possibility of there being any unbiased coin, the result of whose throw (heads or tails) is irrelevant to the ordering Y. For suppose there to be such a coin and denote heads and tails by h and t. Then since the result of the throw is irrelevant to the ordering Y, it follows from (A3.1.4) and (A3.1.8) and P. 2 and P. 3, that

$$\begin{aligned}
&[h\cap A]\cup[h\cap E]\cup[h\cap F]\cup[h\cap I] > [h\cap B]\cup[h\cup D]\cup[h\cap G]\cup[h\cap J]\\
&[h\cap B]\cup[h\cap D]\cup[h\cap H]\cup[t\cap I] > [h\cap C]\cup[h\cap E]\cup[h\cap F]\cup[t\cap J]\\
&[t\cap A]\cup[t\cup D]\cup[h\cup G]\cup[h\cap J] > [t\cap B]\cup[t\cap C]\cup[h\cap H]\cup[h\cap I]\\
&[h\cap C]\cup[t\cap E]\cup[t\cap H]\cup[t\cup J] > [h\cap A]\cup[t\cap F]\cup[t\cap G]\cup[t\cap I]\\
&[t\cap B]\cup[t\cap C]\cup[t\cap F]\cup[t\cap G] > [t\cap A]\cup[t\cap D]\cup[t\cap E]\cup[t\cap H]
\end{aligned} \tag{A3.1.10}$$

But repeated application of (A3.1.1) to the inequalities (A3.1.10) establishes that, where S denotes $A\cup B\cup C\cup D\cup E\cup F\cup G\cup H\cup I\cup J$,

$$[h\cap S]\cup[t\cap S] > [h\cap S]\cup[t\cap S],$$

which contradicts P. 1.

A3.2. Appendix on Numerical Probability

In § **3.19** we introduced the conventional device of calibrating events in probability, by comparing them in probability with the various groups of results of a single spin of a fair roulette-wheel of large scope. The convention usually adopted is to attribute to any $G(t)$ which is the union of t disjoint single results of a spin of a roulette-wheel of scope n, the probability number t/n. In this appendix we shall establish some of the elementary rules governing the probability ordering and probability numbers of such results $G(t)$ and of their unions and intersections. In particular, we shall show that, if the probability ordering is consistent:

(*i*) where $P(G)$ denotes the probability number of G, $P(G) >, <,$ or $= P(G')$ according as $G >, <,$ or $= G'$; (A3.2.1) (*ii*) when G is the (certain) event that the union of all the disjoint single results of the spin occurs, $P(G) = 1$; when G is the (impossible) event that none of the possible single results of the spin occurs, $P(G) = 0$; (A3.2.2)

(*iii*) where G' and G are any two results

$$P(G \cup G') + P(G \cap G') = P(G) + P(G'); \qquad \text{(A3.2.3)}$$

(*iv*) where G and G' are results of two separate spins, and where G is irrelevant to G', then $P(G \cap G') = P(G) \,.\, P(G')$. (A3.2.4)

As a prelude to establishing (A3.2.1) we shall show that if G and G' contain the same number of disjoint single results of a spin and are themselves disjoint, then $G = G'$. (A3.2.5)

It is obvious that (A3.2.5) must hold, if G and G' each contain only one single result; but perhaps if they each contain a sufficiently large number of the single results, then even if each contains the same number, it is still possible for it not to be true that $G = G'$. Let t^* be the smallest number of single events in G for which this can happen. Then we can find G, G' disjoint and each containing only t^* disjoint single events, and yet $G = G'$ is false. Now let g, g' be any pair of single results, g contained in G, and g' contained in G'; let the union of the other single results in G be G_1 and the union of the other single results in G' be $G_1{}'$: then G_1 and $G_1{}'$ each contain only t^*-1 disjoint single results; hence, by the definition of t^*,

$$G_1 = G_1{}'. \qquad \text{(A3.2.6)}$$

Since obviously $\quad g = g', G \equiv G_1 \cup g$ and $G' = G_1{}' \cup g'$,
then if the probability ordering is consistent,

$$G = G_1 \cup g = G_1 \cup g' = G_1{}' \cup g' = G', \qquad \text{(A3.2.7)}$$

which contradicts our assumption that ($G = G'$) was false.

Hence it is not possible to find any smallest value t^*, such that there are two disjoint results G, G' each containing exactly t^* disjoint single results. Hence, whenever two disjoint results G, G' contain the same number of disjoint single results, then $G = G'$. We have thus established (A3.2.5). It follows that for any two disjoint results $P(G) = P(G')$ is a sufficient condition for $G = G'$. (A3.2.8)

This conclusion still holds when G and G' are not necessarily disjoint; for

$$G \equiv (G \cap G') \cup \{G \cap (\sim G')\} \text{ and } G' \equiv (G \cap G') \cup \{(\sim G) \cap G'\}. \qquad \text{(A3.2.9)}$$

Hence, if $P(G) = P(G')$, showing that G and G' contain equal numbers of disjoint single results, it will also be the case that $\{G \cap (\sim G')\}$ and

$\{(\sim G) \cap G'\}$ contain equal numbers of disjoint single results; hence

$$P\{G \cap (\sim G')\} = P\{(\sim G') \cap G'\}. \qquad \text{(A3.2.10)}$$

Hence, by (A3.2.8)

$$G \cap (\sim G') = (\sim G) \cap G'. \qquad \text{(A3.2.11)}$$

It now follows, from (A3.2.9), (A3.2.11) and the fact that the probability ordering is consistent, that

$$G = G', \qquad \text{(A3.2.12)}$$

so that we have extended the result (A3.2.8) to the case where G and G' are not necessarily disjoint.

Now suppose that

$$P(G) > P(G'); \qquad \text{(A3.2.12)}$$

then G must contain more disjoint single results than G', and hence we can find disjoint G_1 and g_1 such that

$$P(G_1) = P(G') \text{ and } P(g_1) > 0, \qquad \text{(A3.2.13)}$$

so that

$$G_1 = G' \text{ and } g_1 > 0, \qquad \text{(A3.2.14)}$$

where, in (A3.2.14), 0 denotes the impossible null-result.
Hence, since the ordering is consistent,

$$G > G' \quad \text{when} \quad P(G) > P(G'). \qquad \text{(A3.2.15)}$$

This completes the proof of (A3.2.1).

The truth of (A3.2.2) is evident from the definition of $P(G)$.

In order to prove (A3.2.3), suppose that G, G' and $(G \cap G')$ contain respectively t, t' and t'' disjoint single results. Then since

$$G \equiv G \cap G' \cup \{G \cap (\sim G')\}, \qquad \text{(A3.2.16)}$$

$G \cap (\sim G')$ contains $(t-t'')$ disjoint single results; and since

$$G \cup G' \equiv G' \cup \{G \cap (\sim G')\},$$

$G \cup G'$ contains $t+t'-t''$ disjoint single results. It then follows from the definition of $P(G)$ that

$$P(G \cup G')+P(G \cap G') = P(G)+P(G'), \qquad \text{(A.3.2.17)}$$

confirming (A3.2.3). In the particular case where G and G' are disjoint, $P(G \cap G') = 0$, so that $P(G \cup G') = P(G)+P(G')$. (A3.2.18)
In the next section, **A3.3**, we shall establish (A3.2.3).

A3.3. The Product Rule for the Probability of the Intersection of a Pair of Irrelevant Events

Let E denote any event which is irrelevant to the spin of a fair roulette-wheel of scope N, and suppose that the probability number of E is $\phi(E)$. Let R be a possible result of spinning the roulette-wheel, which has probability number r/N. We shall first show that the probability number of $[E \cap R]$ is given by

$$\phi[E \cap R] = \phi(E) \,.\, r/N. \qquad \text{(A3.3.1)}$$

Let σ_i denote the event that the spin of the roulette wheel results in the number i ($i = 1, 2, \ldots, N$). Then the event $[\sigma_1 \cup \sigma_2 \ldots \cup \sigma_r]$ has the same probability number r/N as has R. Hence, E being irrelevant to the result of the spin

$$\begin{aligned} \phi[E \cap R] &= \phi[E \cap (\sigma_1 \cup \sigma_2 \ldots \cup \sigma_r)] \\ &= \phi[(E \cap \sigma_1) \cup (E \cap \sigma_2) \ldots \cup (E \cap \sigma_N)] \\ &= \sum_{t=1}^{r} \phi[E \cap \sigma_t] = r\phi[E \cap \sigma_1]. \end{aligned} \qquad \text{(A3.3.2)}$$

Similarly, where I denotes the certain result $[\sigma_1 \cup \sigma_2 \ldots \cup \sigma_N]$,

$$\phi[E] = \phi[E \cap I] = N\phi[E \cap \sigma_1]. \qquad \text{(A3.3.3)}$$

Comparing (A3.3.2) and (A3.3.3), we obtain (A3.3.1).

Now let F be any event which has the probability number θ/N (θ an integer) and to which E is irrelevant. Then let T be a result with the roulette-wheel which has the probability θ/N. Then since

$$\phi[F] = \phi[T] \qquad \text{(A3.3.4)}$$

and since E is irrelevant both to F and to T

$$\phi[E \cap F] = \phi[E \cap T]. \qquad \text{(A3.3.5)}$$

But, by the general argument which confirmed (A3.3.1),

$$\phi[E \cap T] = \phi[E]\, \theta/N. \qquad \text{(A3.3.6)}$$

Hence where E is irrelevant to F, and F has the probability number $\phi[F] = \theta/N$ (θ an integer),

$$\phi[E \cap F] = \phi[E] \,.\, \phi[F]. \qquad \text{(A3.3.7)}$$

A3.4. Asymptotic Irrelevance

In section **3.18** the concept of asymptotic irrelevance was defined, and the paradox was mentioned that if E and E' were two asymptotically irrelevant events, the further knowledge that $E > \sim E$ and that $E' > \sim E'$ would be insufficient to enable one to establish that $E' \cap E \cap \{(\sim E') \cap (\sim E)\} > \{(E' \cap \sim E) \cup (\sim E' \cap E)\}$, namely that the occurrence of both events or neither event is more probable than the occurrence of only one.

The reason that this was called paradoxical is that if numerical probabilities could be assigned to E' and E, then their asymptotic irrelevance would, together with the data that $E' > \sim E'$ and that $E > \sim E$, enable one to prove that both or neither was more probable than only one. For one could argue as follows.

Let p' and p be the probability numbers of E' and E.

Since $$\text{Prob.}\,(E' \cup \sim E') = \text{Prob.}\,(E \cup \sim E) = 1, \quad \text{(A3.4.1)}$$

the probability numbers of $(\sim E')$ and of $(\sim E)$ are given by

$$P(\sim E') = 1-p', \text{ and } P(\sim E) = 1-p. \quad \text{(A3.4.2)}$$

Since $E' > \sim E'$ and $E > \sim E$, it follows that $p' > (1-p')$ and $p > (1-p)$.

So we may put $p' = \frac{1}{2}+d'$ and $p = \frac{1}{2}+d$ where $d' > 0$ and $d > 0$. (A3.4.3)

It is given that E' is asymptotically irrelevant to E, and hence, to any required degree of numerical accuracy,

$$P(E' \cap E) = P(E')P(E) = \tfrac{1}{4}+\tfrac{1}{2}(d+d')+dd'. \quad \text{(A3.4.4)}$$

Since $E' \cap E$ and $E' \cap (\sim E)$ are disjoint,

$$P(E' \cap E)+P(E' \cap \sim E) = P(E') = \tfrac{1}{2}+d'.$$

Hence

$$P(E' \cap \sim E) = \tfrac{3}{4}+\tfrac{1}{2}(d'-d)-dd'.$$

Similarly $P(\sim E' \cap E) = \frac{1}{4}+\frac{1}{2}(d-d')-d'd$

and

$$P(\sim E' \cap \sim E) = \tfrac{1}{4}-\tfrac{1}{2}(d'+d)+d'd.$$

By addition,

$$P(E' \cap E)+P(\sim E' \cap \sim E) = \tfrac{1}{2}+2dd'$$

and

$$P(E' \cap \sim E)+P(\sim E' \cap E) = \tfrac{1}{2}-2d'd,$$

so that

$$P\{(E' \cap E) \cup (\sim E' \cap \sim E)\} > P\{(E' \cap \sim E) \cup E(\sim E' \cup E)\},$$

and hence

$$(E' \cap E) \cup (\sim E' \cap \sim E) > \{(E' \cap \sim E) \cup (\sim E' \cap E)\}. \quad \text{(A3.4.5)}$$

This derivation depended, however, on the assumption that to E' and to E numerical probabilities could be attributed which would exceed $\frac{1}{2}$. The data do not justify this assumption. The following situation is logically possible on the basis of what we are allowed to assume.

For every fair $4N$ roulette-wheel, for every integer N however large,

$$G(2N+1) > E' = E > G(2N)$$

where $G(2N)$ and $G(2N+1)$ are unions of $2N$ and of $(2N+1)$ disjoint single results of the spin. In this case no probability number can be found for E' and E which will always satisfy

$$P\{G(2N+1)\} > P(E') = P(E) > PG(2N)$$

for all positive integers N, since

$$P(G(2N)) = \tfrac{1}{2}, \text{ and } P(G(2N+1) = \tfrac{1}{2}+\frac{1}{4N}.$$

In this case, where we know that $G(2N+1) > E' = E > G(2N)$, there is no valid logical argument to rule out the possibility

$$E' \cap \sim E = \sim E' \cap E > G(N),$$

in which case, choosing $G(N)$ and $g(N)$ as disjoint unions of N disjoint single results each, with $G(N) = g(N)$, it follows that

$$(E' \cap \sim E) \cup (\sim E' \cap E) > G(N) \cup g(N) = G(2N). \quad \text{(A3.4.6)}$$

But $(E' \cap E) \cup (\sim E' \cap \sim E) \cup (E' \cap \sim E) \cup (\sim E' \cap E) = G(4N)$, since both events are certain. Hence by the consistency postulate II (Chapter 2, § **8**), and the corollary (2.8.3),

$$\{(E' \cap E) \cup (\sim' E' \cap \sim E)\} < G(2N) < \{(E' \cap \sim E) \cup (\sim E' \cap E)\}. \quad \text{(A3.4.7)}$$

This possibility cannot be ruled out on the data available.

Exercises 3

3.1. Compare and contrast three different interpretations of the concept 'probability'. Discuss the relevance of each interpretation to economic problems.

3.2. Give an account of Ramsey's method of measuring probability.

3.3. Give examples of reasonable postulates of irrelevance which are used in (a) the natural sciences (b) economics.

3.4. Explain how the concept of a fair roulette-wheel enables the probabilities of a set of events to be calibrated in principle.

3.5. One normally attributes to the union of two disjoint results of an experiment a probability equal to the sum of their separate probabilities. Is this more than simply a convention? Why is it particularly convenient to do so?

In the following exercises the symbol I denotes 'is irrelevant to', and A, B and C denote possible events.

3.6. Does $(A\,I\,C) \cup (B\,I\,C)$ imply $(A \cup B)\,I\,C$? Supply proof or counterexample.

3.7. Does $(A\,I\,C) \cap (B\,I\,C)$ imply $(A \cup B)\,I\,C$? Supply proof or counterexample.

3.8. Does $(A\,I\,C) \cap (B\,I\,C)$ imply $(A \cap B)\,I\,C$? Supply proof or counterexample.

3.9. Investigate the effect, if any, of the further information that A and B are disjoint on the answers to Exercises 6, 7 and 8.

3.10. In a certain probability ordering, $H = T$, $A \geqslant B$ and $C \geqslant D$, and H and T are disjoint and irrelevant to the probability ordering of A, B, C, D, $A \cup C$, and $B \cup D$; moreover, C and D are irrelevant to the probability relation $H = T$. Prove that $A \cup C \geqslant B \cup D$. *Hint*: Show that

$$(A \cap H) \cup (C \cap H) \geqslant (B \cap H) \cup (D \cap T) \geqslant (B \cap H) \cup (D \cap H).$$

Chapter 4

A PROBABILITY MODEL FOR EVENTS

4.1. Events Represented by Vectors or by Sets of Points

In this chapter we shall consider a model for comparing the probabilities of 'events'. In this model events will be represented by column vectors with N elements, each equal either to one or to zero. N will be an integer characterising the model, but we shall not necessarily require that N be finite; for certain purposes it will be convenient to allow the event-vectors to contain an unlimited number of elements.

Since the representation of events by vectors is a novel device, we shall also refer to the analogous, but more familiar, representation of events by sets of points. When the reader has been given time to become familiar with the vector representation, the supplementary representation by sets of points will be abandoned.

4.2. A Model Relating to the Results of an Experiment

Consider, for example, a model relating to the N conceivable mutually exclusive results which may follow an experiment which is to be performed once only. These results, R_t ($t = 1, 2, \ldots, N$) may be represented by N distinct points P_t; alternatively they may be represented by N vectors $\boldsymbol{e}(t)$, where each vector $\boldsymbol{e}(t)$ contains N elements exactly, all of which are zeros except the t-th element, which is a one.

We may refer to these individual results R_t as *elementary events*. We extend the concept of an event to cover sets of alternative elementary events such as (R_3 or R_2 or R_8). This event would be represented by the set of three corresponding points (P_3, P_2 and P_8); alternatively this event may be represented by the *sum* of the three vectors $\boldsymbol{e}(3)$, $\boldsymbol{e}(2)$ and $\boldsymbol{e}(8)$, which represent the three elementary events, R_3, R_2 and R_8, namely the vector whose third, second and eighth elements are ones, and whose other elements are all zeros.

4.3. The Universal Set and the Empty Set

The event (R_1 or R_2 or . . . or R_N) containing all the elementary events R_t may be represented by the set of all the N points P_t; alternatively it may be represented by the vector, all of whose N elements are ones.

The set of all N points will be denoted by U and will be called the universal set; the vector consisting of ones will be called the unit vector and will be denoted by $\mathbf{1}$.

On the other hand the 'impossible' event, that *no* result of type R_t occurs will be represented by the *empty* set of no points: this empty set will be denoted by $\emptyset$. Alternatively the event of 'no result' may be represented by the null vector, all of whose N elements are zeros; this null vector will be denoted by $\mathbf{0}$.

In general, we shall often use the notation that events will be denoted by $E(t)$ $(t = 1, 2, \ldots)$ and the sets of points representing them by $S(t)$, and the vectors representing them by $\mathbf{j}(t)$.

4.4. The Intersection of Two or More Events

The event (both $E(1)$ and $E(2)$) consists of those elementary events R_k which are included both in $E(1)$ and in $E(2)$. This event (both $E(1)$ and $E(2)$) is called the *intersection* of the events $E(1)$ and $E(2)$ and it is denoted by $E(1) \cap E(2)$.

In the point diagram, the event $E(1) \cap E(2)$ is represented by the set of points common to the set $S(1)$, representing $E(1)$, and $S(2)$ representing $E(2)$. In the vector model, the event $E(1) \cap E(2)$ is represented by the product $\hat{\mathbf{j}}(1)\,\mathbf{j}(2)$, where $\mathbf{j}(1)$ and $\mathbf{j}(2)$ are the vectors representing the events $E(1)$ and $E(2)$, respectively and where $\hat{\mathbf{j}}(1)$ is the diagonal $N \times N$ matrix formed from $\mathbf{j}(1)$. Clearly the elements of the product vector $\hat{\mathbf{j}}(1)\,\mathbf{j}(2)$ are the products of the corresponding elements of $\mathbf{j}(1)$ and of $\mathbf{j}(2)$, so that whenever the corresponding elements of both $\mathbf{j}(1)$ and $\mathbf{j}(2)$ are ones, that of $\hat{\mathbf{j}}(1)\,\mathbf{j}(2)$ will also be a one; but in all other cases the element of $\hat{\mathbf{j}}(1)\,\mathbf{j}(2)$ will be a zero.

We may consider the intersection of n events, '$E(1)$ and $E(2)$ and $\ldots$ and $E(n)$', in similar fashion. This event may be denoted by $E(1) \cap E(2) \ldots \cap E(n)$, or more concisely by $\bigcap_{t=1}^{n} E(t)$. The event consists of just those elementary events which enter into every one of the events $E(1), E(2), \ldots, E(n)$. In the point diagram the intersection of $E(1), E(2), \ldots, E(n)$ is represented by the set of points common to all the sets $S(1), S(2), \ldots, S(n)$ representing the events $E(1), E(2), \ldots, E(n)$. In the vector model, the intersection of the n events represented by the n vectors $\mathbf{j}(1), \mathbf{j}(2), \ldots, \mathbf{j}(n)$ is itself represented by the product vector $\hat{\mathbf{j}}(1)\,\hat{\mathbf{j}}(2) \ldots \hat{\mathbf{j}}(n-1)\,\mathbf{j}(n)$ which may without ambiguity be denoted by $\bigcap_{t=1}^{n} \mathbf{j}(t)$. The typical element of this vector will be the product of the corresponding elements of the n vectors $\mathbf{j}(t)$ $(t = 1, 2, \ldots, n)$ and will accordingly be a one if *all* the corresponding elements of the vectors $\boldsymbol{j}(t)$ are ones; otherwise it will be a zero.

4.5. The Negation of an Event or of a Set of Events

The negation of an event $E(t)$ is commonly denoted by $\sim E(t)$ and it consists of all those elementary events which do *not* belong to the event $E(t)$.

In the point diagram, $\sim E(t)$ is represented by $\bar{S}(t)$ the set of points P_k which do *not* belong to the set $S(t)$ which represents the event $E(t)$.

In the vector model, the event $\sim E(t)$ is represented by the vector $(\mathbf{1}-\boldsymbol{j}(t))$ where $\boldsymbol{j}(t)$ is the vector representing the event $E(t)$. It is clear that the vector $(\mathbf{1}-\boldsymbol{j}(t))$ can be obtained from the vector $\boldsymbol{j}(t)$ by altering all its ones into noughts and all its noughts into ones.

Similarly, we may consider the negation of a set of events $E(t)$ $(t = 1, 2, \ldots, n)$. This negation may be denoted by $\bigcap_{t=1}^{n} (\sim E(t))$, and the event consists of all those elementary events which enter into *none* of the events $E(t)$ $(t = 1, 2, \ldots, n)$. In the point diagram $\bigcap_{t=1}^{n} (\sim E(t))$ may be represented by the set of those of the N points P_k which lie outside every one of the sets $S(t)$ $(t = 1, 2, \ldots, n)$ representing the events $E(t)$ $(t = 1, 2, \ldots, n)$. In the vector model, the event $\bigcap_{t=1}^{n} (\sim E(t))$ is represented by the vector $\bigcap_{t=1}^{n} (\mathbf{1}-\boldsymbol{j}(t))$ which is the abbreviated notation for the vector product $\{\hat{\mathbf{1}}-\hat{\boldsymbol{j}}(1)\}\{\hat{\mathbf{1}}-\hat{\boldsymbol{j}}(2)\} \ldots \{\hat{\mathbf{1}}-\hat{\boldsymbol{j}}(n-1)\}\{\mathbf{1}-\boldsymbol{j}(n)\}$. Any element of this vector will be a one if, and only if, the corresponding elements of *all* the vectors $\boldsymbol{j}(t)$ $(t = 1, 2, \ldots, n)$ are zeros: otherwise the element will be a zero.

4.6. The Union of Two or More Events

The event '$E(1)$ and/or $E(2)$' is called the *union* of $E(1)$ and $E(2)$ and is denoted by $E(1) \cup E(2)$. This event consists of all those elementary events which belong to either or both of the events $E(1)$ and $E(2)$.

Clearly the event $E(1) \cup E(2)$ is the negation of the negation of the pair of events $E(1)$, $E(2)$. In the point diagram, $E(1) \cup E(2)$ is represented by the set of all points which belong to either or both the sets $S(1)$, $S(2)$ which represent the events $E(1)$, $E(2)$.

In the vector model, $E(1) \cup E(2)$ is represented by the vector

$$\{\boldsymbol{j}(1) \cup \boldsymbol{j}(2)\} = \mathbf{1}-\{\hat{\mathbf{1}}-\hat{\boldsymbol{j}}(1)\}\{\mathbf{1}-\boldsymbol{j}(2)\}.$$

On reduction

$$\{\boldsymbol{j}(1) \cup \boldsymbol{j}(2)\} = \boldsymbol{j}(1)+\boldsymbol{j}(2)-\hat{\boldsymbol{j}}(1)\,\boldsymbol{j}(2). \tag{4.6.1}$$

It is easy to verify that any element of this vector will be a one if, and only if, one at least of the corresponding elements of the vectors $\boldsymbol{j}(1)$, $\boldsymbol{j}(2)$ is a one, and will be zero if, and only if, both these corresponding elements of $\boldsymbol{j}(1)$, $\boldsymbol{j}(2)$ are zero.

Similarly, we may consider the union of the set of n events $E(t)$ $(t = 1, 2, \ldots, n)$. This union denoted by $E(1) \cup E(2) \ldots \cup E(n)$, or more shortly by $\bigcup_{t=1}^{n} E(t)$, consists of the set of all those elementary events which enter into any one (or more) of the events $E(t)$ $(t = 1, 2, \ldots, n)$. The union, $\bigcup_{t=1}^{n} E(t)$, is the negation of the negation of the set of events $E(t)$ $(t = 1, 2, \ldots, n)$.

In the point diagram the union $\bigcup_{t=1}^{n} E(t)$ consists of all the points which belong to any one (or more) of the sets $S(t)$ $(t = 1, 2, \ldots, n)$ which represents the events $E(t)$.

In the vector model, the union $\bigcup_{t=1}^{n} E(t)$ is represented by the vector

$$\bigcup_{t=1}^{n} \boldsymbol{j}(t) = \mathbf{1}-\bigcap_{t=1}^{n} \{\mathbf{1}-\boldsymbol{j}(t)\} \tag{4.6.2}$$

where the $\boldsymbol{j}(t)$ represent the events $E(t)$. It is clear that the typical element of the vector $\{\bigcup_{t=1}^{n} \boldsymbol{j}(t)\}$ will be a zero if, and only if, the corresponding element of each and every one of the $\boldsymbol{j}(t)$ $(t = 1, 2, \ldots, n)$ is a zero; if any one or more of these corresponding elements is a one, so also will be that element of the vector $\bigcup_{t=1}^{n} \boldsymbol{j}(t)$.

4.7. Implication and Identity

If $E(2)$ contains every one of the elementary events which are contained in $E(1)$, as well possibly as some other elementary events, then $E(1)$ is said to *imply* $E(2)$, and this implication is denoted by $E(1) \subset E(2)$.

In the point diagram such an implication would be represented if every one of the points belonging to the set $S(1)$ representing $E(1)$ belonged also to the set $S(2)$ representing the event $E(2)$.

In the vector model, the implication $E(1) \subset E(2)$ would be represented if the vectors $\boldsymbol{j}(1)$ and $\boldsymbol{j}(2)$ representing $E(1)$ and $E(2)$ satisfied the inequality $\boldsymbol{j}(1) \leqslant \boldsymbol{j}(2)$; for in this case, and only in this case, whenever an element of $\boldsymbol{j}(1)$ is a one, so also is the corresponding element of $\boldsymbol{j}(2)$.

If both $E(1) = E(2)$ and $E(2) \subset E(1)$, then $E(1)$ and $E(2)$ must each consist of the same set of elementary events. This state of affairs is denoted by $E(1) \equiv E(2)$.

In the point diagram, if $E(1) \equiv E(2)$, then the sets $S(1)$ and $S(2)$ representing $E(1)$ and $E(2)$ will be one and the same set of points. In the vector model if $E(1) \equiv E(2)$ then the vectors $\boldsymbol{j}(1)$ and $\boldsymbol{j}(2)$ representing $E(1)$ and $E(2)$ satisfy $\boldsymbol{j}(1) = \boldsymbol{j}(2)$. In other words, if $E(1) \equiv E(2)$, then $E(1)$ and $E(2)$ are represented by the same vector.

4.8. Relations between Events Regarded as Logical Events

It is sometimes convenient to consider relations between events such as $E(1) \subset E(2)$ or $E(1) \equiv E(2)$ even when they *need* not be true. We may in fact regard such a relation as $E(1) \equiv E(2)$ as forming a special kind of event, namely a *logical* event, which like many other events, may or may not occur or turn out to be true.

By the relation $E(1) \equiv E(2)$ we mean that set of elementary events which enter into both or neither of $E(1)$ and $E(2)$. In the point diagram the relation or logical event $E(1) \equiv E(2)$ will be represented by the set of points which lie either wholly outside $S(1)$ and $S(2)$ or else belong both to $S(1)$ and to $S(2)$, the two sets of points representing $E(1)$ and $E(2)$.

In the vector model, the relation $E(1) \equiv E(2)$ will be denoted by

$$\boldsymbol{r}\{\boldsymbol{j}(1) = \boldsymbol{j}(2)\}$$

which stands for the vector defined by

$$\boldsymbol{r}\{\boldsymbol{j}(1) = \boldsymbol{j}(2)\} = \hat{\boldsymbol{j}}(1)\boldsymbol{j}(2) + \{\hat{\mathbf{1}} - \hat{\boldsymbol{j}}(1)\}\{\mathbf{1} - \boldsymbol{j}(2)\} \tag{4.8.1}$$

which is the union of the intersections of $\boldsymbol{j}(1)$ and $\boldsymbol{j}(2)$ and of the intersection of the negations of $\boldsymbol{j}(1)$ and $\boldsymbol{j}(2)$. It will be seen that any element of $\boldsymbol{r}\{\boldsymbol{j}(1) = \boldsymbol{j}(2)\}$ will be a one if the corresponding elements of $\boldsymbol{j}(1)$ and of $\boldsymbol{j}(2)$ are equal, but will be a zero if the corresponding elements of $\boldsymbol{j}(1)$ and of $\boldsymbol{j}(2)$ are unequal.

Similarly, we mean by the relation $E(1) \subset E(2)$ that set of elementary events which either do not enter the event $E(1)$ or else enter both $E(1)$ and $E(2)$. In the point diagram, the relation or logical event $E(1) \subset E(2)$ is represented by those points which either lie outside $S(1)$, the set representing $E(1)$, or else lie both in $S(1)$ and in $S(2)$, the set representing $E(2)$.

In the vector model the relation $E(1) \subset E(2)$ is represented by a vector $\boldsymbol{r}\{\boldsymbol{j}(1) \leqslant \boldsymbol{j}(2)\}$ defined by the equation

$$\boldsymbol{r}\{\boldsymbol{j}(1) \leqslant \boldsymbol{j}(2)\} = \boldsymbol{1} - \boldsymbol{j}(1) + \hat{\boldsymbol{j}}(1)\boldsymbol{j}(2). \tag{4.8.2}$$

Clearly, an element of this vector $\boldsymbol{r}\{\boldsymbol{j}(1) \leqslant \boldsymbol{j}(2)\}$ will be a one whenever the corresponding element of $\boldsymbol{j}(1)$ is a zero, and whenever the corresponding element of $\boldsymbol{j}(2)$ is a one, and will be a zero only when the corresponding element of $\boldsymbol{j}(1)$ is a one but the corresponding element of $\boldsymbol{j}(2)$ is a zero.

In fact the element $r_t\{\boldsymbol{j}(1) \leqslant \boldsymbol{j}(2)\}$ will be a one or a zero according as the relation

$$j_t(1) \leqslant j_t(2)$$

between the corresponding elements of $\boldsymbol{j}(1)$ and $\boldsymbol{j}(2)$ does or does not hold.

The reader should note that the relation $\boldsymbol{r}(\boldsymbol{j} = \boldsymbol{1})$ must be the same vector as $\boldsymbol{j}$, provided $\boldsymbol{j}$ represents any event. This is because

$$\boldsymbol{r}(\boldsymbol{j} = \boldsymbol{1}) = \hat{\boldsymbol{j}}\boldsymbol{1} + (\hat{\boldsymbol{1}} - \hat{\boldsymbol{j}})(\boldsymbol{1} - \boldsymbol{1}) = \hat{\boldsymbol{j}}\boldsymbol{1} = \boldsymbol{j}. \tag{4.8.3}$$

4.9. Disjoint Sets of Events and Exhaustive Sets of Events

A pair of events $E(1)$, $E(2)$, is said to be disjoint if the event $E(1) \cap E(2)$ contains no elementary event, i.e. if no elementary event belongs both to $E(1)$ and to $E(2)$. A set of events $E(t)$ $(t = 1, 2, \ldots, n)$ is said to be disjoint if every pair of the set is disjoint.

In the point diagram a disjoint set of events will be represented by sets of points $S(t)$, no pair of which have any point in common. In the vector model a disjoint set of events $E(t)$ will be represented by vectors $\boldsymbol{j}(t)$ such that for every pair of values t, t' $(0 < t < t' \leqslant n)$, $\hat{\boldsymbol{j}}(t)\boldsymbol{j}(t') = \boldsymbol{0}$. This condition can also be expressed as

$$\sum_{t=1}^{n} \boldsymbol{j}(t) \leqslant \boldsymbol{1}. \tag{4.9.1}$$

A set of events $E(t)$ $(t = 1, 2, \ldots, n)$ is said to be exhaustive if their union includes all the elementary events. We may express this condition as

$$\bigcup_{t=1}^{n} E(t) = U.$$

Such an exhaustive set of events would be represented in the point diagram by sets $S(t)$ of points $(t = 1, 2, \ldots, n)$, such that every point was included in at least one set.

In the vector model, the vectors $\boldsymbol{j}(t)$ $(t = 1, 2, \ldots, n)$ representing an exhaustive set of events $E(t)$ must satisfy

$$\bigcap_{t=1}^{n} \{\mathbf{1}-\boldsymbol{j}(t)\} = \mathbf{0}. \qquad (4.9.2)$$

A necessary and sufficient condition for this is

$$\sum_{t=1}^{n} \boldsymbol{j}(t) \geqslant \mathbf{1}. \qquad (4.9.3)$$

4.10. Definitions for the Vector Model of Events

We may now recapitulate the details of the vector model of events which we have just described, and set them out in the form of the notation and the list of definitions.

N denotes the number of elementary events $\boldsymbol{e}(t)$. These may be thought of as the distinct conceivable results of an experiment which is performed once only.

1 denotes the column vector of N ones.

0 denotes the column vector of N zeros.

$\boldsymbol{e}(t)$ denotes the column vector whose t-th element is a one and whose $N-1$ other elements all are zeros.

D. 1 An elementary event is any of the vectors $\boldsymbol{e}(t)$ $(t = 1, \ldots, N)$.

D. 2 An event is a column vector of N elements each of which is a one or a zero.

D. 3 The intersection of the n events $\boldsymbol{j}(t)$ $(t = 1, 2, \ldots, n)$ is the vector $\bigcap_{t=1}^{n} \boldsymbol{j}(t) = \hat{\boldsymbol{j}}(1)\,\hat{\boldsymbol{j}}(2) \ldots \hat{\boldsymbol{j}}(n-1)\,\boldsymbol{j}(n)$.

D. 4 The negation of the event $\boldsymbol{j}$ is the event $(\mathbf{1}-\boldsymbol{j})$. The negation of the set of events $\boldsymbol{j}(t)$ $(t = 1, 2, \ldots, n)$ is the event

$$\bigcap_{=1}^{n} \{\mathbf{1}-\boldsymbol{j}(t)\}.$$

D. 5 The union of the events $\boldsymbol{j}(1), \boldsymbol{j}(2)$ is the event $\boldsymbol{j}(1) \cup \boldsymbol{j}(2) = \mathbf{1}-\{\hat{\mathbf{1}}-\hat{\boldsymbol{j}}(1)\}\{\mathbf{1}-\boldsymbol{j}(2)\} = \boldsymbol{j}(1)+\boldsymbol{j}(2)-\hat{\boldsymbol{j}}(1)\,\boldsymbol{j}(2)$. The union of the n events $\boldsymbol{j}(t)$ $(t = 1, 2, \ldots, n)$ is the event $\bigcup_{t=1}^{n} \boldsymbol{j}(t) = \mathbf{1}- \bigcap_{t=1}^{n} \{\mathbf{1}-\boldsymbol{j}(t)\}$. This is the negation of the negation of the set of n events.

D. 6 The event $\boldsymbol{j}(1)$ is said to imply the event $\boldsymbol{j}(2)$ if, and only if, $\boldsymbol{j}(1) \leqslant \boldsymbol{j}(2)$.

D. 7 The logical events $\{\boldsymbol{j}(1) \leqslant \boldsymbol{j}(2)$ and $\{\boldsymbol{j}(1) = \boldsymbol{j}(2)\}$ are denoted by $\boldsymbol{r}\{\boldsymbol{j}(1) \leqslant \boldsymbol{j}(2)\}$ and by $\boldsymbol{r}\{\boldsymbol{j}(1) = \boldsymbol{j}(2)\}$. These vectors are defined to satisfy

$$\boldsymbol{r}\{\boldsymbol{j}(1) \leqslant \boldsymbol{j}(2)\} = \mathbf{1} - \boldsymbol{j}(1) + \hat{\boldsymbol{j}}(1)\,\boldsymbol{j}(2)$$
$$\boldsymbol{r}\{\boldsymbol{j}(1) = \boldsymbol{j}(2)\} = \hat{\boldsymbol{j}}(1)\,\boldsymbol{j}(2) + \{\hat{\mathbf{1}} - \hat{\boldsymbol{j}}(1)\}\{\mathbf{1} - \boldsymbol{j}(2)\}.$$

D. 7 implies that ones appear as elements of a logical event vector, whenever the corresponding elements of the vectors involved in the relation satisfy that relation, and noughts appear as the other elements.

D. 8 The set of events $\boldsymbol{j}(t)$ $(t = 1, 2, \ldots, n)$ is said to be disjoint if

$$\sum_{t=1}^{n} \boldsymbol{j}(t) \leqslant \mathbf{1}.$$

D. 9 The set of events $\boldsymbol{j}(t)$ $(t = 1, 2, \ldots, n)$ is said to be exhaustive if

$$\sum_{t=1}^{n} \boldsymbol{j}(t) \geqslant \mathbf{1}.$$

4.11. Probability as a Measure or as a Vector

With each of the elementary events $\boldsymbol{e}(t)$ $(t = 1, 2, \ldots, N)$ we associate a probability p_t. These probabilities must satisfy the conventions

$$p_t \geqslant 0 \quad (t = 1, 2, \ldots, N); \quad \sum_{t=1}^{N} p_t = 1.$$

Where events are represented by points, the probabilities are represented by a measure over the points. The measure associated with each point is the probability p_t of the corresponding elementary event. The probability of any event is the measure over the set of points representing the event. This measure will equal the total of the probabilities of the elementary events of which the event is composed.

Where events are represented by vectors, probability also is represented by a row vector $\boldsymbol{p}'$ whose elements are the N probabilities of the elementary events. The probability of any event $\boldsymbol{j}$ is then found by premultiplying the vector $\boldsymbol{j}$ by the probability vector $\boldsymbol{p}'$. In fact the probability of the event $\boldsymbol{j}$ is simply $\boldsymbol{p}'\boldsymbol{j}$. Clearly, this probability is just the sum of the probabilities of the elementary events of which $\boldsymbol{j}$ is the sum. For example, if $E(1)$ consists of the results R_3, R_2, R_8, then the probability measure over the corresponding set of points $S(1)$, namely the points P_3, P_2 and P_8 is $p_3+p_2+p_8$ and this is the probability of $E(1)$. Moreover if $E(1)$ is represented by a vector $\boldsymbol{j}(1)$, then $\boldsymbol{j}(1) = \boldsymbol{e}(3)+\boldsymbol{e}(2)+\boldsymbol{e}(8)$ and $\boldsymbol{j}(1)$ has its third second and eighth elements as ones and the other elements as zeros. The probability of $j(1)$ is given by

$$p'j(1) = \sum_{t=1}^{N} p_t j_t(1) = p_2 + p_3 + p_8.$$

Suppose, for example, we wish to obtain the probability of $\{j(1) \cup j(2)\}$ the union of two events $j(1), j(2)$. By (6.1)

$$j(1) \cup j(2) = j(1) + j(2) - \hat{j}(1)\, j(2). \qquad (4.11.1)$$

Hence the required probability is given as

$$p'\{j(1) \cup j(2)\} = p'j(1) + p'j(2) - p'\hat{j}(1)\, j(2). \qquad (4.11.2)$$

This equation is sometimes quoted in the form

$$p'\{j(1) \cup j(2)\} + p'\hat{j}(1)\, j(2) = p'j(1) + p'j(2) \qquad (4.11.3)$$

which states that the sum of the probabilities of the union and of the intersection of any two events equals the sum of the probabilities of the two events.

(4.11.2) expresses the same relation as was given in (v) of § **3.20**.

4.12. The Probability of a Logical Relation

The probability of any logical event is to be found in exactly the same way as that of any other event, namely by adding the probabilities of the elementary events of which it is the union, namely those elementary events which satisfy the logical relation.

In the point representation, the probability of a logical event is the probability measure over the set of points which represents that logical event. In the vector model, the probability of any logical event $r(\text{--})$ is found, as usual by premultiplication by the probability row-vector p', as $p'r(\text{--})$.

For example, if $j(1)$ and $j(2)$ are events, then the probability of the logical event $r\{j(1) \leqslant j(2)\}$ is simply $p'r\{j(1) \leqslant j(2)\}$.

4.13. Conditional Probability

One purpose of collecting experimental evidence is to enable probability orderings to be revised in the light of such evidence. It is convenient to make provision for this process by including in our model conditional-probability vectors as well as the main probability row-vector p'.

Corresponding to each logical relation $r(\text{--})$ we therefore define the conditional probability vector $p'_{\mid r(\text{--})}$ which is to replace the probability vector p' whenever the relation $r(\text{--})$ is given or assumed to be true. The most usual case is where the relation $r(\text{--})$ is of the form $r(j = \mathbf{1})$

asserting that some event $\boldsymbol{j}$ is known to have occurred. The rule which we shall adopt for finding any such conditional probability vector is embodied in the following:

$$\boldsymbol{p}'_{|\,\boldsymbol{r}(\text{--})} = \{\boldsymbol{p}'\boldsymbol{r}(\text{--})\}^{-1}\,\boldsymbol{p}'\hat{\boldsymbol{r}}(\text{--}). \tag{4.13.1}$$

The divisor is here the probability of the relation $\boldsymbol{r}(\text{--})$, whereas the numerator is a row vector; $\boldsymbol{p}'_{|\,\boldsymbol{r}(\text{--})}$ is thus a row vector with the properties

$$\boldsymbol{p}'_{|\,\boldsymbol{r}(\text{--})} \geqslant \mathbf{0} \quad \text{and} \quad \boldsymbol{p}'_{|\,\boldsymbol{r}(\text{--})}\mathbf{1} = 1. \tag{4.13.2}$$

These properties enable the conditional probability vector $\boldsymbol{p}'_{|\,\boldsymbol{r}(\text{--})}$ to replace the main probability vector $\boldsymbol{p}'$ when it is given that the relation $\boldsymbol{r}(\text{--})$ holds.

4.14. Probability Conditional on the Occurrence of an Event *k*

Frequently one may wish to consider the conditional probability, given that some specific event $\boldsymbol{k}$ has occurred. The appropriate probability vector to use in this case is the conditional probability vector

$$\boldsymbol{p}'_{|\,\boldsymbol{r}(\boldsymbol{k}=\mathbf{1})} = \{\boldsymbol{p}'\boldsymbol{r}(\boldsymbol{k} = \mathbf{1})\}^{-1}\,\boldsymbol{p}'\hat{\boldsymbol{r}}\{\boldsymbol{k} = 1\} \tag{4.14.1}$$

since $\boldsymbol{r}(\boldsymbol{k} = \mathbf{1})$ is the logical relation that $\boldsymbol{k}$ is true or known. Now the vector $\boldsymbol{r}(\boldsymbol{k} = \mathbf{1})$ is, as we have noted above, identical with the vector $\boldsymbol{k}$, since $\boldsymbol{k}$ is an event. Hence,

$$\boldsymbol{p}'_{\boldsymbol{r}(\boldsymbol{k}=\mathbf{1})} = (\boldsymbol{p}'\boldsymbol{k})^{-1}\,\boldsymbol{p}'\hat{\boldsymbol{k}}. \tag{4.14.2}$$

It is a natural and convenient notation to abbreviate $\boldsymbol{p}'_{|\,\boldsymbol{r}(\boldsymbol{k}=\mathbf{1})}$ to the simpler form $\boldsymbol{p}'_{|\,\boldsymbol{k}}$ and in the following pages we shall adopt this convention. (4.14.2) may then be written

$$\boldsymbol{p}'_{|\,\boldsymbol{k}} = (\boldsymbol{p}'\boldsymbol{k})^{-1}\,\boldsymbol{p}'\hat{\boldsymbol{k}}. \tag{4.14.3}$$

4.15. The Justification of the Formulae for Conditional Probabilities

The justification for the formulae (4.13.1) and (4.14.3) is that they ensure that the occurrence of an event $\boldsymbol{k}$ cannot affect the probability ordering of the two joint events $\hat{\boldsymbol{k}}\boldsymbol{j}$ and $\hat{\boldsymbol{k}}\boldsymbol{m}$, where $\boldsymbol{j}$ and $\boldsymbol{m}$ are any two events.

Originally the probabilities of the two joint events are

$$\boldsymbol{p}'\hat{\boldsymbol{k}}\boldsymbol{j} \quad \text{and} \quad \boldsymbol{p}'\hat{\boldsymbol{k}}\boldsymbol{m}. \tag{4.15.1}$$

Then if $\boldsymbol{k}$ were to occur, the new (conditional) probabilities of the two joint events would be

$$\boldsymbol{p}'_{|\,\boldsymbol{k}}\hat{\boldsymbol{k}}\boldsymbol{j} = (\boldsymbol{p}'\boldsymbol{k})^{-1}\boldsymbol{p}'\hat{\boldsymbol{k}}\hat{\boldsymbol{k}}\boldsymbol{j} \quad \text{and} \quad \boldsymbol{p}'_{|\,\boldsymbol{k}}\hat{\boldsymbol{k}}\boldsymbol{m} = (\boldsymbol{p}'\boldsymbol{k})^{-1}\boldsymbol{p}'\hat{\boldsymbol{k}}\hat{\boldsymbol{k}}\boldsymbol{m}. \tag{415.2.}$$

Since $\boldsymbol{k}$ is an event, $\hat{\boldsymbol{k}}$ is a diagonal matrix composed of zeros and ones. Hence

$$\hat{\boldsymbol{k}}\hat{\boldsymbol{k}} = \hat{\boldsymbol{k}} \tag{4.15.3}$$

and the conditional probabilities (4.15.2) are

$$(\boldsymbol{p}'\boldsymbol{k})^{-1}\boldsymbol{p}'\hat{\boldsymbol{k}}\boldsymbol{j} \quad \text{and} \quad (\boldsymbol{p}'\boldsymbol{k})^{-1}\boldsymbol{p}'\hat{\boldsymbol{k}}\boldsymbol{m}, \tag{4.15.4}$$

so that, provided $\boldsymbol{p}'\boldsymbol{k} > 0$, the conditional probabilities satisfy

$$\boldsymbol{p}'_{|\,\boldsymbol{k}}\hat{\boldsymbol{k}}\boldsymbol{j} >, =, \text{or} < \boldsymbol{p}'_{|\,\boldsymbol{k}}\hat{\boldsymbol{k}}\boldsymbol{m}$$

according as

$$\boldsymbol{p}'\hat{\boldsymbol{k}}\boldsymbol{j} >, =, \text{or} < \boldsymbol{p}'\hat{\boldsymbol{k}}\boldsymbol{m}. \tag{4.15.5}$$

Moreover, the factor $(\boldsymbol{p}'\boldsymbol{k})^{-1}$ in expression (4.14.3) ensures that

$$\boldsymbol{p}'_{|\,\boldsymbol{k}}\boldsymbol{k} = (\boldsymbol{p}'\boldsymbol{k})^{-1}\boldsymbol{p}'\hat{\boldsymbol{k}}\boldsymbol{k} = (\boldsymbol{p}'\boldsymbol{k})^{-1}\boldsymbol{p}'\boldsymbol{k} = \mathbf{1}. \tag{4.15.6}$$

4.16. The Independence of Two Events

In a probability model, independence is the property which is supposed to correspond to irrelevance in the real situations which the model represents. The definition of the independence of two events in the model is chosen, therefore, so as to correspond with the product rule found in the Appendix to Chapter 3, for the probability number of the intersection of an event E with a second event F to which it is irrelevant.

Two events $\boldsymbol{j}(1)$ and $\boldsymbol{j}(2)$ are said to be independent if, and only if, the probability of their intersection (i.e. of their joint occurrence) equals the product of their individual probabilities. This condition may be written as

$$\boldsymbol{p}'\hat{\boldsymbol{j}}(1)\,\boldsymbol{j}(2) = \boldsymbol{p}'\boldsymbol{j}(1)\,\boldsymbol{p}'\boldsymbol{j}(2) \tag{4.16.1}$$

It will be seen that the condition (A3.2.7) found in the Appendix to Chapter 3, is exactly equivalent to this (4.16.1) condition for independence.

Provided $0 < \boldsymbol{p}'\boldsymbol{j}(1) < 1$, this condition will hold if, and only if,

$$\boldsymbol{p}'_{|\,\boldsymbol{j}(1)}\boldsymbol{j}(2) = \boldsymbol{p}'_{|\,\{\mathbf{1}-\boldsymbol{j}(1)\}}\boldsymbol{j}(2) = \boldsymbol{p}'\boldsymbol{j}(2). \tag{4.16.2}$$

It follows that when the event $\boldsymbol{j}(1)$ has a probability strictly between zero and one, the pair of events $\boldsymbol{j}(1)$ and $\boldsymbol{j}(2)$ will be independent if, and only if, the conditional probability for $\boldsymbol{j}(2)$, given that $\boldsymbol{j}(1)$ has occurred, is the same as the conditional probability for $\boldsymbol{j}(2)$, given that $\boldsymbol{j}(1)$ has not occurred, and both equal the ordinary (unconditional) probability of $\boldsymbol{j}(2)$.

One could express the condition for the independence of $\boldsymbol{j}(1)$ and $\boldsymbol{j}(2)$ more loosely by saying that the occurrence or non-occurrence of $\boldsymbol{j}(1)$ must not affect the probability of $\boldsymbol{j}(2)$.

4.17. The Independence of a Pair of Disjoint Exhaustive Sets of Events

The pair of disjoint exhaustive sets of events $\boldsymbol{j}(r)$, $\boldsymbol{k}(s)$ ($r = 1, 2, \ldots, l$; $s = 1, 2, \ldots, n$) are said to be independent if every pair of events containing one from each set is independent.

In view of the definition of independence given in the last section, this means that the two disjoint exhaustive sets are independent if, and only if, for every r, s with $1 \leqslant r \leqslant l$, $1 \leqslant s \leqslant n$;

$$\boldsymbol{p}'\boldsymbol{j}(r)\,\boldsymbol{k}(s) = \boldsymbol{p}'\boldsymbol{j}(r)\,\boldsymbol{p}'\boldsymbol{k}(s). \tag{4.17.1}$$

4.18. The Independence of m Disjoint Exhaustive Sets of Events

The m disjoint exhaustive sets of events $\boldsymbol{j}_t(S_t)$ ($t = 1, 2, \ldots, m$) are independent if, and only if, for every permissible set of values S_t of s_t ($t = 1, 2, \ldots, m$)

$$\boldsymbol{p}'\left\{\bigcap_{t=1}^{m} \boldsymbol{j}_t(S_t)\right\} = \prod_{t=1}^{m} \{\boldsymbol{p}'\boldsymbol{j}_t(S_t)\}. \tag{4.18.1}$$

One may express this condition verbally as follows. The probability of each intersection of m events, one from each of the independent sets, must always equal the product of the separate probabilities of these events.

4.19. Chance Events

The elementary theory of probability has largely been concerned with a very special kind of event, which has been aptly described by Professor Harold Jeffreys as a *chance event*.

A chance event is to be thought of as a sequence of independent events $\boldsymbol{j}(t)$ ($t = 1, 2, \ldots, n$) each with the same numerical probability $\boldsymbol{p}'\boldsymbol{j}(t)$.

In our model, chance events are useful for representing such events as the toss of a coin to get heads or the throw of a die to get a six. Thus $\boldsymbol{j}(t)$ and $(\boldsymbol{1}-\boldsymbol{j}(t))$ might represent 'heads' and 'tails' in the t-th toss of a coin, or they might represent '6' or 'not-6' in the t-th throw of a die.

4.20. Repeatable Experiments

m repeated performances of an experiment with n possible results may be represented in our model by m independent disjoint exhaustive sets of n events $\mathbf{j}_r(t)$ $(t = 1, 2, \ldots, n)$ $(r = 1, 2, \ldots, m)$,

where $\mathbf{p}'\mathbf{j}_r(t) = g(t)$ independent of $\mathbf{r}$. (4.20.1)

A chance event may be regarded as a special case of a repeatable experiment with $n = 2$.

The drawing of one object at random from n objects may be regarded as a repeatable experiment with

$$\mathbf{p}'\mathbf{j}_r(t) = g(t) = (1/n) \tag{4.20.2}$$

for each $t = 1, 2, \ldots, n$ and each $r = 1, 2, \ldots, m$.

The event $\bigcap_{r=1}^{m} \mathbf{j}_r(t_r)$ would represent the draw of the sample of m members t_r $(r = 1, 2, \ldots, m)$.

4.21. Simple Random Sample with Replacement

Suppose that a bag contains balls of m colours, C_t $(t = 1, 2, \ldots, m)$. Let the numbers of balls of each colour be v_t $(t = 1, 2, \ldots, m)$, where $\sum_{t=1}^{m} v_t = v$ and $v_t = vP_t$ $(t = 1, 2, \ldots, m)$.

Then we may regard the drawing of a random sample of k balls, (replacing each ball before drawing the next ball), as a set of k performances of a repeatable experiment which has m possible results corresponding to the colours of the balls.

Let $\mathbf{j}_r(t)$ be the event that colour C_t is drawn at the r-th draw. Then we should require

$$\mathbf{p}'\mathbf{j}_r(t) = P_t \ (t = 1, 2, \ldots, m; \ r = 1, 2, \ldots, k). \tag{4.21.1}$$

We should also require that the sets $\mathbf{j}_r(t)$ $(r = 1, 2, \ldots, k)$ should be independent.

4.22. Definitions

In section **4.10**, we have given nine definitions relating to our vector model for events. In later sections we have set up a model for the *probability of events.* In this section we shall set out the main definitions which we have used in these later sections.

D. 10 The probability row vector $\mathbf{p}'$ contains N elements and satisfies the relations $\mathbf{p}' > 0 \quad (\mathbf{p}'\mathbf{1} = 1)$.

D. 11 The probability of any event $\mathbf{j}$ is $\mathbf{p}'\mathbf{j}$; the probability of any logical event $\mathbf{r}(\text{--})$ is $\mathbf{p}'\mathbf{r}(\text{--})$.

D. 12 Where $\boldsymbol{r}(\text{--})$ is any logical relation, the conditional probability vector $\boldsymbol{p}'_{|\boldsymbol{r}(\text{--})}$ is defined as $(\boldsymbol{p}'\boldsymbol{r})^{-1} \,.\, \boldsymbol{p}'\hat{\boldsymbol{r}}$.

D. 13 Where k is any event $\boldsymbol{p}'_{|k}$ denotes $\boldsymbol{p}'_{|\boldsymbol{r}(k=1)}$.

D. 14 The m disjoint exhaustive sets of events $\boldsymbol{j}_t(s_t)$ $(t = 1, 2, \ldots, m)$ are independent if, and only if, for every permissible set of values S_t of s_t $(t = 1, 2, \ldots, m)$

$$\boldsymbol{p}'\left\{\bigcap_{t=1}^{m} \boldsymbol{j}_t(S_t)\right\} = \prod_{t=1}^{m} \{\boldsymbol{p}'\boldsymbol{j}_t(S_t)\}$$

D. 15 The independent disjoint exhaustive sets of events $\boldsymbol{j}_t(s)$ $(s = 1, 2, \ldots, n;\ t = 1, 2, \ldots, m)$ constitute a repeated experiment if, and only if, for each $s = 1, 2, \ldots, n$, $\boldsymbol{p}'\boldsymbol{j}_t(s)$ is independent of t. A chance event is the special case of such a repeated experiment when $n = 2$.

Appendix to Chapter 4

VECTORS AND MATRICES

A4.1. Introduction

It is expected that most readers of this book will have some former acquaintance with vectors and matrices, and therefore no attempt is made to provide a systematic treatment of them. The reader who wants a systematic treatment is referred to *Determinants and Matrices* by A. C. Aitken (9th edition 1956), Oliver & Boyd, Edinburgh.

For the reader who wishes to revise his knowledge of that part of the subject which we shall use, or who, being previously unacquainted with vectors and matrices, wishes a rough explanation of what they are, the following paragraphs provide a list of definitions and rules of use.

A4.2. Matrices, Vectors and Diagonal Matrices

An $m \times n$ matrix $\boldsymbol{X}$ will denote an array of m rows and n columns of real numbers x_{ij} $(i = 1, 2, \ldots, m;\ j = 1, 2, \ldots, n)$ (A4.2.1)

Thus the 2×3 matrix $\boldsymbol{X}$ will denote the array

$$\begin{pmatrix} x_{11} & x_{12} & x_{13} \\ x_{21} & x_{22} & x_{23} \end{pmatrix} \qquad \text{(A4.2.2)}$$

A column-vector is a matrix with only one column and a row-vector is a matrix with only one row. The m-element column vector $\boldsymbol{x}$ will denote

$$\begin{pmatrix} x_1 \\ x_2 \\ \vdots \\ x_m \end{pmatrix} \tag{A4.2.3}$$

$\boldsymbol{X}'$ will denote the *transpose* of the matrix $\boldsymbol{X}$, namely the matrix obtained by interchanging the rows and the columns of $\boldsymbol{X}$. Thus, where $\boldsymbol{X}$ is the 2×3 matrix (A4.2.2), $\boldsymbol{X}'$ is the 3×2 matrix

$$\begin{pmatrix} x_{11} & x_{21} \\ x_{12} & x_{22} \\ x_{13} & x_{23} \end{pmatrix}; \tag{A4.2.4}$$

and where $\boldsymbol{x}$ is the m-element column vector (A4.2.3), $\boldsymbol{x}'$ will denote the corresponding row-vector $(x_1, x_2, \ldots, x_m)$. (A4.2.5)

Where $\boldsymbol{x}$ is the m-column vector (A4.2.3), $\hat{\boldsymbol{x}}$ will denote the $m \times m$ diagonal matrix with $x_1, x_2, \ldots, x_m$ down the diagonal from top left to bottom right, and with zeros elsewhere. Thus, when $m = 3$ and

$$\boldsymbol{x} = \begin{pmatrix} x_1 \\ x_2 \\ x_3 \end{pmatrix}, \quad \hat{\boldsymbol{x}} = \begin{pmatrix} x_1 & 0 & 0 \\ 0 & x_2 & 0 \\ 0 & 0 & x_3 \end{pmatrix}.$$

A4.3. Sums of Matrices

If $\boldsymbol{X}$ and $\boldsymbol{Y}$ are two $m \times n$ matrices, $[\boldsymbol{X}+\boldsymbol{Y}]$ will denote the $m \times n$ matrix $\boldsymbol{Z}$ whose typical element z_{ij} is given by

$$z_{ij} = x_{ij}+y_{ij}, \tag{A4.3.1}$$

as the sum of the corresponding elements of $\boldsymbol{X}$ and of $\boldsymbol{Y}$. The rule may be extended to allow

$$[\boldsymbol{X}+\boldsymbol{Y}+\boldsymbol{W}] = [\boldsymbol{X}+\boldsymbol{Y}]+\boldsymbol{W} \text{ etc.}$$

so that, where $\boldsymbol{X}+\boldsymbol{Y}+\boldsymbol{W} = \boldsymbol{V}$, the typical element of $\boldsymbol{V}$ is given by

$$v_{ij} = x_{ij}+y_{ij}+w_{ij} \quad \text{etc.} \tag{A4.3.2}$$

A4.4. Matrix Equations and Inequalities

Equations and inequalities, such as $\boldsymbol{X} = \boldsymbol{Y}$ or $\boldsymbol{X} \geqslant \boldsymbol{Z}$ or $\boldsymbol{X} > \boldsymbol{W}$, between two matrices or matrix expressions, each of m rows and n columns, will assert the mn corresponding relations between the elements, namely,

$$x_{ij} = y_{ij} \text{ or } x_{ij} \geqslant z_{ij} \text{ or } x_{ij} > w_{ij}$$
$$(i = 1, 2, \ldots, m;\ j = 1, 2, \ldots, n). \tag{A4.4.1}$$

F

A4.5. Products of Matrices

If $\boldsymbol{X}$ is an $m \times n$ matrix and $\boldsymbol{Y}$ is an $n \times p$ matrix, then the product $\boldsymbol{XY}$ is the $m \times p$ matrix $\boldsymbol{Z}$ whose typical element z_{ik} is given by

$$z_{ik} = \sum_{j=1}^{n} x_{ij} y_{jk} \quad (i = 1, 2, \ldots, m;\ k = 1, 2, \ldots, p). \quad \text{(A4.5.1)}$$

In particular, where $\boldsymbol{x}$ is a column vector and λ is a single number, $\boldsymbol{x}\lambda$ is the column vector obtained by multiplying each element of $\boldsymbol{x}$ by λ.

The following properties of certain products of matrices follow directly from the various definitions, and they are used so often in the main part of the book that the reader is advised to confirm them as exercises, if they are not already familiar to him.

(*i*) If $\boldsymbol{x}$ and $\boldsymbol{y}$ are two m-element column vectors,

$$\boldsymbol{x}'\boldsymbol{y} = \boldsymbol{y}'\boldsymbol{x} = \lambda,$$

where λ is the real number obtained by summing the products of corresponding elements of $\boldsymbol{x}$ and $\boldsymbol{y}$.

(*ii*) If $\boldsymbol{X}$ is an $m \times n$ matrix and $\boldsymbol{Y}$ is an $n \times p$ matrix,

$$(\boldsymbol{XY})' = \boldsymbol{Y}'\boldsymbol{X}'.$$

(*iii*) If $\boldsymbol{x}$ and $\boldsymbol{y}$ are two m-element column vectors

$$\hat{\boldsymbol{x}}\boldsymbol{y} = \boldsymbol{z} = \hat{\boldsymbol{y}}\boldsymbol{x}, \quad \text{(A4.5.2)}$$

where $\boldsymbol{z}$ is the m-element column vector whose typical element z_i is given by

$$z_i = x_i y_i. \quad \text{(A4.5.3)}$$

(*iv*) If $\boldsymbol{x}, \boldsymbol{y}, \ldots, \boldsymbol{u}, \boldsymbol{v}$ are m-element column vectors, then

$$\hat{\boldsymbol{x}}\hat{\boldsymbol{y}} \ldots \hat{\boldsymbol{u}}\boldsymbol{v} = \boldsymbol{z} \quad \text{(A4.5.4)}$$

where $\boldsymbol{z}$ is the m-element column vector whose typical element z_i is given by

$$z_i = x_i y_i \ldots u_i v_i \quad \text{(A4.5.5)}$$

as the product of the corresponding elements of all the column vectors $\boldsymbol{x}, \boldsymbol{y}, \ldots, \boldsymbol{u}$ and $\boldsymbol{v}$.

A4.6. Trace and the Unit Vector

If $\boldsymbol{X}$ is an $m \times m$ matrix, the *Trace* of $\boldsymbol{X}$ is defined by

$$\text{tr}\,[\boldsymbol{X}] = \sum_{i=1}^{m} x_{ii} \quad \text{(A4.6.1)}$$

as the sum of the diagonal elements of $\boldsymbol{X}$.

The symbol **1** will always denote a column vector composed of ones. It should always be clear from the context, how many unit elements the vector is to be composed of. **1** is known as the *Unit vector*.

Unless he is already familiar with the following results the reader is advised to check them.

(*i*) The sum of the elements of any column vector $\boldsymbol{x}$ is given by the expression $\mathbf{1}'\boldsymbol{x}$ and their arithmetic mean average is given by $\mathbf{1}'\boldsymbol{x} / \mathbf{1}'\mathbf{1}$ it being understood that **1** has the same number of elements when used in the denominator as in the numerator.

(*ii*) The sum of the elements of any matrix $\boldsymbol{X}$ are given by $\mathbf{1}'\boldsymbol{X}\mathbf{1}$.

(*iii*) The sum of the squares of the elements of any matrix $\boldsymbol{X}$ are given by tr $[\boldsymbol{X}'\boldsymbol{X}]$ or alternatively by tr $[\boldsymbol{X}\boldsymbol{X}']$.

(*iv*) If $\boldsymbol{X}$, $\boldsymbol{Y}$ and $\boldsymbol{Z}$ are $m \times n$, $n \times p$ and $p \times m$ matrices

$$\text{tr}\,[\boldsymbol{XYZ}] = \text{tr}\,[\boldsymbol{YZX}] = \text{tr}\,[\boldsymbol{ZXY}].$$

(*v*) If $\boldsymbol{X}$ is any matrix $\hat{\mathbf{1}}\boldsymbol{X} = \boldsymbol{X}$ and $\boldsymbol{X}\hat{\mathbf{1}} = \boldsymbol{X}$.

A4.7. Inverse of a Square Matrix

If $\boldsymbol{X}$ is an $m \times m$ matrix, $|\boldsymbol{X}|$ denotes the determinant of the array of numbers x_{ij} which constitutes $\boldsymbol{X}$. Thus, if $m = 2$,

$$|\boldsymbol{X}| = \begin{vmatrix} x_{11} & x_{12} \\ x_{21} & x_{22} \end{vmatrix} = x_{11}x_{22} - x_{12}x_{21}. \qquad \text{(A4.7.1)}$$

For an account of determinants see Aitken *op. cit.* Any $m \times m$ matrix $\boldsymbol{X}$ whose determinant $|\boldsymbol{X}|$ does not equal zero has an inverse matrix denoted $\boldsymbol{X}^{-1}$ which is defined by the property

$$\boldsymbol{X}\boldsymbol{X}^{-1} = \hat{\mathbf{1}}. \qquad \text{(A4.7.2)}$$

It is known that (provided $|\boldsymbol{X}| \neq 0$) the inverse matrix $\boldsymbol{X}^{-1}$ exists and is unique, and moreover that $\boldsymbol{X}^{-1}\boldsymbol{X} = \hat{\mathbf{1}}$. (A4.7.3)
Inverse matrices are useful for solving sets of linear equations. Thus if x_1, x_2 and x_3 solve the equations

$$\begin{aligned} a_{11}x_1 + a_{12}x_2 + a_{13}x_3 &= c_1 \\ a_{21}x_1 + a_{22}x_2 + a_{23}x_3 &= c_2 \\ a_{31}x_1 + a_{32}x_2 + a_{33}x_3 &= c_3 \end{aligned}$$

then these equations may be written as the single matrix equation

$$\boldsymbol{A}\boldsymbol{x} = \boldsymbol{c}. \qquad \text{(A4.7.4)}$$

Assuming that $|\boldsymbol{A}|$ is not zero, we may premultiply each side of (A4.7.4) by the inverse matrix $\boldsymbol{A}^{-1}$ to obtain

$$A^{-1}Ax = A^{-1}c$$
$$\therefore \quad \hat{1}x = A^{-1}c$$
$$\therefore \quad x = A^{-1}c \tag{A4.7.5}$$

so that the solutions x_1, x_2 and x_3 will be given as the three elements of the column-vector $A^{-1}c$. The same method extends to provide the solutions to any number, m, of simultaneous linear equations in m variables.

A4.8. Determinants

Reference has already been made to a text-book on determinants. In this paragraph certain of the main properties of the determinant will be noted without proof.

(*i*) If a_{ij} $(i, j = 1, 2, \ldots, m)$ are the elements of $|A|$, the determinant of the $m \times m$ matrix A, the minor, A_{ij}, of any element a_{ij} is the determinant obtained by omitting the i-th row and j-th column from $|A|$.

(*ii*)
$$\sum_{j=1}^{m} (-1)^{i+j} a_{ij} A_{ij} = |A| \quad (i = 1, 2, \ldots, m) \tag{A4.8.1}$$
$$\sum_{j=1}^{m} (-1)^{j} a_{hj} A_{ij} = 0 \quad (h \neq i,\ h, i = 1, 2, \ldots, m) \tag{A4.8.2}$$

(*iii*) The typical element α_{ij} of the inverse matrix A^{-1} is given by

$$\alpha_{ij} = (-1)^{i+j} A_{ji} / |A|, \text{ provided } |A| \neq 0. \tag{A4.8.3}$$

(*iv*) If A and B are two $m \times m$ matrices

$$|AB| = |A| \,.\, |B|. \tag{A4.8.4}$$

It follows that if $|A| \neq 0$,

$$|A^{-1}| = |A|^{-1}. \tag{A4.8.5}$$

A4.9. Orthogonal Matrix

An $m \times m$ matrix A is said to be orthogonal if

$$A'A = \hat{1}. \tag{A4.9.1}$$

It follows from (A4.8.4) that if A is orthogonal

$$|A'||A| = 1 \tag{A4.9.2}$$

and hence, since $|A'| = |A|$,

$$|A| = \pm 1. \tag{A4.9.3}$$

Since $|A|$ is not zero, A will have an inverse A^{-1}, and when we postmultiply (A4.9.1) by this inverse, we obtain

$$A' = A^{-1} \quad \text{and} \quad A = (A')^{-1}. \tag{A4.9.4}$$

Where $\boldsymbol{a}_1$ is any vector of m elements such that

$$\boldsymbol{a}_1'\boldsymbol{a}_1 = 1, \tag{A4.9.5}$$

it is always possible to find an orthogonal $m \times m$ matrix A, of which $\boldsymbol{a}_1$ is the first column: for example, if $m = 5$, and a_{ij} ($j = 1, 2, 3, 4, 5$) are the 5 elements of $\boldsymbol{a}_1$, we may expand (A4.9.5) to the form

$$a_{11}^2 + a_{12}^2 + a_{13}^2 + a_{14}^2 + a_{15}^2 = 1. \tag{A4.9.6}$$

It will then be possible to find angles θ_i ($i = 1, 2, 3, 4$) such that, where s_i and c_i denote $\sin\theta_i$ and $\cos\theta_i$ ($i = 1, 2, 3, 4$),

$$a_{11} = s_1,\ a_{21} = c_1 s_2,\ a_{31} = c_1 c_2 s_3,\ a_{41} = c_1 c_2 c_3 s_4,\ a_{51} = c_1 c_2 c_3 c_4. \tag{A4.9.7}$$

One way of completing the matrix A so as to ensure its orthogonality would then be to form it as

$$A = \begin{Bmatrix} s_1 & 0 & 0 & 0 & -c_1 \\ c_1 s_2 & 0 & 0 & -c_2 & s_1 s_2 \\ c_1 c_2 s_3 & 0 & -c_3 & s_2 s_3 & s_1 c_2 s_3 \\ c_1 c_2 c_3 s_4 & -c_4 & s_3 s_4 & s_2 c_3 s_4 & s_1 c_2 c_3 s_4 \\ c_1 c_2 c_3 c_4 & s_4 & s_3 c_4 & s_2 c_3 c_4 & s_1 c_2 c_3 c_4 \end{Bmatrix} \tag{A4.9.8}$$

The method may easily be generalised from the case $m = 5$ to the general case where m is any positive integer, to show that subject to (A4.9.5) it is always possible to find an orthogonal matrix A of which $\boldsymbol{a}_1$ is the first column. This result will be used in Appendix 8.

Exercises 4

(In these Exercises vectors are printed in italics and not as elsewhere in bold type.)

4.1. A pack of four cards consists of the two red kings, the queen of clubs and the knave of hearts. An experiment consists of shuffling these cards and drawing first one and then another.

Construct a probability model with $N = 12$ on the assumption that at each draw each card in the pack has an equal probability of being drawn.

4.2. Let $j(1), j(2), j(3), j(4), j(5)$ represent the events

(*i*) the first card drawn is red,
(*ii*) the second card drawn is red,
(*iii*) the first card drawn is a king,
(*iv*) the second card drawn is a king,
(*v*) at least one king is drawn.

Give the vectors for each $j(r)$ $(r = 1, 2, 3, 4, 5)$ in the model you provided in question 1.

4.3. Find the probabilities of (*i*) $j(1) \cap j(2)$; (*ii*) $j(3) \cup j(4)$; (*iii*) $\bigcap_{r=1}^{5} j(r)$, and (*iv*) $j(1) \cup j(2)$.

4.4. If $j(6)$ is the event that the first card drawn is a heart, to which of the events $j(r)$ $(r = 1, 2, 3, 4, 5)$ is $j(6)$ relevant? Verify this numerically.

4.5. Give an example of a disjoint exhaustive set of events in your model.

4.6. Find $p'r\{j(1) \leqslant j(3)\}$, and explain what it measures in your model.

4.7. Find the probability that the second card drawn is a king if the first card drawn is black.

4.8. Explain what is meant in the model of the earlier question by

(*i*) $p'r\{j(1) = j(2)\}$
(*ii*) $p'r\{j(5) \leqslant j(3)\}$
(*iii*) $p'r\{(1-j(1)) \leqslant j(3)\}$,

and write down the vectors for $r\{j(1) = j(2)\}$, $r\{j(5) \leqslant j(3)\}$ and $r[\{1-j(1)\} \leqslant j(3)]$ in your model. Hence calculate the three probabilities.

4.9. Modify the model constructed in question 1, so as to allow the first card drawn to be replaced before the second card is drawn. How would you represent the event of drawing a black card in each draw?

4.10. In your model for question 9, represent the modification that in the second draw there was a probability of 0·4 for drawing again the first card drawn but a probability of 0·2 for drawing each other card.

4.11. j and k are events and m is their union. If their probabilities are $p'j = 0{\cdot}25$, $p'k = 0{\cdot}\dot{6}$, and $p'm = 0{\cdot}8\dot{3}$, find $p'jk$, the probability of the intersection of j and k.

4.12. In the notation of question 11, find the conditional probabilities of j given k, and of k given j, namely, $p'_{|k}j$ and $p'_{|j}k$.

4.13. m is the union of the events j and k. If $p'j = 0{\cdot}5$ and $p'k = 0{\cdot}\dot{3}$, and if j and k are independent, find $p'm$.

4.14. Two fair dice are rolled simultaneously. j and k denote the events 'the numbers shown differ by one' and 'one number shown is twice the other'. Find $p'j$, $p'k$, $p'\hat{j}k$, $p'_{|k}j$ and $p'_{|j}k$.

4.15. r represents the intersection of three events j, k and m. Prove that

$$p'r = p'm\,p'_{|m}k\,p'_{|\hat{k}m}j.$$

4.16. Prove that, where j, k and m represent three events and where $p'\hat{j}m > 0$ and $p'\hat{k}m > 0$,

$$\frac{p'_{|m}j}{p'_{|m}k} = \frac{p'_{|j}m\;p'j}{p'_{|k}m\;p'k}.$$

(Bayes' theorem)

4.17. u is the union of three events, j, k and m. Show that

$$p'u = p'j+p'k+p'm-p'\hat{k}m-p'\hat{m}j-p'\hat{j}k+p'\hat{j}\hat{k}m.$$

4.18. u represents the union of n events $j(1), j(2), \ldots, j(n)$. Obtain the result

$$p'u = \sum_{t} p'j(t) - \sum_{s\,t} p'\hat{j}(s)j(t) + \sum_{r\,s\,t} p'\hat{j}(r)\hat{j}(s)j(t) - \ldots + (-)^{n-1} \bigcap_{i=1}^{n} [j(i)].$$

4.19. Smith, Jones, Brown and Carruthers each admire one only of four girls at a social function and no two admire the same one of these girls. If the hostess at random introduces one of these girls to each admirer, what is the probability that she is introducing at least one girl to that girl's own admirer?

Chapter 5

A MODEL FOR DISCRETE VARIATES

5.1. Discrete Variates Introduced

In this chapter, we extend the probability model which dealt with events and their probabilities, to cover discrete variates as well. Discrete variates are variables which can take any one of a sequence of different real values with various probabilities. The occurrence of an event may be thought of as the observation of a particularly simple type of variate limited to two values only, namely one and zero, since the event either occurs (value one), or does not occur (value zero).

5.2. Representation of Variates by Vectors

D. 16 In our probability model, any vector with the usual number N of elements will stand for a *variate*. As in the case of event-vectors, it will for many purposes be convenient to allow each variate vector to contain an unlimited number of elements.

At this stage we may pick out for notice two particularly simple types of variate. The first is any variate of the form

$$\boldsymbol{x} = \mathbf{1}x$$

where x is a real number. Such a vector will have each of its elements equal to x; in this case the variate $\boldsymbol{x}$ is one which is constrained to take the real value x, and it may accordingly be regarded as a constant. The second is any vector $\boldsymbol{x}$ whose elements each are equal either to one or to zero; in this case the variate can take one of the two values one or zero, and it reduces to an 'event' of the class which we examined in Chapter 4.

5.3. Logical Relations Involving Variates

The introduction of variates greatly increases the possible scope of logical relations. These may now include relations of the type $\boldsymbol{r}(\boldsymbol{x} = \boldsymbol{y})$, $\boldsymbol{r}(\boldsymbol{x} \leqslant \boldsymbol{y})$ etc. where $\boldsymbol{x}$ and $\boldsymbol{y}$ represent variates.

Any such logical relation is represented by a vector according to the terms of the following definition.

D. 17 The logical event $\boldsymbol{r}(L)$, where L stands for some relation between variates, is a column vector of N elements whose t-th element is a one or a zero according as the relation L does or does not hold between the t-th elements of the two vectors representing the variates.

In particular the relation L may contain variates of the type $\mathbf{1}x$ where x is a real number: these variates, we have seen, are really constants. We shall adopt the convention of omitting the vector $\mathbf{1}$ in all cases where variates of type $\mathbf{1}x$ appear in a logical relation. For example, instead of writing $\boldsymbol{r}(\mathbf{1}2 \leqslant \boldsymbol{x} \leqslant \mathbf{1}3)$ we should write $\boldsymbol{r}(2 \leqslant \boldsymbol{x} \leqslant 3)$.

5.4. The Expansion of a Function of a Variate in Terms of the Corresponding Function of the Values which it may Take

Let $\boldsymbol{x}$ be the vector representing a variate and $\boldsymbol{\phi}$ denote the vector obtained by putting each element ϕ_t equal to $\phi(x_t)$ where x_t is the corresponding element of $\boldsymbol{x}$ and where ϕ is a specified real function.

Then the vector $\boldsymbol{\phi}$ may be expanded as

$$\boldsymbol{\phi} = \sum_x \boldsymbol{r}(\boldsymbol{x} = x)\,\phi(x), \tag{5.4.1}$$

where the summation extends over all values x which it is possible for $\boldsymbol{x}$ to take.

To prove (4.1), let ϕ_t be any element of $\boldsymbol{\phi}$ and let x_t be the corresponding element of $\boldsymbol{x}$, and y_t be the corresponding element of $\phi(x)\boldsymbol{r}(\boldsymbol{x} = x)$. Then

$$y_t = 0 \quad \text{or} \quad y_t = \phi_t,$$

according as $\qquad x_t \neq x \quad \text{or} \quad x_t = x. \qquad$ (5.4.2)

Hence $\qquad \phi_t = \sum_x y_t \quad$ for each value of t. $\qquad$ (5.4.3)

Hence $\qquad \boldsymbol{\phi} = \sum_x \boldsymbol{r}(\boldsymbol{x} = x)\,\phi(x) \quad$ as in (5.4.1).

In particular $\qquad \boldsymbol{x} = \sum_x \boldsymbol{r}(\boldsymbol{x} = x)x. \qquad$ (5.4.4)

5.5. Multivariate Relations

If $\boldsymbol{x}$ and $\boldsymbol{y}$ are a pair of variates, then the joint relation $\boldsymbol{r}\{(\boldsymbol{x} = x)$ and $(\boldsymbol{y} = y)\}$ may be represented by the product

$$\hat{\boldsymbol{r}}(\boldsymbol{x} = x)\,\boldsymbol{r}(\boldsymbol{y} = y)$$

whose typical element will be a one when both the corresponding element of $\boldsymbol{x}$ is x and the corresponding element of $\boldsymbol{y}$ is y, but otherwise it will be a zero.

Similarly, if $\boldsymbol{x}(j)$ $(j = 1, 2, \ldots, m)$ are any m variates, the joint relation

$$\boldsymbol{r}[\{\boldsymbol{x}(1) = x(1)\} \text{ and } \{\boldsymbol{x}(2) = x(2)\} \text{ and} \ldots \text{and } \{\boldsymbol{x}(m) = x(m)\}]$$

may be represented by the product

$$\bigcap_{j=1}^{m} \boldsymbol{r}\{\boldsymbol{x}(j) = x(j)\}$$

giving the intersection of the m events

$$\boldsymbol{r}\{\boldsymbol{x}(j) = x(j)\} \quad (j = 1, 2, \ldots, m).$$

5.6. The Expansion of a Function of Many Variates in Terms of the Corresponding Function of the Values which they may Take

Let $\boldsymbol{x}(j)$ $(j = 1, 2, \ldots, m)$ be a set of m variates and let $\boldsymbol{\phi}$ denote the vector obtained by putting each element ϕ_t equal to

$$\phi\{x_t(1), x_t(2), \ldots, x_t(m)\}$$

where $x_t(j)$ $(j = 1, 2, \ldots, m)$ are the corresponding elements of the m vectors $\boldsymbol{x}(j)$, and where ϕ is a specified real function of m real variables.

Then the expansion of $\boldsymbol{\phi}$ corresponding to (5.4.1) the expansion discussed above is

$$\boldsymbol{\phi} = \sum_{x(1)} \sum_{x(2)} \cdots \sum_{x(m)} \bigcap_{j=1}^{m} \boldsymbol{r}\{\boldsymbol{x}(j) = x(j)\}\, \phi\{x(1), x(2), \ldots, x(m)\}. \tag{5.6.1}$$

The proof is very similar to that given above for (5.4.1) and it is left as an exercise for the reader.

5.7. The Probability Function of a Variate

Consider the probability of the event that a variate $\boldsymbol{x}$ takes a particular value x. This probability is

$$\boldsymbol{p}'\boldsymbol{r}(\boldsymbol{x} = x)$$

and it is a real function of the real variable x. It will be called the 'probability function of the variate $\boldsymbol{x}$', and it will be convenient to denote it by $f(x)$ when the context makes it clear that $\boldsymbol{x}$ is the variate to which the probability function relates.

D. 18 The probability function for the variate $\boldsymbol{x}$ is

$$f(x) = \boldsymbol{p}'\boldsymbol{r}(\boldsymbol{x} = x). \tag{5.7.1}$$

This notation will often be found useful in discussing the estimation of properties of probability functions of variates from a set of observations of those variates or relevant to them.

5.8. Expected Values

If $\boldsymbol{x}$ is an event then the product $\boldsymbol{p}'\boldsymbol{x}$ is simply the probability of that event. But if $\boldsymbol{x}$ is a variate and not an event, there is no sense in speaking of the probability of $\boldsymbol{x}$. However $\boldsymbol{p}'\boldsymbol{x}$ remains an important property of $\boldsymbol{x}$ and

D. 19 $\boldsymbol{p}'\boldsymbol{x}$ is called the *expected value* of the variate $\boldsymbol{x}$. Similarly if $\boldsymbol{\phi}$ is any function of $\boldsymbol{x}$, then $\boldsymbol{p}'\boldsymbol{\phi}$ is called the expected value of $\boldsymbol{\phi}$.

Now we have seen above (5.4.1) that

$$\boldsymbol{\phi} = \sum_x \boldsymbol{r}(\boldsymbol{x} = x)\,\phi(x). \tag{5.8.1}$$

where the summation extends over all values x which $\boldsymbol{x}$ may take. Premultiplying (5.8.1) by $\boldsymbol{p}'$ we find that the expected value of $\boldsymbol{\phi}$ is

$$\boldsymbol{p}'\boldsymbol{\phi} = \sum_x \boldsymbol{p}'\boldsymbol{r}(\boldsymbol{x} = x)\,\phi(x). \tag{5.8.2}$$

But the probability function for $\boldsymbol{x}$ was defined in D. 18 as

$$f(x) = \boldsymbol{p}'\boldsymbol{r}(\boldsymbol{x} = x). \tag{5.8.3}$$

Substituting (5.8.3) in (5.8.2) we find that the expected value of any function $\boldsymbol{\phi}$ of x may be expanded as

$$\boldsymbol{p}'\boldsymbol{\phi} = \sum_x f(x)\,\phi(x) \tag{5.8.4}$$

where $f(x)$ is the probability function for $\boldsymbol{x}$.

In particular, the expected value for $\boldsymbol{x}$ is

$$\boldsymbol{p}'\boldsymbol{x} = \sum_x f(x)\,.\,x. \tag{5.8.5}$$

It will be found convenient in later chapters to use the abbreviation $E(\boldsymbol{x})$ to denote 'the expected value of' so that

$$E(\boldsymbol{x}) = \sum_x f(x)x$$

and

$$E(\boldsymbol{\phi}) = \sum_x f(x)\,\phi(x). \tag{5.8.6}$$

We shall see below, in § **27**, that for some variates $\boldsymbol{x}$, and for some $\phi(\boldsymbol{x})$, the expressions (5.8.6) may fail to converge to finite sums: this failure can only occur when, as foreshadowed in § **2**, we allow the variate vectors to contain an unlimited number of elements.

5.9. Illustrative Example of a Variate

A true six-faced die has inscribed on its six faces the number pairs (6,1); (6,2); (6,3); (5,3); (5,4); (4,4). As an adequate probability model for discussing the probabilities of various possible outcomes of throwing the die once, we might take the following one:

$$\begin{aligned} N &= 6, \\ \boldsymbol{p}' &= (1/6,\ 1/6,\ 1/6,\ 1/6,\ 1/6,\ 1/6) \\ \boldsymbol{x}' &= (6,\ 6,\ 6,\ 5,\ 5,\ 4) \\ \boldsymbol{y}' &= (1,\ 2,\ 3,\ 3,\ 4,\ 4) \end{aligned} \tag{5.9.1}$$

$\boldsymbol{x}'$, $\boldsymbol{y}'$ are row vectors; and $\boldsymbol{x}$, $\boldsymbol{y}$, the corresponding column vectors are variates representing the first and second number on the uppermost face of the thrown die.

One may consider various non-trivial logical events such as $\boldsymbol{r}(\boldsymbol{y} = 1)$, $\boldsymbol{r}(\boldsymbol{y} = 2)$, $\boldsymbol{r}(\boldsymbol{y} = 3)$, $\boldsymbol{r}(\boldsymbol{y} = 4)$, $\boldsymbol{r}(\boldsymbol{x} = 4)$, $\boldsymbol{r}(\boldsymbol{x} = 5)$ and $\boldsymbol{r}(\boldsymbol{x} = 6)$. These all are represented by column vectors but it is more convenient to write out their transposes, $\boldsymbol{r}'(\boldsymbol{y} = 1)$, etc. For example

$$\boldsymbol{r}'(\boldsymbol{y} = 3) = (0,\ 0,\ 1,\ 1,\ 0,\ 0). \tag{5.9.2}$$

In the same way, it is easy to write down the row vectors for the other relations and to derive probabilities such as, for example,

$$\boldsymbol{p}'\boldsymbol{r}(\boldsymbol{y} = 3) = 1/3 \quad \text{and} \quad \boldsymbol{p}'\boldsymbol{r}(\boldsymbol{x} = 5) = 1/3$$

and

$$\boldsymbol{p}'\hat{\boldsymbol{r}}(\boldsymbol{x} = 5)\,\boldsymbol{r}(\boldsymbol{y} = 3) = 1/6 \tag{5.9.3}$$

by matrix or vector multiplication.

The probability function for $\boldsymbol{x}$, namely $f(x) = \boldsymbol{p}'\boldsymbol{r}(\boldsymbol{x} = x)$ takes the values

$$f(4) = 1/6, \quad f(5) = 1/3, \quad f(6) = 1/2, \quad f(x) = 0 \tag{5.9.4}$$

for other values of x, as may be verified by multiplying the vectors $\boldsymbol{p}'$ and $\boldsymbol{r}(\boldsymbol{x} = x)$ for the various values of x.

The reader may care to derive in the same way the probability function for $\boldsymbol{y}$.

The expected values of $\boldsymbol{x}$ and $\boldsymbol{y}$ may be found by forming the scalar products:

$$E(\boldsymbol{x}) = \boldsymbol{p}'\boldsymbol{x} = 16/3; \quad E(\boldsymbol{y}) = \boldsymbol{p}'\boldsymbol{y} = 17/6. \tag{5.9.5}$$

5.10. Bivariate Probability Functions

If $\boldsymbol{x}$ and $\boldsymbol{y}$ are any pair of variates, then the probability that they take any specified pair (x, y) of real values may be written as

D. 20 $$f(x, y) = \boldsymbol{p}'\hat{\boldsymbol{r}}(\boldsymbol{x} = x)\boldsymbol{r}(\boldsymbol{y} = y) \quad (5.10.1)$$

and this function of the pair of variables (x, y) is called the probability function for the pair of variates $(\boldsymbol{x}, \boldsymbol{y})$.

The expected value $E(\boldsymbol{\phi})$ of any function $\boldsymbol{\phi}$ of the two variates $\boldsymbol{x}$ and $\boldsymbol{y}$ may, on account of (5.6.1) be expanded as

$$E(\boldsymbol{\phi}) = \boldsymbol{p}'\boldsymbol{\phi} = \sum_x \sum_y \boldsymbol{p}'\hat{\boldsymbol{r}}(\boldsymbol{x} = x)\boldsymbol{r}(\boldsymbol{y} = y)\phi(x, y). \quad (5.10.2)$$

Substituting from (5.10.1) in (5.10.2) we obtain

$$E(\boldsymbol{\phi}) = \boldsymbol{p}'\boldsymbol{\phi} = \sum_x \sum_y f(x, y)\phi(x, y). \quad (5.10.3)$$

In particular, the expected value of the product $\hat{\boldsymbol{x}}\boldsymbol{y}$ of two variates $\boldsymbol{x}$ and $\boldsymbol{y}$ is given by

$$E(\hat{\boldsymbol{x}}\boldsymbol{y}) = \boldsymbol{p}'\hat{\boldsymbol{x}}\boldsymbol{y} = \sum_x \sum_y f(x, y)xy. \quad (5.10.4)$$

5.11. Example of a Bivariate Probability Function

The bivariate probability function $f(x, y)$ for the variate pair $(\boldsymbol{x}, \boldsymbol{y})$ in the example (5.9.1) above can be set out as in the following table.

TABLE 5.11.1

	x	4	5	6	*other*	*Total*
	y		$f(x \ \ y)$			$\boldsymbol{p}'\boldsymbol{r}\ (\boldsymbol{y} = y)$
	1	0	0	1/6	0	1/6
	2	0	0	1/6	0	1/6
	3	0	1/6	1/6	0	1/3
	4	1/6	1/6	0	0	1/3
	other	0	0	0	0	
Total	$\boldsymbol{p}'\boldsymbol{r}\ (\boldsymbol{x} = x)$	1/6	1/3	1/2		

The expected value of the product $\hat{\boldsymbol{x}}\boldsymbol{y}$ can be read off by summing $x\,y\,f(x, y)$ over all cells of the table: it is

$$\begin{aligned} \boldsymbol{p}'\hat{\boldsymbol{x}}\boldsymbol{y} &= \tfrac{1}{6}\{6\times1+6\times2+5\times3+6\times3+4\times4+5\times4\} \\ &= \tfrac{1}{6}\times87 = 14\tfrac{1}{2}. \end{aligned} \quad (5.11.1)$$

It will be noted that the probability functions $\boldsymbol{p}'\boldsymbol{r}(\boldsymbol{x} = x)$ and $\boldsymbol{p}'\boldsymbol{r}(\boldsymbol{y} = y)$ can be obtained by summing the columns and rows of the table. This is in general possible when such a table is available for a bivariate probability function, and for this reason the functions

$\boldsymbol{p}'\boldsymbol{r}(\boldsymbol{x} = x)$ and $\boldsymbol{p}'\boldsymbol{r}(\boldsymbol{y} = y)$ are commonly known as the *marginal probability functions* for the variates $\boldsymbol{x}$ and $\boldsymbol{y}$.

5.12. The Interpretation of Expected Values in Gambling Situations

Suppose that a wager may yield prizes of X_1, X_2 and X_3 dollars with probabilities P_1 P_2 P_3, and that there are no other possibilities, so that

$$P_1+P_2+P_3 = 1.$$

In common speech, the worth of the wager would then be said to be

$$W = P_1X_1+P_2X_2+P_3X_3. \tag{5.12.1}$$

Now let $\boldsymbol{x}$ be the variate representing the prize yielded by the wager. Then, by equation (5.8.6), the expected value of $\boldsymbol{x}$ is

$$E(\boldsymbol{x}) = \sum_{j=1}^{3} f(X_j)\,X_j \tag{5.12.2}$$

where $f(x)$ is the probability function for $\boldsymbol{x}$.

The probabilities of the three values of $\boldsymbol{x}$ have been given as

$$f(X_j) = P_j\ (j = 1, 2, 3). \tag{5.12.3}$$

Hence, substituting in (5.12.2)

$$E(\boldsymbol{x}) = P_1X_1+P_2X_2+P_3X_3 = W \text{ by (5.12.1).} \tag{5.12.4}$$

Thus the expected value of the gain from the wager corresponds to what is commonly meant by the *worth* of the wager.

5.13. Conditional Probability Functions for Variates

Conditional probabilities for discrete variates to take particular values may be defined in exactly the same way as conditional probabilities for other events.

The conditional probability vector $\boldsymbol{p}'_{|\,\boldsymbol{r}(\boldsymbol{x}=x)}$ is obtained by regarding the relation $\boldsymbol{r}(\boldsymbol{x} = x)$ as an event. We obtain

$$\boldsymbol{p}'_{|\,\boldsymbol{r}(\boldsymbol{x}=x)} = \{\boldsymbol{p}'\boldsymbol{r}(\boldsymbol{x} = x)\}^{-1}\,\boldsymbol{p}'\hat{\boldsymbol{r}}(\boldsymbol{x} = x), \tag{5.13.1}$$

provided

$$\boldsymbol{p}'\boldsymbol{r}(\boldsymbol{x} = x) > 0. \tag{5.13.2}$$

This rule implies that the conditional probability, given that $\boldsymbol{x} = x$, for $\boldsymbol{y} = y$ is

$$\boldsymbol{p}'_{|\,\boldsymbol{r}(\boldsymbol{x}=x)}\boldsymbol{r}(\boldsymbol{y} = y) = \{\boldsymbol{p}'\boldsymbol{r}(\boldsymbol{x} = x)\}^{-1}\boldsymbol{p}'\hat{\boldsymbol{r}}(\boldsymbol{x} = x)\boldsymbol{r}(\boldsymbol{y} = y). \tag{5.13.3}$$

If we write $h(y\,|\,\boldsymbol{x} = x)$ for the conditional probability function,

$f(x, y)$ for the bivariate probability function and $g(x)$ for the marginal probability function for $\mathbf{x}$, then the result may be put in the convenient form:

$$h(y \mid \mathbf{x} = x) = f(x, y) \,/\, g(x) \text{ provided } g(x) > 0. \qquad (5.13.4)$$

Accordingly we adopt the definition,

D. 21 The conditional probability function for $\mathbf{y}$, given that $\mathbf{x} = x$, is obtained by dividing the bivariate probability function for $(\mathbf{x}, \mathbf{y})$ at (x, y) by the marginal probability function for $\mathbf{x}$ at x, provided the latter is not zero.

5.14. Independence of Variates

We may define the independence of variates in terms of the independence of the logical events that they take specific values.

Thus a pair of variates $\mathbf{x}$, $\mathbf{y}$ are by definition independent, if for all possible values x, y the events $\mathbf{r}(\mathbf{x} = x)$ and $\mathbf{r}(\mathbf{y} = y)$ are independent.

More formally

D. 22 The discrete variates $\mathbf{x}$ and $\mathbf{y}$ are independent if, for all real values (x, y),

$$\mathbf{p'\hat{r}}(\mathbf{x} = x)\mathbf{r}(\mathbf{y} = y) = \mathbf{p'r}(\mathbf{x} = x)\mathbf{p'r}(\mathbf{y} = y). \qquad (5.14.1)$$

In other words, the bivariate probability function for $(\mathbf{x}, \mathbf{y})$ must be the product of the two marginal probability functions for $\mathbf{x}$ and for $\mathbf{y}$. The condition in fact may be written

D. 22*a* The discrete variates $\mathbf{x}$ and $\mathbf{y}$ are independent if, and only if for all real values (x, y),

$$f(x, y) = g(x)\, h(y), \qquad (5.14.2)$$

where $f(x, y)$ is the bivariate probability function and $g(x)$ and $h(y)$ satisfy

$$\sum_{x} g(x) = \sum_{y} h(y) = 1. \qquad (5.14.3)$$

It is then easily shown by summation that $g(x)$ and $h(y)$ are the marginal probability functions for $\mathbf{x}$ and $\mathbf{y}$.

It follows from this definition that if $(\mathbf{x}, \mathbf{y})$ are independent and if x is any possible value of $\mathbf{x}$, then

$$f(y \mid \mathbf{x} = x) = h(y). \qquad (5.14.4)$$

Thus, if $\mathbf{x}$, $\mathbf{y}$ are independent, then the conditional probability function for $\boldsymbol{y}$, given that $\boldsymbol{x}$ takes any specified possible value x, is the same as the (unconditional) marginal probability function for $\boldsymbol{y}$.

This property of the independence of two variates illustrates the correspondence between the notion of the independence of $\boldsymbol{x}$ and $\boldsymbol{y}$ and the notion that the value of $\boldsymbol{x}$ is irrelevant to the probabilities of getting various values for $\boldsymbol{y}$.

Needless to say, these conditions for the independence of a pair of variates are entirely symmetrical between the two variates. If $\boldsymbol{x}$ and $\boldsymbol{y}$ are independent, then assuredly $\boldsymbol{y}$ and $\boldsymbol{x}$ are independent also.

5.15. The Expected Value of the Product of Two Independent Variates

If $\boldsymbol{x}$ and $\boldsymbol{y}$ are a pair of independent variates, then, by (5.10.2), the expected value of their product is

$$E(\hat{\boldsymbol{x}}\boldsymbol{y}) = \sum_x \sum_y \boldsymbol{p'\hat{r}}(\boldsymbol{x} = x)\boldsymbol{r}(\boldsymbol{y} = y)xy \tag{5.15.1}$$

$$= \sum_x \sum_y \boldsymbol{p'r}(\boldsymbol{x} = x)\boldsymbol{p'r}(\boldsymbol{y} = y)xy \tag{5.15.2}$$

by virtue of the definition of independence of $\boldsymbol{x}$ and $\boldsymbol{y}$.

$$\therefore \quad E(\hat{\boldsymbol{x}}\boldsymbol{y}) = \sum_x \boldsymbol{p'r}(\boldsymbol{x} = x)x \sum_y \boldsymbol{p'r}(\boldsymbol{y} = y)y$$

$$= \sum_x f(x)x \sum_y g(y)y \tag{5.15.3}$$

where $f(x)$ and $g(y)$ are the (marginal) probability functions for the variates $\boldsymbol{x}$ and $\boldsymbol{y}$.

Hence, by (5.8.6), the expected value of the product of the two independent variates $\boldsymbol{x}$ and $\boldsymbol{y}$ is

$$E(\hat{\boldsymbol{x}}\boldsymbol{y}) = E(\boldsymbol{x})\, E(\boldsymbol{y}), \tag{5.15.4}$$

which is the product of the expected values of the two independent variates.

5.16. Distribution Functions

D. 23 The probability $\boldsymbol{p'r}(\boldsymbol{x} \leqslant x)$ that a discrete variate $\boldsymbol{x}$ does not exceed the variable x is known as the *distribution function* for $\boldsymbol{x}$. It is often convenient to denote it by $F(x)$ when it is clear from the context to what variate it refers.

The following obvious relation exists between the probability function $f(x)$ and the distribution function $F(x)$ for a variate $\boldsymbol{x}$:

$$F(x) = \sum_{X \leqslant x} f(X). \tag{5.16.1}$$

D. 23*a* The distribution function for a pair of variates $(\boldsymbol{x}, \boldsymbol{y})$ may similarly be defined as

$$F(x, y) = \boldsymbol{p}'\hat{\boldsymbol{r}}(\boldsymbol{x} \leqslant x)\boldsymbol{r}(\boldsymbol{y} \leqslant y) \qquad (5.16.2)$$

and clearly this is related to the bivariate probability function $f(x, y)$ by

$$F(x, y) = \sum_{X \leqslant x} \sum_{Y = y} f(X, Y). \qquad (5.16.3)$$

Owing to these relations between the distribution functions and the probability functions, the conditions for the independence of two variates $\boldsymbol{x}$ and $\boldsymbol{y}$ may be expressed in the form

$$F(x, y) = G(x)\,H(y) \qquad (5.16.4)$$

where F is the bivariate distribution function for $(\boldsymbol{x}, \boldsymbol{y})$ and G and H are the separate distribution functions for $\boldsymbol{x}$ and for $\boldsymbol{y}$.

We shall occasionally refer to the function $F(y \mid \boldsymbol{x} \leqslant x)$ defined as

$$F(y \mid \boldsymbol{x} \leqslant x) = \boldsymbol{p}'_{\mid \boldsymbol{r}(\boldsymbol{x} \leqslant x)}\boldsymbol{r}(\boldsymbol{y} \leqslant y) \qquad (5.16.5)$$

as the conditional distribution function for $\boldsymbol{y}$, given $(\boldsymbol{x} \leqslant x)$.

5.17. Multivariate Probability Functions

Where $\boldsymbol{x}(t)$ $(t = 1, 2, \ldots, m)$ are any m discrete variates, the probability that they will take any particular set of values $x(t)$ $(t = 1, 2, \ldots, m)$ will be given by their probability function

$$f[x(1), x(2), \ldots, x(m)] = \boldsymbol{p}'[\bigcap_{t=1}^{m} \boldsymbol{r}\{\boldsymbol{x}(t) = x(t)\}]. \qquad (5.17.1)$$

The expected value of any function $\phi[\boldsymbol{x}(1), \boldsymbol{x}(2), \ldots, \boldsymbol{x}(m)]$ of the $\boldsymbol{x}(t)$ will then be given by

$$E(\phi) = \sum_{x(1)} \sum_{x(2)} \cdots \sum_{x(m)} f[x(1), \ldots, x(m)]\,\phi[x(1), \ldots, x(m)]. \qquad (5.17.2)$$

The definition D. 14 (§ **4.22**) of the independence of m disjoint exhaustive events may now be used to provide the definition of the independence of m variates $\boldsymbol{x}(t)$ $(t = 1, 2, \ldots, m)$.

D. 24 The m variates $\boldsymbol{x}(t)$ $(t = 1, 2, \ldots, m)$ are independent, if and only if, for all possible sets of values $x(t)$, their joint probability function is equal to the product of m functions, one of each of the variables $x(t)$; these can then be chosen so that they are the individual (marginal) probability functions of the $\boldsymbol{x}(t)$.

This condition can be written

$$\mathbf{p}'[\bigcap_{t=1}^{m} \mathbf{r}\{\mathbf{x}(t) = x(t)\}] = \prod_{t=1}^{m} [\mathbf{p}'\mathbf{r}\{\mathbf{x}(t) = x(t)\}]$$

for all possible sets of values $x(t)$ $(t = 1, 2, \ldots, m)$. (5.17.3)

The special case where the independent variates $\mathbf{x}(t)$ $(t = 1, 2, \ldots, m)$ all have the *same* marginal probability function $g(x)$ is of particular interest. We may then regard them as m independent observations of a particular kind of variate to be called a *chance variate*. We accordingly adopt the definition

D. 25 The set of variates $\mathbf{x}(t)$ $(t = 1, 2, \ldots, m)$ constitute m observations of a *chance variate* $\mathbf{x}$ with probability function $g(x)$ if they are an independent set and each has the probability function $g(x)$.

The probability that the m observations of a chance variate $\mathbf{x}$ will take the specified values $x(t)$ $(t = 1, 2, \ldots, m)$ is then $\prod_{t=1}^{m} g\{x(t)\}$, where $g(x)$ is the probability function for $\mathbf{x}$.

5.18. Deviations

If $\mathbf{y}$ is any variate, one may consider the deviation $\{\mathbf{y}-\mathbf{1}E(\mathbf{y})\}$ of $\mathbf{y}$ from its own expected value as being itself a variate. We shall refer to this variate more shortly as the *deviation* of $\mathbf{y}$ and we shall denote it by $\mathbf{y} \cdot \mathbf{1}$. The reason for using this notation will appear in § **5.23** below. The formal definition of the deviation of $\mathbf{y}$ is as follows.

D. 26 The deviation of any variate $\mathbf{y}$ is given by

$$\mathbf{y} \cdot \mathbf{1} = \mathbf{y} - \mathbf{1}E(\mathbf{y}).$$

EXAMPLE 5.18.1. Prove that if $\mathbf{y}$ has an expected value $E(y)$ then

$$E(\mathbf{y} \cdot \mathbf{1}) = 0.$$

Solution. $E(\mathbf{y} \cdot \mathbf{1}) = E\{\mathbf{y}-\mathbf{1}E(\mathbf{y})\} \cdot \mathbf{p}'\{\mathbf{y}-\mathbf{1}\mathbf{p}'\mathbf{y}\}$

$\therefore$ $E(\mathbf{y} \cdot \mathbf{1}) = \mathbf{p}'\mathbf{y}-\mathbf{p}'\mathbf{y} = 0.$

5.19. Covariance and Variance

If $\mathbf{x}$ and $\mathbf{y}$ are any two variates, the expected value of the product of their deviations is called their *covariance*. Thus

D. 27 The covariance of two variates $\mathbf{x}$ and $\mathbf{y}$ is given by

$$\text{cov}(\mathbf{x}, \mathbf{y}) = (\mathbf{x} \cdot \mathbf{1})' \hat{\mathbf{p}} (\mathbf{y} \cdot \mathbf{1}).$$

EXAMPLE 5.19.1. Prove that if $\mathbf{x}$ and $\mathbf{y}$ have expected values and are independent, then cov $(\mathbf{x}, \mathbf{y}) = 0$.

Solution. If $\mathbf{x}$ $\mathbf{y}$ are independent, it follows from (5.15.4) that

$$\mathbf{x}'\hat{\mathbf{p}}\mathbf{y} = \mathbf{p}'\mathbf{x}\ \mathbf{p}'\mathbf{y}$$

$$\begin{aligned} \therefore \quad \text{cov}(\mathbf{x}, \mathbf{y}) &= (\mathbf{x} \cdot \mathbf{1})' \hat{\mathbf{p}} (\mathbf{y} \cdot \mathbf{1}) = \{\mathbf{x} - \mathbf{1}\mathbf{p}'\mathbf{x}\}' \hat{\mathbf{p}} \{\mathbf{y} - \mathbf{1}\mathbf{p}'\mathbf{y}\} \\ &= \mathbf{x}'\hat{\mathbf{p}}\mathbf{y} - \mathbf{x}'\mathbf{p}\mathbf{1}'\hat{\mathbf{p}}\mathbf{y} - \mathbf{x}'\hat{\mathbf{p}}\mathbf{1}\mathbf{p}'\mathbf{y} + \mathbf{x}'\mathbf{p}\mathbf{1}'\hat{\mathbf{p}}\mathbf{1}\mathbf{p}'\mathbf{y} \\ &= \mathbf{p}'\mathbf{x}\mathbf{p}'\mathbf{y} - \mathbf{p}'\mathbf{x}\mathbf{p}'\mathbf{y} - \mathbf{p}'\mathbf{x}\mathbf{p}'\mathbf{y} + \mathbf{p}'\mathbf{x}\mathbf{p}'\mathbf{y} \end{aligned}$$

$$\therefore \quad \text{cov}(\mathbf{x}, \mathbf{y}) = 0.$$

EXAMPLE 5.19.2. Show that, if $x_0 = \mathbf{p}'\mathbf{x}$, $y_0 = \mathbf{p}'\mathbf{y}$ and $f(x, y)$ is the probability function for $(\mathbf{x}, \mathbf{y})$, then

$$\text{cov}(\mathbf{x}, \mathbf{y}) = \mathbf{p}'\mathbf{x}\mathbf{y} - x_0 y_0 = \sum_x \sum_y f(x, y)(x - x_0) y.$$

Solution

$$\begin{aligned} \text{cov}(\mathbf{x}, \mathbf{y}) &= (\mathbf{x} - \mathbf{1}x_0)' \hat{\mathbf{p}} (\mathbf{y} - \mathbf{1}y_0) = \mathbf{x}'\hat{\mathbf{p}}\mathbf{y} - \mathbf{x}'\hat{\mathbf{p}}\mathbf{1}y_0 - x_0\mathbf{1}'\hat{\mathbf{p}}\mathbf{y} + x_0\mathbf{1}'\hat{\mathbf{p}}\mathbf{1}y_0 \\ &= \mathbf{p}'\hat{\mathbf{x}}\mathbf{y} - \mathbf{x}'\mathbf{p}y_0 - x_0\mathbf{p}'\hat{\mathbf{1}}\mathbf{y} + x_0\mathbf{1}'\mathbf{p}y_0 \\ &= \mathbf{p}'\hat{\mathbf{x}}\mathbf{y} - x_0 y_0 - x_0 y_0 + x_0 y_0 = (\mathbf{p}'\hat{\mathbf{x}}\mathbf{y}) - x_0 y_0 \end{aligned}$$

$\therefore$ cov $(\mathbf{x}, \mathbf{y}) = E(\hat{\mathbf{x}}\mathbf{y}) - x_0 E(\mathbf{y}) = E(\phi(\mathbf{x}, \mathbf{y}))$ where $\phi(x, y) = (x - x_0)y$

Hence, by (5.10.3), cov $(\mathbf{x}, \mathbf{y}) = \sum_x \sum_y f(x, y)(x - x_0) y$.

EXAMPLE 5.19.3. Let $g(x)$ be the marginal probability function for $\mathbf{x}$, and let $\bar{y}(x)$ denote the expected value of $\mathbf{y}$, given that $\mathbf{x} = x$. Prove that $g(x)\bar{y}(x) = \sum_y f(x, y) y$ where $f(x, y)$ is the probability function for $(\mathbf{x}, \mathbf{y})$.

Solution $\quad \bar{y}(x) = \mathbf{p}'_{|\mathbf{r}(\mathbf{x} = x)}\mathbf{y}, \quad g(x) = \mathbf{p}'\mathbf{r}(\mathbf{x} = x)$, and hence

$$\begin{aligned} g(x)\bar{y}(x) &= \mathbf{p}'\hat{\mathbf{r}}(\mathbf{x} = x)\mathbf{y} = \mathbf{p}'\hat{\mathbf{r}}(\mathbf{x} = x) \sum_y \mathbf{r}(\mathbf{y} = y) y \\ &= \sum_y \mathbf{p}'\hat{\mathbf{r}}(\mathbf{x} = x)\mathbf{r}(\mathbf{y} = y) y = \sum_y f(x, y) y \end{aligned}$$

EXAMPLE 5.19.4. Show that if $\bar{y}(x)$ (as defined in Example 5.19.3) is a monotonic increasing function of x, then the covariance of $\mathbf{x}$ and $\mathbf{y}$ is positive, provided $\bar{y}(x)$ does not take the same value for all possible values of x.

Solution. By example 5.19.2 $\text{cov}(\mathbf{x}, \mathbf{y}) = \sum_x \sum_y f(x, y)(x - x_0)y$.

Hence $\text{cov}(\mathbf{x}, \mathbf{y}) = \sum_x (x - x_0)g(x)y(x)$ by example 5.19.3.

Let $\bar{y}(x_0) = Y_0$; then since $\sum_x (x - x_0)g(x) = \sum_x xg(x) - x_0 = 0$,

$$\text{cov}(\mathbf{x}, \mathbf{y}) = \sum_x (x - x_0)g(x)\bar{y}(x) = \sum_x (x - x_0)\{\bar{y}(x) - Y_0\}g(x) > 0,$$

since, for all x, $(x - x_0)\{\bar{y}(x) - Y_0\} \geqslant 0$,
because $\bar{y}(x)$ is a monotonic increasing function of x, and $\bar{y}(x) = Y_0$ when $x - x_0 = 0$. Moreover, since $\bar{y}(x)$ does not take the same value for all values of x, there must be at least one value of x for which $(x - x_0)\{\bar{y}(x) - Y_0\}g(x)$ is not zero.

The expected value of the square of the deviation of any variate is called its *variance*. Thus the variance of $\mathbf{y}$ may be thought of as the covariance of $\mathbf{y}$ and $\mathbf{y}$. Thus

D. 28 The variance of any variate $\mathbf{y}$ is given by

$$\text{var}(\mathbf{y}) = \text{cov}(\mathbf{y}, \mathbf{y}) = E(\mathbf{y} \cdot 1)^2 = (\mathbf{y} \cdot \mathbf{1})'\hat{\mathbf{p}}(\mathbf{y} \cdot 1).$$

The variance of $\mathbf{y}$ is of course to be measured in units which are the squares of the units in which $\mathbf{y}$ is measured. For this reason, it is often convenient to use a related measure, *the standard error* of $\mathbf{y}$, which is simply the positive square root of the variance of $\mathbf{y}$.

D. 29 The standard error of any variate $\mathbf{y}$ is given by

$$\sigma(\mathbf{y}) = \sqrt{\{\text{var}(\mathbf{y})\}} \geqslant 0.$$

The standard error of any variate is measured in the same units as the variate itself is measured; it may be regarded as giving some indication of the extent to which the variate is likely to differ from its own expected value.

EXAMPLE 5.19.5. Prove that $\text{var}(\mathbf{x} + \mathbf{y}) = \text{var}\,\mathbf{x} + 2\,\text{cov}(\mathbf{x}, \mathbf{y}) + \text{var}\,\mathbf{y}$.

Solution
$$\begin{aligned} \text{var}(\mathbf{x} + \mathbf{y}) &= \{(\mathbf{x} + \mathbf{y}) \cdot \mathbf{1}\}'\hat{\mathbf{p}}\{(\mathbf{x} + \mathbf{y}) \cdot \mathbf{1}\} \\ &= (\mathbf{x} \cdot \mathbf{1})'\hat{\mathbf{p}}(\mathbf{x} \cdot \mathbf{1}) + 2(\mathbf{x} \cdot \mathbf{1})'\hat{\mathbf{p}}(\mathbf{y} \cdot \mathbf{1}) + (\mathbf{y} \cdot \mathbf{1})'\hat{\mathbf{p}}(\mathbf{y} \cdot \mathbf{1}) \\ \text{var}(\mathbf{x} + \mathbf{y}) &= \text{var}(\mathbf{x}) + 2\,\text{cov}(\mathbf{x}, \mathbf{y}) + \text{var}\,\mathbf{y}. \end{aligned}$$

5.20. Other Expressions for Variance and Covariance

It follows from definitions D. 26 and D. 27 that

$$\text{cov}(\mathbf{x}, \mathbf{y}) = \{\mathbf{x} - \mathbf{1p'x}\}'\hat{\mathbf{p}}\{\mathbf{y} - \mathbf{1p'y}\} = \mathbf{x}'\hat{\mathbf{p}}\mathbf{y} - \mathbf{p'xp'y} \tag{5.20.1}$$

$$\therefore \quad \text{cov}(\boldsymbol{x}, \boldsymbol{y}) = E(\boldsymbol{\hat{x}y}) - E(\boldsymbol{x})E(\boldsymbol{y}), \tag{5.20.2}$$

a form which we shall use in later chapters.

Again, we may use (5.8.6), (5.10.4) and (5.20.2) to obtain

$$\text{cov}(\boldsymbol{x}, \boldsymbol{y}) = \sum_x \sum_y f(x, y)xy - \sum_x g(x)x \sum_y h(y)y, \tag{5.20.3}$$

where $f(x, y)$ is the bivariate probability function and $g(x)$, $h(y)$ are the marginal probability functions for $\boldsymbol{x}$ and $\boldsymbol{y}$.

Then, putting $\boldsymbol{y} = \boldsymbol{x}$ in (5.20.2) and using D. 28, we find

$$\text{var}\,\boldsymbol{y} = E(\boldsymbol{y}^2) - \{E(\boldsymbol{y})\}^2 \tag{5.20.4}$$

and hence, by (5.8.6)

$$\text{var}\,\boldsymbol{y} = \sum_y h(y)y^2 - \{\sum_y h(y)y\}^2. \tag{5.20.5}$$

5.21. Correlation Coefficient

It is often convenient to use a measure reflecting covariance, which is independent of the units in which the variates are measured. For this purpose one may employ the *correlation coefficient* which is defined as follows.

D. 30 The correlation coefficient between two variates, $\boldsymbol{x}$ and $\boldsymbol{y}$ is given by

$$r(\boldsymbol{x}, \boldsymbol{y}) = \text{cov}(\boldsymbol{x}, \boldsymbol{y}) \,/\, \{\sigma(\boldsymbol{x})\sigma(\boldsymbol{y})\}.$$

EXAMPLE 5.21.1. Prove that $-1 \leqslant r(\boldsymbol{x}, \boldsymbol{y}) \leqslant 1$.

Solution. It follows from Example 5.19.3 that

$$\text{var}\left\{\frac{\boldsymbol{x}}{\sigma(\boldsymbol{x})} + \frac{\boldsymbol{y}}{\sigma(\boldsymbol{y})}\right\} = \text{var}\left\{\frac{\boldsymbol{x}}{\sigma(\boldsymbol{x})}\right\} + 2\,\text{cov}\left\{\frac{\boldsymbol{x}}{\sigma(\boldsymbol{x})}, \frac{\boldsymbol{y}}{\sigma(\boldsymbol{y})}\right\} + \text{var}\left\{\frac{\boldsymbol{y}}{\sigma(\boldsymbol{y})}\right\}$$

$$\therefore \quad 0 \leqslant 1 + 2\frac{\text{cov}(\boldsymbol{x}, \boldsymbol{y})}{\sigma(\boldsymbol{x})\sigma(\boldsymbol{y})} + 1 = 2\{1 + r(\boldsymbol{x}, \boldsymbol{y})\}.$$

$$\therefore \quad -1 \leqslant r(\boldsymbol{x}, \boldsymbol{y}).$$

Similarly, since $0 \leqslant \text{var}\left\{\frac{\boldsymbol{x}}{\sigma(\boldsymbol{x})} - \frac{\boldsymbol{y}}{\sigma(\boldsymbol{y})}\right\}$, it follows from Example 5.19.3 that $0 \leqslant 2\{1 - r(\boldsymbol{x}, \boldsymbol{y})\}$.

Hence $$-1 \leqslant r(\boldsymbol{x}, \boldsymbol{y}) \leqslant 1.$$

If the correlation coefficient $r(\boldsymbol{x}, \boldsymbol{y})$ is zero, then the variates are said to be *uncorrelated.* Provided $\boldsymbol{x}$ and $\boldsymbol{y}$ have finite variances, they will be uncorrelated if, and only if, $\text{cov}(\boldsymbol{x}, \boldsymbol{y}) = 0$.

EXAMPLE 5.21.2. Show that a pair of uncorrelated variates need not be independent, but that a pair of independent variates with finite variance must be uncorrelated.

Solution. Consider the pair of variates $\boldsymbol{x}' = (1, 1, 1, -3)$ and $\boldsymbol{y}' = (0, 1, 2, 1)$ with probability vector $\boldsymbol{p}' = (1/4, 1/4, 1/4, 1/4)$.

$$\text{cov}\,(\boldsymbol{x}, \boldsymbol{y}) = (\boldsymbol{x}\cdot\mathbf{1})'\hat{\boldsymbol{p}}(\boldsymbol{y}\cdot\mathbf{1}) = \boldsymbol{x}'\hat{\boldsymbol{p}}\boldsymbol{y} - \boldsymbol{p}'\boldsymbol{x}\boldsymbol{p}'\boldsymbol{y} = \boldsymbol{x}'\hat{\boldsymbol{p}}\boldsymbol{y}, \quad \text{since} \quad \boldsymbol{p}'\boldsymbol{x} = 0.$$

$\therefore$ cov $(\boldsymbol{x}, \boldsymbol{y}) = 0$ showing that $\boldsymbol{x}$ and $\boldsymbol{y}$ are uncorrelated.

But
$$\boldsymbol{p}'\hat{\boldsymbol{r}}(\boldsymbol{x} = -3)\boldsymbol{r}(\boldsymbol{y} = 0) = 0,$$

whereas
$$\boldsymbol{p}'\boldsymbol{r}(\boldsymbol{x} = -3)\boldsymbol{p}'\boldsymbol{r}(\boldsymbol{y} = 0) = 1/16.$$

Hence the events $\boldsymbol{r}(\boldsymbol{x} = -3)$ and $\boldsymbol{r}(\boldsymbol{y} = 0)$ are not independent; and hence $\boldsymbol{x}$ and $\boldsymbol{y}$ are not independent.

We thus have an example of a pair of uncorrelated variates which are not independent.

But if $\boldsymbol{x}$ and $\boldsymbol{y}$ have finite variance and are independent then by (5.15.4)

$$E(\hat{\boldsymbol{x}}\boldsymbol{y}) = E(\boldsymbol{x})E(\boldsymbol{y}).$$

Hence, by (5.20.2), $\qquad$ cov $(\boldsymbol{x}, \boldsymbol{y}) = 0$
and $\boldsymbol{x}$ $\boldsymbol{y}$ are uncorrelated.

5.22. The Variance-Covariance Matrix

We may generalise the concept of the covariance of a pair of variates $\boldsymbol{x}$, $\boldsymbol{y}$ to apply to a set of k variates $\boldsymbol{x}(1), \boldsymbol{x}(2), \ldots, \boldsymbol{x}(k)$.

Suppose that, as usual, these variates are represented by k column vectors: then these k vectors form a k-column matrix which we denote by $\boldsymbol{X}$. Similarly we may denote by $\boldsymbol{X}\cdot\mathbf{1}$ the k-column matrix whose columns are $\boldsymbol{x}(t)\cdot\mathbf{1}$ $(t = 1, 2, \ldots, k)$ namely the deviations of the k variates.

It will be recalled that for each (s, t),

$$\{\boldsymbol{x}(s)\cdot\mathbf{1}\}'\hat{\boldsymbol{p}}\{\boldsymbol{x}(t)\cdot\mathbf{1}\} = \text{cov}\,\{\boldsymbol{x}(s), \boldsymbol{x}(t)\}, \tag{5.22.1}$$

hence the matrix

$$V = \{\boldsymbol{X}\cdot\mathbf{1}\}'\hat{\boldsymbol{p}}\{\boldsymbol{X}\cdot\mathbf{1}) \tag{5.22.2}$$

will have elements V_{st} $(s = 1, 2, \ldots, k;\ t = 1, 2, \ldots, k)$ such that

$$V_{st} = \text{cov}\,\{\boldsymbol{x}(s), \boldsymbol{x}(t)\} \tag{5.22.3}$$

and
$$V_{ss} = \text{var}\,\{\boldsymbol{x}(s)\}. \tag{5.22.4}$$

The matrix V is known as the variance-covariance matrix for the set of variates $\boldsymbol{X}$.

The formal definition is as follows.

D. 31 If the columns of the matrix $\boldsymbol{X}$ represent variates, and if the columns of the matrix $\boldsymbol{X} \cdot \mathbf{1}$ represent the deviations of these variates, then the variance-covariance matrix for $\boldsymbol{X}$ is given by

$$V(X) = \{\boldsymbol{X} \cdot \mathbf{1}\}' \hat{p} \{\boldsymbol{X} \cdot \mathbf{1}\}$$

and its typical element V_{st} gives the covariance between the variates represented by columns s and t of $\boldsymbol{X}$.

In the particular case where the variates represented by the columns of $\boldsymbol{X}$ are uncorrelated, all the off-diagonal elements of the variance-covariance matrix of $\boldsymbol{X}$ must be zero.

If k linearly independent variates are represented by a k-column matrix $\boldsymbol{X}$, it is always possible, and often convenient, to apply a linear transformation

$$\boldsymbol{Y} = \boldsymbol{X}\boldsymbol{B} \tag{5.22.5}$$

where $\boldsymbol{B}$ is a non-singular $k \times k$ matrix, to obtain a new set of k uncorrelated variates represented by the matrix $\boldsymbol{Y}$.

In discussing such transformations it will be found helpful to use the concepts of orthogonality, and of corrected variates, as developed in the following section.

5.23. Orthogonality and Corrected Variates

D. 32 Two variates $\boldsymbol{x}$ and $\boldsymbol{y}$ are said to be *orthogonal* if the expected value of their product is zero, i.e. if $\boldsymbol{x}'\hat{\boldsymbol{p}}\boldsymbol{y} = 0$.

The importance of this concept of orthogonality arises from the fact that a necessary and sufficient condition for two variates to be uncorrelated is that their deviations should be orthogonal.

EXAMPLE 5.23.1. Show that two variates $\boldsymbol{x}$, $\boldsymbol{y}$ are uncorrelated
(*i*) if, and only if, $\boldsymbol{x} \cdot \mathbf{1}$ and $\boldsymbol{y} \cdot \mathbf{1}$ are orthogonal
(*ii*) if, and only if, $\boldsymbol{x}$ and $\boldsymbol{y} \cdot \mathbf{1}$ are orthogonal.

Solution (*i*) follows immediately from D. 27 and D. 32.
(*ii*) follows from (*i*) and from the fact that

$$\begin{aligned}(\boldsymbol{x} \cdot \mathbf{1})' \hat{\boldsymbol{p}} (\boldsymbol{y} \cdot \mathbf{1}) &= \boldsymbol{x}' \hat{\boldsymbol{p}} (\boldsymbol{y} \cdot \mathbf{1}) - (\boldsymbol{p}'\boldsymbol{x}) \mathbf{1}' \hat{\boldsymbol{p}} (\boldsymbol{y} \cdot \mathbf{1}) \\ &= \boldsymbol{x}' \hat{\boldsymbol{p}} (\boldsymbol{y} \cdot \mathbf{1}).\end{aligned}$$

Suppose $\boldsymbol{x}$ and $\boldsymbol{y}$ are any two variates. Then it is always possible to find a new variate $\boldsymbol{z}$, given by

$$\boldsymbol{z} = \boldsymbol{y} - \boldsymbol{x}\beta \qquad (\beta \text{ is a scalar}) \tag{5.23.1}$$

which is orthogonal to the variate $\boldsymbol{x}$. This variate, $\boldsymbol{z}$ will be called '$\boldsymbol{y}$ *corrected for* $\boldsymbol{x}$' and denoted $\boldsymbol{y} \cdot \boldsymbol{x}$. We adopt the following definition:

D. 33. Where $\boldsymbol{x}$ and $\boldsymbol{y}$ are variates, '$\boldsymbol{y}$ corrected for $\boldsymbol{x}$' is given by

$$\boldsymbol{y} \cdot \boldsymbol{x} = \boldsymbol{y} - \boldsymbol{x}\beta$$

where the scalar β is so chosen that

$$\boldsymbol{x}'\hat{\boldsymbol{p}}(\boldsymbol{y} \cdot \boldsymbol{x}) = 0$$

(this being the condition that $\boldsymbol{y} \cdot \boldsymbol{x}$ and $\boldsymbol{x}$ should be orthogonal).

The value of the coefficient β such that

$$\boldsymbol{y} \cdot \boldsymbol{x} = \boldsymbol{y} - \boldsymbol{x}\beta \tag{5.23.2}$$

is easily found; for, since

$$\boldsymbol{x}'\hat{\boldsymbol{p}}(\boldsymbol{y} \cdot \boldsymbol{x}) = 0, \tag{5.23.3}$$

$$\boldsymbol{x}'\hat{\boldsymbol{p}}\boldsymbol{y} = (\boldsymbol{x}'\hat{\boldsymbol{p}}\boldsymbol{x})\beta.$$

$$\therefore \quad \beta = (\boldsymbol{x}'\hat{\boldsymbol{p}}\boldsymbol{x})^{-1}\boldsymbol{x}'\hat{\boldsymbol{p}}\boldsymbol{y}. \tag{5.23.4}$$

In the particular case where $\boldsymbol{x} = \boldsymbol{1}$ we find

$$\boldsymbol{y} \cdot \boldsymbol{1} = \boldsymbol{y} - \boldsymbol{1}\beta, \tag{5.23.5}$$

where

$$\beta = (\boldsymbol{1}'\hat{\boldsymbol{p}}\boldsymbol{1})^{-1}\boldsymbol{1}'\hat{\boldsymbol{p}}\boldsymbol{y} = \boldsymbol{p}'\boldsymbol{y}. \tag{5.23.6}$$

$$\therefore \quad \boldsymbol{y} \cdot \boldsymbol{1} = \boldsymbol{y} - \boldsymbol{1}(\boldsymbol{p}'\boldsymbol{y}) = \boldsymbol{y} - \boldsymbol{1}E(\boldsymbol{y}). \tag{5.23.7}$$

and the right-hand side is the deviation of $\boldsymbol{y}$ from its expected value $\boldsymbol{p}'\boldsymbol{y}$. Equation (23.7) underlies the notation employed in the above definition, D. 26, for the deviation of a variate.

The definitions of orthogonality and of corrected variates may be simply extended to allow correction for a set of variates $\boldsymbol{X}$ instead of for a single variate $\boldsymbol{x}$ only.

D. 34 The variate $\boldsymbol{y}$ is said to be orthogonal to the set of variates represented by the columns of the matrix $\boldsymbol{X}$ if

$$\boldsymbol{y}'\hat{\boldsymbol{p}}\boldsymbol{X} = \boldsymbol{0}.$$

Similarly, if a set of variates are represented by the columns of the matrix, we may speak of the variate '$\boldsymbol{y}$ corrected for the variates $\boldsymbol{X}$', and define it as follows:

D. 35 The variate '$\boldsymbol{y}$ corrected for the variates $\boldsymbol{X}$' is given by

$$\boldsymbol{y} \cdot \boldsymbol{X} = \boldsymbol{y} - \boldsymbol{X\beta},$$

where the vector $\boldsymbol{\beta}$ is so chosen that

$$\boldsymbol{X}'\hat{\boldsymbol{p}}(\boldsymbol{y} \cdot \boldsymbol{X}) = \boldsymbol{0}$$

(this being the condition that $\boldsymbol{y} \cdot \boldsymbol{X}$ and $\boldsymbol{X}$ should be orthogonal).

EXAMPLE 5.23.2. Show that, if $\boldsymbol{y}$ is orthogonal to the variates represented by the k-column matrix $\boldsymbol{X}$, then it is orthogonal also to the m variates represented by the k-column matrix $\boldsymbol{XB}$ where $\boldsymbol{B}$ is any $k \times m$ matrix.

Solution. We need only note that, where

$$\boldsymbol{y}'\hat{\boldsymbol{p}}\boldsymbol{X} = \boldsymbol{0} \quad \text{and } \boldsymbol{B} \text{ is a } k \times m \text{ matrix}$$

$$\boldsymbol{y}'\hat{\boldsymbol{p}}\boldsymbol{XB} = \boldsymbol{0}.$$

5.24. The Pair of Uncorrelated Variates $\boldsymbol{x}$ and $\boldsymbol{y} \cdot (\boldsymbol{1}, \boldsymbol{x})$

By correcting any variate $\boldsymbol{y}$ both for the 'variate' $\boldsymbol{1}$ and for the variate $\boldsymbol{x}$ we obtain a new variate $\boldsymbol{y} \cdot (\boldsymbol{1}, \boldsymbol{x})$ orthogonal both to $\boldsymbol{1}$ and to $\boldsymbol{x}$, and hence orthogonal also to $\boldsymbol{x} \cdot \boldsymbol{1}$.

Since $\boldsymbol{y} \cdot (\boldsymbol{1}, \boldsymbol{x})$ is orthogonal to $\boldsymbol{x} \cdot \boldsymbol{1}$, it is uncorrelated with $\boldsymbol{x}$. Since $\boldsymbol{y} \cdot (\boldsymbol{1}, \boldsymbol{x})$ is orthogonal to $\boldsymbol{1}$, it has expected value zero.

Thus,

$$(\boldsymbol{x} \cdot \boldsymbol{1})'\hat{\boldsymbol{p}}\boldsymbol{y} \cdot (\boldsymbol{1}, \boldsymbol{x}) = \boldsymbol{1}'\hat{\boldsymbol{p}}\boldsymbol{y} \cdot (\boldsymbol{1}, \boldsymbol{x}) = 0. \tag{5.24.1}$$

It follows from the definition of $(\boldsymbol{y} \cdot \boldsymbol{X})$ (D. 35) that

$$\boldsymbol{y} \cdot (\boldsymbol{1}, \boldsymbol{x}) = \boldsymbol{y} - \boldsymbol{1}\beta_0 - \boldsymbol{x}\beta_1 \tag{5.24.2}$$

where β_0, β_1 are scalars. The values of β_0 and β_1 are easily found from the conditions

$$\boldsymbol{1}'\hat{\boldsymbol{p}}\boldsymbol{y} \cdot (\boldsymbol{1}, \boldsymbol{x}) = \boldsymbol{x}'\hat{\boldsymbol{p}}\boldsymbol{y} \cdot (\boldsymbol{1}, \boldsymbol{x}) = 0. \tag{5.24.3}$$

It follows from (5.24.3) that (5.24.2) may be written as

$$\boldsymbol{y} \cdot (\boldsymbol{1}, \boldsymbol{x}) = (\boldsymbol{y} \cdot \boldsymbol{1}) - \beta_1(\boldsymbol{x} \cdot \boldsymbol{1}). \tag{5.24.4}$$

Hence by (5.24.3)

$$(\boldsymbol{x} \cdot \boldsymbol{1})'\hat{\boldsymbol{p}}\boldsymbol{y} \cdot (\boldsymbol{1}, \boldsymbol{x}) = 0 \tag{5.24.5}$$

and by (5.24.4)

$$\beta_1 = (\boldsymbol{x} \cdot \boldsymbol{1})'\hat{\boldsymbol{p}}(\boldsymbol{y} \cdot \boldsymbol{1}) \,/\, (\boldsymbol{x} \cdot \boldsymbol{1})'\hat{\boldsymbol{p}}(\boldsymbol{x} \cdot \boldsymbol{1}) \tag{5.24.6}$$

$$\beta_1 = \operatorname{cov}(\boldsymbol{x}, \boldsymbol{y}) \,/\, \operatorname{var} \boldsymbol{x}. \tag{5.24.7}$$

Finally from (5.24.2) (5.24.3)

$$\mathbf{1}'\hat{\boldsymbol{p}}\boldsymbol{y} = \mathbf{1}'\hat{\boldsymbol{p}}\mathbf{1}\beta_0 + \mathbf{1}'\hat{\boldsymbol{p}}\boldsymbol{x}\beta_1, \tag{5.24.8}$$

$$\therefore \quad \beta_0 = E(\boldsymbol{y}) - \beta_1 E(\boldsymbol{x}), \tag{5.24.9}$$

where β_1 is given by (5.24.7).

5.25. The Linear Transformation of a Set of Variates to an Uncorrelated Set

We have seen that any pair of variates $\boldsymbol{x}$, $\boldsymbol{y}$ may be transformed to the uncorrelated pair $\boldsymbol{x}$, $\boldsymbol{y} \cdot (\mathbf{1}, \boldsymbol{x})$.

By a similar treatment we may transform the variates represented by the columns of a matrix $\boldsymbol{X}$ (provided they are linearly independent), to an uncorrelated set.

Let the columns be $\boldsymbol{x}(1), \boldsymbol{x}(2), \ldots, \boldsymbol{x}(k)$. Then we may show by applications of the definitions D. 32 and D. 35 that every pair of the set of variates

$$\mathbf{1},\ \boldsymbol{x}(1) \cdot \mathbf{1},\ \boldsymbol{x}(2) \cdot \{\mathbf{1}, x(1)\},\ \boldsymbol{x}(3) \cdot \{\mathbf{1}, \boldsymbol{x}(1), \boldsymbol{x}(2)\}, \ldots,$$
$$\boldsymbol{x}(k) \cdot \{\mathbf{1}, \boldsymbol{x}(1), \boldsymbol{x}(2), \ldots, \boldsymbol{x}(k-1)\} \text{ is orthogonal.}$$

It follows that the set of variates

$$\boldsymbol{x}(1),\quad \boldsymbol{x}(2) \cdot \{\mathbf{1}, \boldsymbol{x}(1)\}, \ldots, \boldsymbol{x}(k) \cdot \{\mathbf{1}, \boldsymbol{x}(1), \boldsymbol{x}(2), \ldots, \boldsymbol{x}(k-1)\}$$

is uncorrelated.

5.26. Moments

Expected values, variance and covariance are special cases of a class of measures known as *moments*. These may be used to summarise the properties of the probability function of a variate or pair of variates.

The expected values of the various powers of a variate are given by the *moments about the origin*, whereas the expected values of the various powers of the deviations of the variate are given by the *moments about the mean*. These are defined as follows:

D. 36 For any variate $\boldsymbol{x}$ the 'moments about the origin' are given by

$$\mu_r' = E(\boldsymbol{x}^r) \quad (r = 0, 1, 2, \ldots)$$

and the 'moments about the mean' by

$$\mu_r = E\{(\boldsymbol{x} \cdot \mathbf{1})^r\} \quad (r = 0, 1, 2, \ldots),$$

where $\boldsymbol{x}^r$ denotes $\hat{\boldsymbol{x}}^r\mathbf{1}$ and $(\boldsymbol{x} \cdot 1)^r$ denotes $\widehat{(\boldsymbol{x} \cdot \mathbf{1})}^r\mathbf{1}$.

Similarly, if $\boldsymbol{x}\ \boldsymbol{y}$ are a variate-pair, the bivariate moments are defined as follows:

D. 37 For any variate pair $(\boldsymbol{x}, \boldsymbol{y})$ the bivariate moments about the origin are given by

$$\mu'_{rs} = E\{\hat{\boldsymbol{x}}^r \boldsymbol{y}^s\} \quad (r = 0, 1, 2, \ldots;\ s = 0, 1, 2, \ldots),$$

and the bivariate moments about the mean are given by

$$\mu_{rs} = E\{\widehat{(\boldsymbol{x}\cdot\mathbf{1})^r}(\boldsymbol{y}\cdot\mathbf{1})^s\} \quad (r = 0, 1, 2, \ldots;\ s = 0, 1, 2, \ldots)$$

where $\boldsymbol{y}^s$ denotes $\hat{\boldsymbol{y}}^s\mathbf{1}$ and $(\boldsymbol{y}\cdot\mathbf{1})^s$ denotes $\widehat{(\boldsymbol{y}\cdot\mathbf{1})^s}\mathbf{1}$.

It follows from (5.8.6) and D. 36 that

$$\mu'_r = \sum_x f(x)x^r \quad \text{and} \quad \mu_r = \sum_x f(x)(x-\mu'_1)^r, \qquad (5.26.1)$$

where $f(x)$ is the probability function for $\boldsymbol{x}$; and it follows from (5.10.3) and D. 37 that

$$\mu'_{rs} = \sum_x \sum_y f(x, y)x^r y^s \quad \text{and}$$

$$\mu_{rs} = \sum_x \sum_y f(x, y)(x-\mu'_{10}))^r(y-\mu'_{01})^s \qquad (5.26.2)$$

where $f(x, y)$ is the probability function for $(\boldsymbol{x}, \boldsymbol{y})$.

EXAMPLE 5.26.1. The reader should check from the various definitions that for any variate $\boldsymbol{y}$,

$$\mu_1 = 0; \quad \mu_2 = \text{var}\ \boldsymbol{y}$$

and that for any variate-pair $(\boldsymbol{x}, \boldsymbol{y})$

$$\mu'_{01} = E(\boldsymbol{y});\ \mu_{02} = \text{var}\ \boldsymbol{y};\ \mu_{11} = \text{cov}\,(\boldsymbol{x}, \boldsymbol{y}) \quad \text{and}$$

$$\mu_{11} / \sqrt{(\mu_{20}\mu_{02})} = r(\boldsymbol{x}, \boldsymbol{y}).$$

5.27. Failure of Moments to Converge

At the end of § **8** it was mentioned that the sum $\boldsymbol{p}'\boldsymbol{x}$ need not converge to a limit for all variates $\boldsymbol{x}$.

EXAMPLE 5.27.1. Show that, if $\boldsymbol{p}' = \left(\frac{1}{1\,.\,2}, \frac{1}{2\,.\,3}, \frac{1}{3\,.\,4}, \ldots\right)$, although $\boldsymbol{p}'\mathbf{1} = 1$, $\boldsymbol{p}'\boldsymbol{x}$ fails to converge,

(*i*) if $\boldsymbol{x} = (1, 2, 3, 4, \ldots)$,
(*ii*) if $\boldsymbol{x} = (1, -1, 2, -2, 3, -3, \ldots)$.

Solution

$$p'\mathbf{1} = \frac{1}{1.2}+\frac{1}{2.3}+\frac{1}{3.4}\dots = (\tfrac{1}{1}-\tfrac{1}{2})+(\tfrac{1}{2}-\tfrac{1}{3})+(\tfrac{1}{3}-\tfrac{1}{4})+\dots=1. \tag{5.27.1}$$

$$(i)\ p'x = \tfrac{1}{2}+\tfrac{1}{3}+\tfrac{1}{4}\dots > \tfrac{1}{2}+(\tfrac{1}{4}+\tfrac{1}{4})+(\tfrac{1}{8}+\tfrac{1}{8}+\tfrac{1}{8}+\tfrac{1}{8})+\dots$$
$$= \tfrac{1}{2}+\tfrac{1}{2}+\tfrac{1}{2}+\dots,\quad \text{which is unbounded.} \tag{5.27.2}$$

(*ii*) $p'x$ may be arranged into its positive and negative terms; then

$$p'x = \left\{\frac{1}{1.2}+\frac{2}{3.4}+\frac{3}{5.6}+\frac{4}{7.8}\dots\right\}$$

$$-\left\{\frac{1}{2.3}+\frac{2}{4.5}+\frac{3}{6.7}+\frac{4}{8.9}\dots\right\}$$

$$\therefore\ p'x = \tfrac{1}{2}\{\tfrac{1}{1}+\tfrac{1}{3}+\tfrac{1}{5}+\tfrac{1}{7}\dots\}-\tfrac{1}{2}\{\tfrac{1}{3}+\tfrac{1}{5}+\tfrac{1}{7}+\tfrac{1}{9}\dots\}. \tag{5.27.3}$$

It is clear that each of the two terms is unbounded. Hence $p'x$ can be made to converge only if we prescribe the order in which the various products are to be taken during summation. The series for $p'x$ is, in this instance, only 'conditionally convergent'.

We have seen that all moments are expected values of powers of variates or of products of powers of variates. Moments may thus be regarded as expected values of variates, and it is understandable that moments may often be indeterminate or undefined.

EXAMPLE 5.27.2. x has the probability function

$$g(x) = x^{-\alpha} / \sum_{t=1}^{\infty} t^{-\alpha} \quad (\alpha > 1) \quad (x = 1, 2, 3, \dots).$$

Investigate which of the moments about the origin have finite values.

Solution. Whereas when $\alpha \leqslant +1$ the series $\sum_{t=1}^{\infty} t^{-\alpha}$ has no finite value

since
$$\sum_{t=1}^{T} t^{-\alpha} \geqslant \sum_{t=1}^{T} t^{-1} > \int_1^{T+1} t^{-1}dt = \log_e(T+1),$$

the series $\sum_{t=1}^{\infty} t^{-\alpha}$ has a finite sum when $\alpha > 1$ since

$$\sum_{t=1}^{T} t^{-\alpha} < 1+\int_1^T t^{-\alpha}dt = 1+\{1-T^{1-\alpha}\}/(\alpha-1).$$

Now, for this variate x,

$$\mu_r' = \sum_{x=1}^{\infty} x^{r-\alpha} \Big/ \sum_{t=1}^{\infty} t^{-\alpha}; \tag{5.27.4}$$

hence μ_r' is finite if, and only if, $r < (\alpha - 1)$.

When one is interested in the whole series of moments for a probability function $f(x)$, it may be convenient to regard them as generated by the moment-generating function

$$M'(\alpha) = E(e^{\alpha x}) = \sum_x f(x)e^{\alpha x} \tag{5.27.5}$$

provided this function converges. When it does so, we may expand $e^{\alpha x}$ as a power series in αx to obtain

$$M'(\alpha) = \sum_{r=0}^{\infty} \mu_r'\alpha^r / r! \tag{5.27.6}$$

A fuller discussion of the moment-generating function is postponed until §§ **6.24-5** and §§ **8.19-21**.

EXAMPLE 5.27.3. x has the probability function

$$g(x) = (1+x^4)^{-1} / \sum_{t=-\infty}^{\infty} (1+t^4)^{-1}$$

$$x = 0, 1, -1, 2, -2, \ldots$$

Show that, although μ_r and μ_r' are finite for $r \leqslant 2$, they are not so for $r > 2$.

Solution. $\sum_{t=1}^{\infty} (1+t^4)^{-1}$ is convergent, since each term is less than the corresponding term of the convergent series $\sum_{t=1}^{\infty} t^{-4}$.

$\sum_{t=0}^{\infty} t(1+t^4)^{-1}$ may similarly be proved convergent.

$$\therefore \quad \sum_{t=-\infty}^{\infty} t(1+t^4)^{-1} = \sum_{t=1}^{\infty} t(1+t^4)^{-1} + \sum_{t=-\infty}^{-1} t(1+t^4)^{-1} = 0.$$

$$\therefore \quad E(x) = \mu_1' = 0.$$

$\therefore \quad \mu_r' = \mu_r$ whenever either is finite.

$$\mu_2' = \sum_{t=-\infty}^{\infty} t^2(1+t^4)^{-1} / 2S \quad \text{where} \quad S = \sum_{t=0}^{\infty} (1+t^4)^{-1};$$

$$\therefore \quad \mu_2' = \sum_{t=0}^{\infty} t^2(1+t^4)^{-1} / S.$$

But $\sum_{t=0}^{T} t^2(1+t^4)^{-1} < \sum_{t=1}^{T} t^{-2}$;

hence μ_2' is finite.

But $S\mu_3' = \sum_{t=1}^{\infty} t^3(1+t^4)^{-1} - \sum_{t=1}^{\infty} t^3(1+t^4)^{-1}$.

Now, $\sum_{t=1}^{\infty} t^3(1+t^4)^{-1}$ is not finite, since for $t > 1$ $t^3 / (1+t^4) > 1/2t$

and $\sum_{t=1}^{T} (1/2t) > \frac{1}{2} \log_e (T+1)$

$\therefore$ μ_3' is not determinate. *A fortiori* μ_r' is not determinate for $r > 3$. Moreover, if μ_r is finite $\mu_r' = \mu_r$.

Thus for $r \leqslant 2$, μ_r and μ_r' are determinate, but, for $r > 2$, μ_r and μ_r' are not determinate.

5.28. Other Descriptive Measures: Skewness, Kurtosis, Median, Quartiles and Other Quantiles

It was mentioned in § **5.21** that it is often found convenient to use a measure of covariance which is independent of the units in which the variates are measured. Such a measure, the correlation coefficient, was obtained by dividing the covariance by the product of the standard errors of the variates. In the same way it is often found convenient to obtain measure of the third and fourth moments μ_3 and μ_4, which are independent of the units in which the variate is measured. The measures of this kind usually employed are

$$\gamma_1 = \frac{\mu_3}{\mu_2^{3/2}} \text{ (skewness), and } \beta_2 = \frac{\mu_4}{\mu_2^2} \text{ (kurtosis).} \qquad (5.28.1)$$

Variates which cannot take negative values but can take values which are high multiples of the expected value, usually have considerable positive skewness: such variates are very common in economic applications. Variates which represent the arithmetic means of large samples taken at random from a fixed population usually have a kurtosis near to three. It is a common practice to measure kurtosis, not by the formula (5.28.1) but instead by

$$\gamma_2 = \beta_2 - 3 = \frac{\mu^4 - 3\mu_2^2}{\mu_2^2}. \qquad (5.28.2)$$

Illustrations. The values of γ_1 and γ_2 are given below for a few illustrative probability functions:

(*i*) $f(x) = 0{\cdot}01 \quad (x = 0, 1, 2, \ldots, 99) \quad \gamma_1 = 0 \quad \gamma_2 = 1{\cdot}20006$

(*ii*) $f(x) = \dfrac{x^2}{60} \quad (x = -4, -3, -2, -1, 0, 1, 2, 3, 4) \quad \gamma_1 = 0$

$\gamma_2 = -1{\cdot}83$

(*iii*) $f(x) = \dfrac{1}{\mathrm{e}.x!} \quad (x = 0, 1, 2, \ldots) \quad \gamma_1 = 1 \quad \gamma_2 = 1$

(*iv*) $f(x) = \binom{8}{x}(\tfrac{2}{3})^8\tfrac{1}{2}^x \quad (x = 0, 1, 2, 3, 4, 5, 6, 7, 8) \quad \gamma_1 = 1{\cdot}25$

$\gamma_2 = -0{\cdot}1875$

(*v*) $f(x) = \tfrac{1}{2}\,2^{-x} \quad (x = 0, 1, 2, \ldots) \quad \gamma_1 = 2{\cdot}121 \quad \gamma_2 = 6{\cdot}5$

Another set of measures for summarising the properties of a probability function are the *quantiles.* The simplest of these is the *Median,* which is the value of x such that $\boldsymbol{p'r}(\boldsymbol{x} \leqslant x) \geqslant \frac{1}{2}$ and that $\boldsymbol{p'r}(\boldsymbol{x} \geqslant x) \geqslant \frac{1}{2}$: but in cases where the value is not unique and there is a range of such values, the mid-point of the range is to be taken. Similarly the *lower quartile* is the value of x such that $\boldsymbol{p'r}(\boldsymbol{x} \leqslant x) \geqslant \frac{1}{4}$ and that $\boldsymbol{p'r}(\boldsymbol{x} \geqslant x) \geqslant \frac{3}{4}$, and the *upper quartile* is the value of x such that $\boldsymbol{p'r}(\boldsymbol{x} \leqslant x) \geqslant \frac{3}{4}$ and that $\boldsymbol{p'r}(\boldsymbol{x} \geqslant x) \geqslant \frac{1}{4}$. In each case, if there is a range of such values, the mid-point of the range is to be taken. Similarly other quantiles such as deciles and percentiles may be defined: thus the third decile is that value of x for which $\boldsymbol{p'r}(\boldsymbol{x} \leqslant x) \geqslant 0{\cdot}3$ and $\boldsymbol{p'r}(\boldsymbol{x} \geqslant x) \geqslant 0{\cdot}7$, etc. etc.

EXAMPLE 5.28.1. Find the median and the quartiles for the five probability functions given as illustrations earlier in this section.

Solutions

	Lower quartile	*Median*	*Upper quartile*
(*i*)	24·5	49·5	74·5
(*ii*)	−4	0	4
(*iii*)	0	1	2
(*iv*)	2	3	4
(*v*)	0	0·5	1·5

5.29. Examples of Probability Functions

Certain probability functions crop up so often in probability models of experimental situations that they have been intensively studied and given special names.

In this section we shall consider the following four: the rectangular distribution, the Poisson distribution, the binomial distribution and the trinomial distribution.

(*i*) *Rectangular distribution.* Consider a fair roulette-wheel where the slots are given the $(n+1)$ numbers in the arithmetic progression

$a, a+h, \ldots, a+nh$ $(h>0)$. If x denotes the number of the slot into which the ball drops, then x has the probability function

$$g(x) = 1/(n+1); \; (x = a, a+h, \ldots, a+nh) \qquad (5.29.1)$$

as illustrated in Fig. 3.

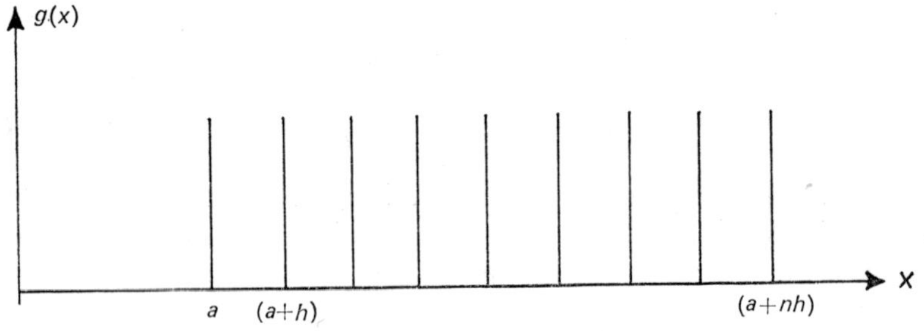

Fig. 3

By use of the formulae

$$\mu_r' = (n+1)^{-1} \sum_{t=0}^{n} (a+th)^r$$

$$\mu_r = (n+1)^{-1} \sum_{t=0}^{n} (a+th-\mu_1')^r \qquad (5.29.2)$$

the first few moments may be found easily, as

$$\mu_1' = a+(nh/2) \qquad \mu_2' = a^2+nah+\frac{n}{6}(2n+1)h^2$$

$$\mu_2 = n(n+2)h^2/12 \qquad \mu_3 = 0$$

$$\mu_3' = a^3+\tfrac{3}{2}na^2h+\frac{an(2n+1)h^2}{2}+\frac{n^2(n+1)h^3}{4}$$

(*ii*) *Poisson distribution.* A Poisson variate with parameter λ has the probability function

$$g(x) = e^{-\lambda}\lambda^x/x! \quad (x = 0, 1, 2, \ldots). \qquad (5.29.3)$$

This is illustrated in Fig. 4*a*, *b*, *c* for the cases $\lambda = 0{\cdot}5$, $\lambda = 3$ and $\lambda = 10$.

The first few moments for this distribution are

$\mu_1' = \lambda$, $\mu_2' = \lambda(\lambda+1)$, $\mu_3' = \lambda(\lambda^2+3\lambda+1)$, $\mu_4' = \lambda(\lambda^3+4\lambda^2+6\lambda+1)$
$\mu_2 = \lambda$, $\mu_3 = \lambda$, $\mu_4 = \lambda(1+3\lambda)$.

This function (5.29.3), with $\lambda = \pi n$, gives a good approximation for the binomial distribution, discussed below, see (5.29.4), when n is large but πn is not large.

If the chance of an event happening in any time-interval of given short length is independent of earlier happenings of events, then the probability function for the number x of events in any period of length T is of the form (5.29.3) with λ equal to T times the mean number of events happening per unit time.

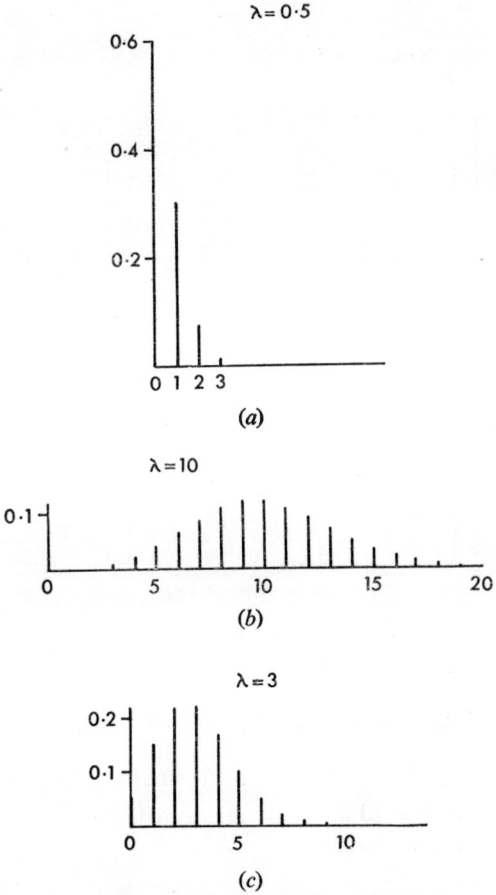

Fig. 4*a*, *b*, *c*. Poisson variates

We shall see later (§ **8.23**), that the sum of k independent Poisson variates with parameters $\lambda_1, \lambda_2 \ldots \lambda_k$ is itself a Poisson variate with parameter $(\lambda_1+\lambda_2+\ \ldots\ +\lambda_k)$.

(*iii*) *Binomial distribution.* A binomial variate has the probability function

$$g(x) = \binom{n}{x} \pi^x (1-\pi)^{n-x} \quad (x = 0, 1, 2, \ldots, n), \tag{5.29.4}$$

where n is a positive integer, and where $0 < \pi < 1$, n and π are parameters.

The distribution is illustrated for various cases in Figs. 5*a*, *b*, *c*, *d*.

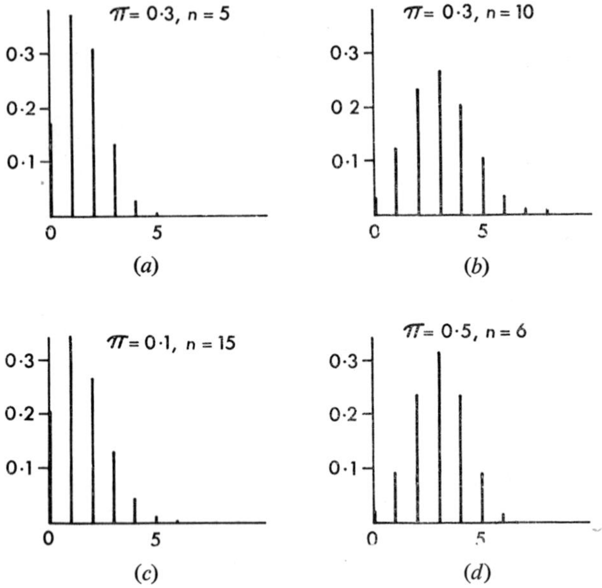

Fig. 5*a*, *b*, *c*, *d*. Binomial variates

By use of the formulae of D. 36 the first few moments may be found as

$$\mu_1' = n\pi \qquad \mu_2' = n\pi\{(n-1)\pi+1\}$$
$$\mu_2 = n\pi(1-\pi) \qquad \mu_3 = n\pi(1-\pi)(1-2\pi)$$
$$\mu_4 = n\pi(1-\pi)\{1-3\pi+3\pi^2\}$$

The following properties of the binomial variate are established in the Appendix to this Chapter.

(*a*) The number of successes in n independent trials, each with chance π of success, is a binomial variate with parameters (n, π).

(*b*) The greatest value of $g(x)$ as given by (5.28.4) is for that value x which satisfies

$$\pi(n+1) > x > \pi(n+1)-1,$$

or, if $\pi(n+1)$ is an integer, then the values $g\{\pi(n+1)\}$ and $g\{\pi(n+1)-1\}$ are jointly greatest.

(*c*) If $\boldsymbol{x}$ and $\boldsymbol{y}$ are independent binomial variates with parameters (m, π) and (n, π), and if $\boldsymbol{x}+\boldsymbol{y} = \boldsymbol{z}$, then $\boldsymbol{z}$ is a binomial variate with parameters $(m+n, \pi)$.

(*iv*) *Trinomial distribution.* Let $(\boldsymbol{x}, \boldsymbol{y})$ have the bivariate probability function

$$f(x, y) = n!\, \pi^x \phi^y (1-\pi-\phi)^{n-x-y} \,/\, x!\, y!\, (n-x-y)! \quad (5.29.5)$$

Then the three related variates $\boldsymbol{x}$, $\boldsymbol{y}$ and $(n-\boldsymbol{x}-\boldsymbol{y})$ may be said to follow the trinomial distribution.

This probability function arises when we carry out n independent performances of an experiment which has only three possible results which have probabilities π, ϕ and $1-\pi-\phi$. (5.29.5) gives the probability of getting x, y and $(n-x-y)$ results of the three kinds.

By applying the formulae of D. 37 and (5.10.3)

$$\mu'_{rs} = \sum_x \sum_y f(x\ y) x^r y^s$$

$$\mu_{rs} = \sum_x \sum_y f(x\ y)(x-\mu'_{10})^r (y-\mu'_{01})^s$$

and the following values may be found for the earlier moments.

$$\mu'_{10} = n\pi \qquad \mu'_{01} = n\phi$$

$$\mu_{20} = n\pi(1-\pi) \qquad \mu_{02} = n\phi(1-\phi) \qquad \mu_{11} = -n\pi\phi$$

$$\mu'_{20} = n\pi\{(n-1)\pi+1\} \quad \mu'_{02} = n\phi\{(n-1)\phi+1\} \quad \mu'_{11} = n(n-1)\pi\phi.$$

5.30. The Applications of the Probability Functions of §5.29

The probability function of the rectangular variate (5.29.1) arises naturally in games of chance. If, for example, we deal one card from a shuffled pack of orthodox playing cards, complete without jokers, and if we count knaves, queens, kings and aces respectively as 11, 12, 13 and 1, then the probability function for the number on the card dealt is

$$g(x) = 1/13 \quad (x = 1, 2, 3, \ldots, 12, 13). \qquad (5.30.1)$$

The same probability function (5.29.1) arises also where x_i $(i = 1, 2, \ldots, n)$ denotes the rank of the i-th member of a sample or set of n independent observations. It underlies a number of the distribution-free tests to which allusion will be made in Chapter 11.

The probability function (5.29.2) for a Poisson variate crops up in many situations where the chance of (at least one of) a particular kind of occurrence in I_θ, an interval of time of given short length θ, is approximately proportional to θ, but in no way depends either on the moment at which the interval starts or on the number or timing of such occurrences in any time-interval before the start of I_θ. It can be shown that in such situations, if x denotes the number of occurrences in an interval of length T (not necessarily small), then x is a Poisson variate of parameter λ, where λ/T is the expected number of occurrences per unit of time.

Situations approximately of this kind arise in the context of motor traffic problems, doctors' waiting-rooms (numbers of arrivals or departures in a given interval), warehouse storage (number of items ordered or delivered during week), and in many other practical and theoretical contexts.

The binomial probability function (5.29.3) arises wherever a sample is taken from a population, or a set of independent observations are taken of a variate. Where y denotes the number of n observations of a variate x, which fall into a given range R, then y is a binomial variate with parameters n and π, where

$$\pi = p'r[x \subset R]. \tag{5.30.2}$$

For example, the probability function for the number y out of n observations of a 'continuous' variate x which exceed the median value of x is a binomial variate with parameters n and $1/2$. This result depends upon the fact that for continuous variates (introduced in Chapter 8) the probability of exceeding the median value is exactly one-half. The reason that the same is not always true of discrete variates is that there may be a non-zero probability that a discrete variate will equal its median value.

The binomial probability function is useful also for establishing a number of the significance tests and confidence intervals which will be explained in Chapters 11 and 13 of Volume 2.

The trinomial probability function (5.29.5) is a special case of the multinomial probability function which gives the probability of getting any particular frequency distribution for a set of n observations of a given discrete variate. It can also be used to find the probability of getting any particular grouped frequency distribution (see Chapter 6, § **26**) of a continuous variate (see Chapter 8). Its use arises naturally in finding the probability functions for various measures of 'badness of fit' calculated from observed grouped frequency distributions. These measures of badness of fit can then be used for conducting significance tests (see Chapter 11 of Volume 2).

Appendix to Chapter 5

THE BINOMIAL VARIATE

A5.1. The Derivation of the Probability Function for a Binomial Variate

Let x be the number of successes in n independent trials, where in each trial the probability of success is π. Let the event $j(s)$ denote success in the s-th trial ($s = 1, 2, \ldots, n$), and let $k(s)$ denote failure in the s-th trial. Then

$$\pi = p'j(s) = 1-p'k(s) \quad (s = 1, 2, \ldots, n). \qquad \text{(A5.1.1)}$$

Since the trials are independent, the probability of success in each of the first x trials and of failure in the remaining $(n-x)$ trials is given by

$$p'\left\{\bigcap_{s=1}^{x} j(s)\right\} \cap \left\{\bigcap_{j=x+1}^{n} p'k(s)\right\} = \prod_{s=1}^{x}\{p'j(s)\} \prod_{s=x+1}^{n}\{p'k(s)\}$$

$$= \pi^{x}(1-\pi)^{n-x}. \qquad \text{(A5.1.2)}$$

Similarly the probability of the event of any other sequence composed of 'x successes and $(n-x)$ failures' whatever the order, is also

$$\pi^{x}(1-\pi)^{(n-x)}. \qquad \text{(A5.1.2)}$$

The total number of the disjoint events each representing x successes and $(n-x)$ failures is equal to the number of ways of selecting a group of x objects out of a group of n objects, and this number is given by

$$\binom{n}{x} = \frac{n!}{x!(n-x)!} \qquad \text{(A5.1.3)}$$

Hence the probability of the union of these $\binom{n}{x}$ disjoint equiprobable events, each with probability $\pi^{x}(1-\pi)^{n-x}$ is given as the sum of their individual probabilities by

$$g(x) = \binom{n}{x}\pi^{x}(1-\pi)^{n-x}, \qquad \text{(A5.1.4)}$$

and this must be the probability of getting exactly x successes in n independent trials where π is the probability of success in each trial. This formula for $g(x)$ agrees with formula (5.29.4) in the text for a binomial variate with parameters (π, n).

A5.2. The Most Probable Value of x

The greatest value of $g(x)$ with x an integer will be achieved for that integer x which satisfies

$$\pi(n+1) > x > \pi(n+1)-1 \qquad \text{(A5.2.1)}$$

or if $\pi(n+1)$ is an integer, it will be achieved for each of the values of x

$$\pi(n+1) \quad \text{and} \quad \pi(n+1)-1. \qquad \text{(A5.2.2)}$$

Proof: Let

$$q(x) = g(x+1)/g(x). \qquad \text{(A5.2.3)}$$

Then if X is the value of x for which $g(x)$ is greatest,

$$q(X) < 1 > q(X-1). \qquad \text{(A5.2.4)}$$

But

$$q(X) = \frac{n-X}{X+1}\frac{\pi}{1-\pi} \quad \text{and} \quad q(X-1) = \frac{n-X+1}{X}\frac{\pi}{1-\pi}$$

$$\therefore \quad (n-X)\pi < (X+1)(1-\pi) \quad \text{and} \quad (n-X+1)\pi > X(1-\pi)$$

so that

$$(n+1)\pi-1 < X < (n+1)\pi. \qquad \text{(A5.2.5)}$$

If X and $X+1$ give equal greatest values of $g(x)$ then $q(X) = 1$, so that

$$(n-X)\pi = (X+1)(1-\pi),$$

whence

$$X = (n+1)\pi-1, \text{ so that} \qquad \text{(A5.2.6)}$$

$$x = (n+1)\pi-1 \quad \text{and} \quad x = (n+1)\pi \text{ give equal greatest values of } g(x). \qquad \text{(A5.2.7)}$$

A5.3. The Reproductive Property of Binomial Variates

If $\boldsymbol{x}$ and $\boldsymbol{y}$ are independent binomial variates with parameters (m, π) and (n, π) and if $\boldsymbol{z} = \boldsymbol{x}+\boldsymbol{y}$, then $\boldsymbol{z}$ is a binomial variate with parameters $(m+n, \pi)$. (A5.3.1)

Proof: Since $\boldsymbol{x}$ and $\boldsymbol{y}$ are independent binomial variates with parameters (m, π) and (n, π), $\boldsymbol{x}$ has the same probability function as the variate $\boldsymbol{X}$ which is the number of successes in m independent trials, each with probability π of success: and $\boldsymbol{y}$ has the same probability function as has a variate $\boldsymbol{Y}$ which is the number of successes in n independent trials, each with probability π of success. Hence $\boldsymbol{z}$ has the same probability function as has the variate $\boldsymbol{Z} = \boldsymbol{X}+\boldsymbol{Y}$. But $\boldsymbol{Z}$ is then the number of successes in $(m+n)$ independent trials each with probability π of success, and $\boldsymbol{Z}$ is therefore a binomial variate with parameters $(m+n, \pi)$, and hence also $\boldsymbol{z}$ must be a binomial variate with parameters $(m+n, \pi)$.

Exercises 5

5.1. The expected values of any two variates $\boldsymbol{x}$ and $\boldsymbol{y}$ are $\boldsymbol{p}'\boldsymbol{x} = a$ and $\boldsymbol{p}'\boldsymbol{y} = b$. Prove that where $\boldsymbol{u} = \boldsymbol{x} - \mathbf{1}a$ and $\boldsymbol{v} = \boldsymbol{y} - \mathbf{1}b$, and $\boldsymbol{z} = \boldsymbol{x}c + \boldsymbol{y}k + \mathbf{1}m$
(*i*) $\boldsymbol{p}'\boldsymbol{u} = \boldsymbol{p}'\boldsymbol{v} = 0$ (*ii*) $\boldsymbol{p}'\boldsymbol{z} = c\,\boldsymbol{p}'\boldsymbol{x} + k\,\boldsymbol{p}'\boldsymbol{y} + m$.

5.2. Retaining the notation of question 1, the covariance of $\boldsymbol{x}$ and $\boldsymbol{y}$ may be defined as

$$\text{cov}\,(\boldsymbol{x}, \boldsymbol{y}) = \boldsymbol{p}'\hat{\boldsymbol{u}}\boldsymbol{v} = E[\boldsymbol{uv}].$$

Show that $\text{cov}\,(\boldsymbol{x}, \boldsymbol{y}) = \boldsymbol{p}'\hat{\boldsymbol{x}}\boldsymbol{y} - ab.$

5.3. Where $\text{var}\,(\boldsymbol{x}) = \text{cov}\,(\boldsymbol{x}, \boldsymbol{x})$ prove that, in the notation of question 1,

$$\text{var}\,(\boldsymbol{x}) = \boldsymbol{p}'\hat{\boldsymbol{x}}\boldsymbol{x} - a^2 = E[\boldsymbol{x}^2] - (E[\boldsymbol{x}])^2.$$

5.4. When $\boldsymbol{y} = \mathbf{1}\alpha + \boldsymbol{x}\beta$, prove that $\text{var}\,\boldsymbol{y} = \beta^2\,\text{var}\,\boldsymbol{x}$.

5.5. When $\boldsymbol{u} = \boldsymbol{x} + \boldsymbol{y}$, prove that

$$\text{var}\,\boldsymbol{u} = \text{var}\,[\boldsymbol{x}] + \text{var}\,[\boldsymbol{y}] + 2\,\text{cov}\,[\boldsymbol{x}, \boldsymbol{y}].$$

5.6. When $\boldsymbol{z} = \mathbf{1}\alpha + \boldsymbol{x}\beta + \boldsymbol{y}\gamma$, obtain a formula for $\text{var}\,[\boldsymbol{z}]$ in terms of the variances and covariance of $\boldsymbol{x}$ and $\boldsymbol{y}$.

5.7. $\boldsymbol{x}$ is a binomial (π, n) variate, having the probability function

$$f(x) = \frac{n!}{x!\,(n-x)!}\pi^x(1-\pi)^{n-x} \quad (x = 0, 1, \ldots, n;\ 0 < \pi < 1).$$

Prove that $E(\boldsymbol{x}) = n\pi$ and that $\text{var}\,[\boldsymbol{x}] = n\pi(1-\pi)$.

Hint. Use $E[\phi] = \sum_{x=0}^{n} f(x)\phi(x)$, first with $\phi(x) = x$ and then with $\phi(x) = x(x-1)$.

5.8. $\boldsymbol{x}$ is a Poisson variate with probability function

$$f(x) = \mathrm{e}^{-\lambda}\lambda^x/x! \quad (x = 0, 1, 2, \ldots, \lambda > 0).$$

Prove that $E(\boldsymbol{x}) = \lambda$, and find $E[\boldsymbol{x}(\boldsymbol{x}-\mathbf{1})]$ and $\text{var}\,[\boldsymbol{x}]$.

5.9. $(\boldsymbol{x}, \boldsymbol{y})$ have the bivariate probability function

$$f(x, y) = \binom{n}{x}\binom{m}{y-x}\pi^x(1-\pi)^{n-x}\phi^{y-x}(1-\phi)^{m-y+x}$$ for x and y integers, $(0 \leqslant x \leqslant n,\ x \leqslant y \leqslant x+m,\ 0 < \pi < 1,\ 0 < \phi < 1)$.

Find the variance-covariance matrix

$$\boldsymbol{M} = \begin{pmatrix} \text{var}\,[\boldsymbol{x}] & \text{cov}\,[\boldsymbol{x}, \boldsymbol{y}] \\ \text{cov}\,[\boldsymbol{x}, \boldsymbol{y}] & \text{var}\,[\boldsymbol{y}] \end{pmatrix}.$$

5.10. Find the modes of the probability functions of questions 7 and 8, namely the values of x for which $f(x)$ is greatest.

5.11. Where $\boldsymbol{x}$ and $\boldsymbol{y}$ are independent events each having probability π, write down the probability function $f(x, y)$ for the variate pair $(\boldsymbol{x}, \boldsymbol{y})$ and obtain the probability function for their sum, $\boldsymbol{z} = \boldsymbol{x}+\boldsymbol{y}$ and find the expected value and variance of $\boldsymbol{z}$.

5.12. Two variates $\boldsymbol{x}$ and $\boldsymbol{y}$ are said to be independent if for all possible pairs of values (x, y), the events $\boldsymbol{r}(\boldsymbol{x} = x)$ and $\boldsymbol{r}(\boldsymbol{y} = y)$ are independent (see D. 22 of § **5.14**). Prove that for any pair of independent variates $(\boldsymbol{x}, \boldsymbol{y})$ $\boldsymbol{p}'\boldsymbol{xy} = \boldsymbol{p}'\boldsymbol{x}\boldsymbol{p}'\boldsymbol{y}$ (provided the expected values exist) and hence that cov $(\boldsymbol{x}, \boldsymbol{y}) = 0$. Where $\boldsymbol{z} = \boldsymbol{x}+\boldsymbol{y}$ and where $\boldsymbol{x}$ and $\boldsymbol{y}$ are independent, prove that

$$\text{var}\,[\boldsymbol{z}] = \text{var}\,[\boldsymbol{x}]+\text{var}\,[\boldsymbol{y}].$$

5.13. $f(x)$ is defined as in question 7 above and $\pi = \lambda/n$. Keeping λ constant, prove that

$$\lim_{n \to \infty} [f(x)/f(0)] = \lambda^x/x!$$

Using the fact that $f(0) \sum_{x=0}^{\infty} [f(x)/f(0)] = 1$

find $\lim_{n \to \infty} f(x)$ for $x = 0, 1, 2, 3, \ldots$

Verify that $\lim_{n = \infty} f(x) = g(x)$ where $g(x)$ is the probability function for a Poisson variate with parameter λ. (See q. 8.)

5.14. Find the moment generating function $M'(\alpha) = \sum_{x=0}^{\infty} e^{\alpha x} f(x)$ for a Poisson variate with parameter λ. Hence obtain μ_1' and the moment generating function $M(\alpha) = e^{-\mu_1'\alpha} M'(\alpha)$.

5.15. Use the solutions to question 14 to obtain the first 4 moments μ_1, μ_2, μ_3 and μ_4 for a Poisson variate of parameter λ.

5.16. Use the solution of question 15 to find the skewness and kurtosis for a Poisson variate of parameter λ. (See § **5.28**.)

5.17. In order to test whether a large batch of components contains as many as 1% of defectives, Ambrose will test $\boldsymbol{n}$ components selected at random and will accept the batch if none of these $\boldsymbol{n}$ components is found defective. How large must $\boldsymbol{n}$ be to ensure that if the batch contains 1% or more of defectives, the risk of accepting it is 5% or less.

Hint: Regard the number found defective as a Poisson variate and use tables of e^{-x}.

5.18. If in question 17 Ambrose were to test m components and accept the batch if the number found defective were 0 or 1, how large would m have to be to ensure the same safety?

5.19. Under the schemes of questions 17 and 18, if in fact only 1 in 500 of the batch were defective, what are the risks that Ambrose will reject the batch?
How might one modify the scheme so as to diminish the risk of rejecting good batches without increasing the risk of accepting bad batches?

5.20. Four discs are numbered -6, $+1$, $+2$, $+3$. I draw 2 discs at random and $\boldsymbol{x}$ and $\boldsymbol{y}$ denote the numbers on the first and second of the discs drawn. Obtain the variance-covariance matrix for the variate-pair $(\boldsymbol{x}, \boldsymbol{y})$.

5.21. $\boldsymbol{x}$ and $\boldsymbol{y}$ have the bivariate probability function $f(x, y) = 0{\cdot}1$ when $y > x$, $(x = 0, 1, 2, 3, 4;\ y = 0, 1, 2, 3, 4)$. Find the correlation coefficient between $\boldsymbol{x}$ and $\boldsymbol{y}$.

5.22. $\boldsymbol{x}$ has the probability function

$$f(-2) = 0{\cdot}5, \quad f(1) = 0{\cdot}4, \quad f(6) = 0{\cdot}1.$$

Find the skewness γ_1 and kurtosis γ_2 for $\boldsymbol{x}$.

5.23. Find the corrected variate $\boldsymbol{y} \cdot \boldsymbol{x}$ under the conditions (*i*) of question 20, (*ii*) of question 21.

Chapter 6

DESCRIPTION OF MEASUREMENTS

6.1. Attributes and Measurements

Suppose that various observations are taken concerning each of n entities. What these entities are (whether persons, households, objects, firms or nations) need not for the moment concern us.

Suppose that for each of these n entities we are provided with just one measurement $x(j)$. Such a set of n measurements, $x(j)$ $(j = 1, 2, \ldots, n)$, may be represented by a vector $\boldsymbol{x}$.

The simplest case is where $x(j)$ always takes the value 0 or 1: it takes the value 1 if the corresponding entity does possess some specified quality, and the value 0 if it does not. Such a vector will be termed an *attribute vector*. In the more general case, where the $x(j)$ may take values other than zero or one, the vector will be called a *measurement vector*.

Attribute vectors and measurement vectors may be described and handled by methods very closely analogous to those by which events and discrete variates were described and handled in the last two chapters.

6.2. The Scaling Vector

The description is in fact simpler in the case of attribute vectors and measurement vectors since the probability vector $\boldsymbol{p}'$ may be replaced by the scaling vector $\boldsymbol{f}'$

D. 36 The scaling vector $\boldsymbol{f}'$ is defined as the row vector of elements each equal to $1/n$, where n is the number of terms in the attribute vector or measurement vector $\boldsymbol{x}$.

6.3. Measurement Matrices

There is no need to suppose that only one measurement is given for each of the n entities. In case there are m such measurements, $x_k(j)$ $(k = 1, 2, \ldots, m)$ for each entity $(j = 1, 2, \ldots, n)$, then the whole set of measurements may be represented by an $n \times m$ matrix $\boldsymbol{X}$ whose k-th column $(k = 1, 2, \ldots, m)$ is the column vector $\boldsymbol{x}_k$ with elements $x_k(j)$ $(j = 1, 2, \ldots, n)$. The column vectors $\boldsymbol{x}_k$ may be all measurement vectors, or all attribute vectors, or there may be some of each.

6.4. Arithmetical Relations between Measurements

As in the case of variates, if $\boldsymbol{x}_k$, $\boldsymbol{x}_l$ are any two measurement vectors, we may denote by $\boldsymbol{r}(\boldsymbol{x}_k = \boldsymbol{x}_l)$ a vector whose j-th element will be a one or a zero according as the equality $x_k(j) = x_l(j)$ is or is not satisfied. $\boldsymbol{r}(\boldsymbol{x}_k = \boldsymbol{x}_l)$ may be regarded as an attribute vector, the attribute in question being equality of the measurements x_k and x_j. Similarly we may denote by $\boldsymbol{r}(\boldsymbol{x}_k \leqslant \boldsymbol{x}_l)$ the vector whose j-th element will be a one or a zero according as $x_k(j) \leqslant x_l(j)$ or $x_k(j) > x_l(j)$.

As in the case of variates, we shall suppress the unit vector $\mathbf{1}$ in such relations as $\boldsymbol{r}(\boldsymbol{x} = \mathbf{1}x)$ and write $\boldsymbol{r}(\boldsymbol{x} = x)$ where x is a scalar.

Thus $\boldsymbol{r}(\boldsymbol{x} = x)$ denotes the vector whose j-th element is a 1 if $x(j) = x$ but is a zero if $x(j) \neq x$.

6.5. Frequency Functions

In the case of variates the probability function $\boldsymbol{p}'\boldsymbol{r}(\boldsymbol{x} = x)$ gave the probability that a variate $\boldsymbol{x}$ took various specified values x. In the case of measurements, we similarly define the frequency function.

D. 37 The frequency function for the measurement vector $\boldsymbol{x}$ is $\boldsymbol{f}'\boldsymbol{r}(\boldsymbol{x} = x)$ (where, as usual $\boldsymbol{f}' = (1/n, 1/n, \ldots, 1/n)$).

The frequency function for $\boldsymbol{x}$ tells us what *proportion* of the measurements $x(j)$ equal the specified value x. The reason for this is that each entity for which the measurement $x(j) = x$ will contribute $1/n$ to the scalar product $\boldsymbol{f}'\boldsymbol{r}(\boldsymbol{x} = x)$, whereas each other entity will contribute zero.

In the particular case where $\boldsymbol{x}$ is an attribute vector,

$$\boldsymbol{f}'\boldsymbol{r}(\boldsymbol{x} = 1)$$

will be the proportion of the n entities which possess the attribute, $\boldsymbol{f}'r(\boldsymbol{x} = 0)$ will be the proportion which do not, and $\boldsymbol{f}'\boldsymbol{r}(\boldsymbol{x} = x) = 0$ for all values of x other than 0 or 1.

6.6. Arithmetic Means and Proportions

The simple average of n quantities, obtained by dividing their total by n, is known as their arithmetic mean. If $\boldsymbol{x}$ is a measurement vector, we shall denote the arithmetic mean of its n elements $x(j)$ $(j = 1, 2, \ldots, n)$ by $\bar{x}$. Since $\boldsymbol{f}'$ is the row vector $(1/n, 1/n, \ldots, 1/n)$, it follows that where $\boldsymbol{x}$ is any measurement vector its arithmetic mean is given by

$$\bar{x} = \boldsymbol{f}'\boldsymbol{x} = \sum_{j=1}^{n} x(j) \,/\, n. \qquad (6.6.1)$$

This corresponds to the fact that the expected value of any variate is given by

$$E(\boldsymbol{x}) = \boldsymbol{p}'x. \tag{6.6.2}$$

In the special case where $\boldsymbol{x}$ is an attribute vector, the number of entities which possess the attribute (and for which $x(j) = 1$) will be $\sum_{j=1}^{n} x(j)$, so that the proportion of entities which possess the attribute will be

$$P = \boldsymbol{f}'\boldsymbol{x}. \tag{6.6.3}$$

This corresponds to the fact that the probability of any event $\boldsymbol{j}$ is $\boldsymbol{p}'\boldsymbol{j}$.

6.7. Expression of the Arithmetic Mean in Terms of the Frequency Distribution

It is often convenient to express the arithmetic mean $\bar{x}$ of a measurement vector $\boldsymbol{x}$ in the form

$$\bar{x} = \boldsymbol{f}'\boldsymbol{x} = \sum_{x} [\boldsymbol{f}'\boldsymbol{r}(\boldsymbol{x} = x)]x, \tag{6.7.1}$$

where the summation extends over all values x taken by $\boldsymbol{x}$.

The formula (6.7.1) corresponds to the expansion of the expected value of a variate in the form

$$E(\boldsymbol{x}) = \boldsymbol{p}'\boldsymbol{x} = \sum_{x} [\boldsymbol{p}'\boldsymbol{r}(\boldsymbol{x} = x)]x, \tag{6.7.2}$$

and (6.7.1) can be proved in the same manner as (6.7.2) was proved in § **8** of Chapter 5.

6.8. Orthogonality

Where $\boldsymbol{x}$ and $\boldsymbol{y}$ represent two sets of measurements relating to the same entities, $\boldsymbol{x}'\boldsymbol{y}$ must represent the sum of the products of the two measurements. We term the two sets $\boldsymbol{x}$ and $\boldsymbol{y}$ *orthogonal*, if this sum of products is zero. Thus

D. 38 Two measurement vectors $\boldsymbol{x}$, $\boldsymbol{y}$ are orthogonal if, and only if,

$$\boldsymbol{x}'\boldsymbol{y} = 0. \tag{6.8.1}$$

D. 38a Two measurement matrices $\boldsymbol{X}$, $\boldsymbol{Y}$ are orthogonal if, and only if

$$\boldsymbol{X}'\boldsymbol{Y} = \boldsymbol{0} \qquad \text{(6.8.1a,}$$

The correspondence of D. 38 with D. 32, the definition of orthogonality of two variates $\boldsymbol{x}$ and $\boldsymbol{y}$, namely

$$\boldsymbol{x}'\hat{\boldsymbol{p}}\boldsymbol{y} = 0 \tag{6.8.2}$$

is very close. For the condition $x'y = 0$ is equivalent to

$$x'\hat{f}y = 0, \tag{6.8.3}$$

since

$$x'\hat{f}y = x'y/n. \tag{6.8.4}$$

6.9. Projection and Correction of Measurement Vectors

If y and x are two measurement-vectors, it is sometimes useful to divide one of them, y, into two portions, '*the projection of y on x*', denoted by $y \perp x$ and '*y* corrected for *x*', denoted by $y \cdot x$. In this division $y \perp x$ is a scalar multiple of x, so chosen that $y \cdot x$ is orthogonal to x.

We may then define $y \perp x$ and $y \cdot x$ as follows:

D. 39 If y, x are any two measurement vectors,

$$y = y \perp x + y \cdot x,$$

where $\quad y \perp x = xb, \quad$ and b is a scalar,

and $\quad x'(y \cdot x) = 0.$

The definition extends naturally to measurement matrices in the following form:

D. 40 If Y and X are measurement matrices,

$$Y = Y \perp X + Y \cdot X,$$

where $\quad Y \perp X = XB \quad (B \text{ is a matrix})$

and $\quad X'(Y \cdot X) = 0.$

EXAMPLE 6.9.1. Show that if X, Y and Z are measurement matrices, then we may put

$$(Xr + Ys) \perp Z = (X \perp Z)r + (Y \perp Z)s.$$

Solution. Let $X \perp Z = ZB$ and $Y \perp Z = ZC$.

Then $\quad (X \perp Z)r + (Y + Z)s = Z\{Br + Cs\} = ZH$, say.

The only further condition which ZH must satisfy in order to qualify as $(Xr + Ys) \perp Z$, is that

$$Z'ZH = Z'\{Xr + Ys\}. \tag{A}$$

Now, since

$$X \perp Z = ZB, \quad Y \perp Z = ZC$$

and

$$Z'(X \cdot Z) = Z'(Y \cdot Z) = 0,$$

we have

$$Z'X = Z'ZB \quad \text{and} \quad Z'Y = Z'ZC.$$

Hence

$$Z'ZH = Z'Z\{Br+Cs\} = Z'\{Xr+Ys\}$$

so that (A) is satisfied. Thus we may put

$$(Xr+Ys) \perp Z = (X \perp Z)r+(Y \perp Z)s.$$

EXAMPLE 6.9.2. Show that if X and Y are measurement matrices, then $Y \perp X$ is uniquely determined although the matrix B such that $Y \perp X = XB$ is not always uniquely determined.

Solution. Suppose that XB and XC are possible values of $Y \perp X$. Then

$$X'Y = X'XB = X'XC$$

$$\therefore \quad X'X(B-C) = 0$$

$$\therefore \quad \{X(B-C)\}'\{X(B-C)\} = 0$$

$$\therefore \quad X(B-C) = 0$$

$$\therefore \quad XB = XC \text{ and } Y \perp X \text{ is uniquely determined.}$$

But suppose X to be a matrix whose rank is less than its number of columns. Then we can find a vector b such that $Xb = 0$ and $b \neq 0$.
Let $Y \perp X = XB$.
Let E be a matrix conformable with B whose first column is b and whose remaining columns consist of zeros.
Then $\quad B+E \neq B.$
But $\quad X(B+E) = XB = Y \perp X,$ since $Xb = 0$
and hence $\quad XE = 0.$
Hence the matrix B such that $Y \perp X = XB$ is not unique.

6.10. The Geometrical Interpretation of $y \perp x$ and $y \cdot x$

Let x and y be represented (Fig. 6) by rays OX, OY from the origin to points X, Y in n-dimensional space whose coordinates are the n elements of x and the n elements of y. Then the usual condition for perpendicularity is such that OX and OY will be perpendicular if $x'y = 0$, i.e. if x and y are orthogonal.

Now, let x, y and $y \perp x$ be represented by rays OX, OY and OZ from the origin to the three points whose coordinates are the elements of x, y and $y \perp x$. Then Z must lie on OX since

$$y \perp x = xb. \qquad (b \text{ being a scalar}). \tag{6.10.1}$$

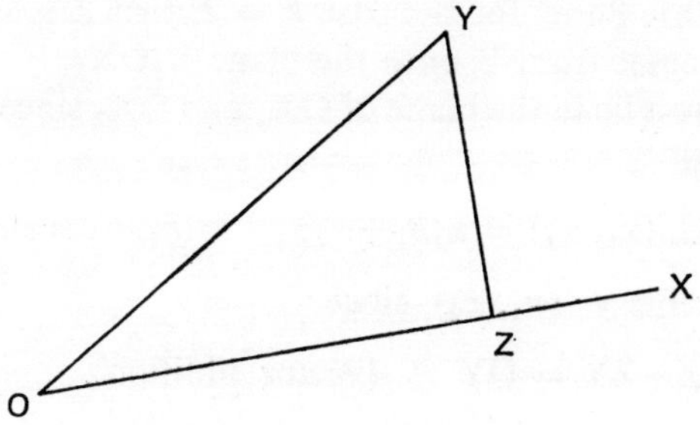

Fig. 6

Also, since

$$y - y \perp x = y \cdot x \tag{6.10.2}$$

and $\quad y \cdot x$ is orthogonal to x. (6.10.3)

The ray ZY must be perpendicular to the ray OX.

Hence Z is the foot of the perpendicular from Y onto OX.

In the diagram, x, y, $y \cdot x$ and $y \perp x$ are represented by OX, OY, ZY and OZ.

A similar geometrical illustration (Fig. 7) can be given for the case of the projection of a vector y onto a matrix X composed of column-vectors $x_1 \ldots x_k$. Let the vector y be represented by OY and the vectors $x_1 \ldots x_k$, by $OX_1, OX_2, \ldots, OX_k$ where, as before, the coordinates of the points Y, X_1, etc. are given by the elements of the corresponding vectors $y_1 x_1$, etc. Let the projection $y \perp X$ be represented by OZ in the same fashion. Let H be the hyperplane generated by the rays $OX_1, OX_2, \ldots, OX_k$; then it can be shown that Z is the foot of the perpendicular from Y onto H.

The diagram illustrates the case $k = 2$.

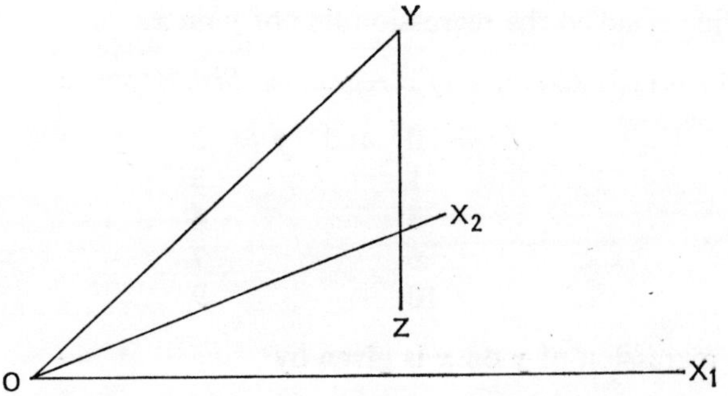

Fig. 7

EXAMPLE 6.10.1. Show for the case $k = 2$, that Z is indeed the foot of the perpendicular from Y onto the plane X_1OX_2.

Solution. Z must lie in the plane of OX_1 and OX_2 since OZ represents $\boldsymbol{y} \perp (\boldsymbol{x}_1, \boldsymbol{x}_2)$ and

$$\boldsymbol{y} \perp (\boldsymbol{x}_1, \boldsymbol{x}_2) = \boldsymbol{x}_1 b_1 + \boldsymbol{x}_2 b_2 \qquad (b_1, b_2 \text{ scalars}).$$

Also ZY represents $\boldsymbol{y} \cdot (\boldsymbol{x}_1, \boldsymbol{x}_2)$, since

$$\text{OZ}+\text{ZY} = \text{OY} \qquad \text{(vector addition)}.$$

Hence ZY is perpendicular both to OX_1 and OX_2, since $\boldsymbol{y} \cdot (\boldsymbol{x}_1, \boldsymbol{x}_2)$ is orthogonal both to $\boldsymbol{x}_1$ and $\boldsymbol{x}_2$.
Hence ZY is perpendicular to the plane X_1OX_2.
Hence Z is the foot of the perpendicular from Y onto the plane X_1OX_2.

6.11. The Regression of *y* on *x*, and the Regression of *Y* on *X*

If $\boldsymbol{x}$ and $\boldsymbol{y}$ are any two measurement vectors, the projection of $\boldsymbol{y}$ on the 2-column measurement matrix $(\boldsymbol{1}, \boldsymbol{x})$ is called the regression of $\boldsymbol{y}$ on $\boldsymbol{x}$. Thus

D. 41 The regression of $\boldsymbol{y}$ on $\boldsymbol{x}$ is given by

$$\boldsymbol{y} \perp (\boldsymbol{1}, \boldsymbol{x}) = (\boldsymbol{1}, \boldsymbol{x})\boldsymbol{b} = \boldsymbol{1}b_1 + \boldsymbol{x}b_2 = \boldsymbol{y}^* \text{ (say)}. \qquad (6.11.1)$$

We may represent (Fig. 8) the two vectors $\boldsymbol{x}$ and $\boldsymbol{y}$ by the n points with coordinates (x_i, y_i) $(i = 1, 2, \ldots, n)$, and we may represent the two vectors $\boldsymbol{x}$ and $\boldsymbol{y}^*$ by the n points with coordinates (x_i, y_i^*) $(i = 1, 2, \ldots, n)$, where, for each value of i,

$$y_i^* = b_1 + b_2 x_i. \qquad (6.11.2)$$

These points (x_i, y_i^*) thus lie all on the line

$$y = b_1 + b_2 x, \qquad (6.11.3)$$

and this line is called the regression line of $\boldsymbol{y}$ on $\boldsymbol{x}$.

EXAMPLE 6.11.1. *Example of a regression line.* Suppose

$$\boldsymbol{x} = \begin{matrix} 0 \\ 1 \\ 3 \\ 6 \\ 10 \end{matrix} \quad \text{and} \quad \boldsymbol{y} = \begin{matrix} 2 \\ 3 \\ 4 \\ 7 \\ 9 \end{matrix}$$

Then the regression of $\boldsymbol{y}$ on $\boldsymbol{x}$ is given by

$$\boldsymbol{y}^* = \boldsymbol{y} \perp (\boldsymbol{1}, \boldsymbol{x}) = (\boldsymbol{1}, \boldsymbol{x})\boldsymbol{b}, \qquad (6.11.4)$$

where $\quad (\mathbf{1}, \mathbf{x})'\mathbf{y}^* = (\mathbf{1}, \mathbf{x})'\mathbf{y} \quad$ (by D. 41a). $\quad$ (6.11.5)

$$\therefore \quad (\mathbf{1}, \mathbf{x})'(\mathbf{1}, \mathbf{x})\mathbf{b} = (\mathbf{1}, \mathbf{x})'\mathbf{y}$$

$$\therefore \quad \mathbf{b} = [(\mathbf{1}, \mathbf{x})'(\mathbf{1}, \mathbf{x})]^{-1}(\mathbf{1}, \mathbf{x})'\mathbf{y}. \qquad (6.11.6)$$

Now $\quad (\mathbf{1}, \mathbf{x})'(\mathbf{1}, \mathbf{x}) = \begin{pmatrix} 5 & 20 \\ 20 & 146 \end{pmatrix} \quad (\mathbf{1}, \mathbf{x})'\mathbf{y} = \begin{pmatrix} 25 \\ 147 \end{pmatrix}$

and $\quad [(\mathbf{1}, \mathbf{x})'(\mathbf{1}, \mathbf{x})]^{-1} = (1/330)\begin{pmatrix} 146 & -20 \\ -20 & 5 \end{pmatrix}, \quad$ (6.11.7)

so that $\quad \mathbf{b} = \begin{pmatrix} 71/33 \\ 23/33 \end{pmatrix},$

and $\quad \mathbf{y}^* = \mathbf{1}(71/33) + \mathbf{x}(23/33) = \begin{pmatrix} 71/33 \\ 94/33 \\ 140/33 \\ 209/33 \\ 301/33 \end{pmatrix}, \quad$ (6.11.8)

and the regression line of $\mathbf{y}$ on $\mathbf{x}$ is

$$y = (71/33) + (23/33)x,$$

or, approximately, in decimals

$$y = 2{\cdot}15 + 0{\cdot}70x. \qquad (6.11.9)$$

The diagram shows the points (x_i, y_i) $(i = 1, 2, 3, 4, 5)$ scattered about the regression line (6.11.9) passing through the five points (x_i, y_i^*).

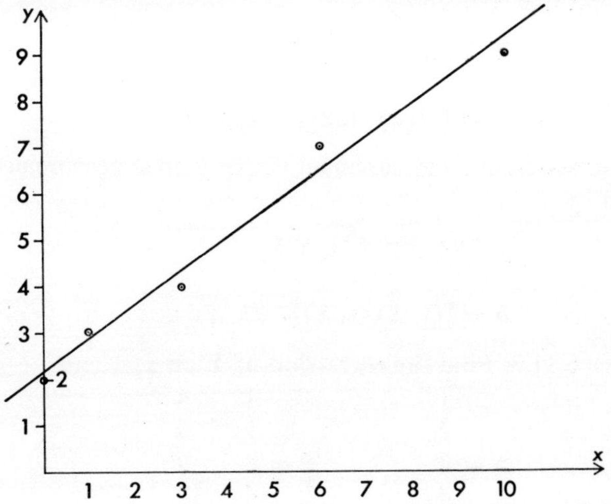

Fig. 8

It will be noted that the regression line follows closely the lie of the five points. This is of course no accident but is due to the 'least-squares' property of regression lines which will be explained in § **6.12** following.

We now give the formal definition of the regression of one measurement matrix $\boldsymbol{Y}$ on another measurement matrix $\boldsymbol{X}$. This will include, as a special case, the definition of the regression of a measurement vector $\boldsymbol{y}$ on another regression vector $\boldsymbol{x}$.

D. 42 The regression of any measurement matrix $\boldsymbol{Y}$ on any measurement matrix $\boldsymbol{X}$ is the projection $\boldsymbol{Y} \perp (\mathbf{1}, \boldsymbol{X})$ where $(\mathbf{1}, \boldsymbol{X})$ is the matrix obtained by adding a column of ones to the matrix $\boldsymbol{X}$.

EXAMPLE 6.11.2. Show that if the determinant of the matrix $(\mathbf{1}, \boldsymbol{X})'(\mathbf{1}, \boldsymbol{X})$ is non-zero, the regression of $\boldsymbol{Y}$ on $\boldsymbol{X}$ is given by

$$\boldsymbol{Y} \perp (\mathbf{1}, \boldsymbol{X}) = (\mathbf{1}, \boldsymbol{X})\boldsymbol{B}, \tag{6.11.10}$$

with

$$\boldsymbol{B} = [(\mathbf{1}, \boldsymbol{X})'(\mathbf{1}, \boldsymbol{X})]^{-1}(\mathbf{1}, \boldsymbol{X})'\boldsymbol{Y}. \tag{6.11.11}$$

Solution. It follows from D. 41a that $\boldsymbol{Y} \perp (\mathbf{1}, \boldsymbol{X})$ satisfies (6.11.10) with $\boldsymbol{B}$ satisfying

$$(\mathbf{1}, \boldsymbol{X})'\boldsymbol{Y} = (\mathbf{1}, \boldsymbol{X})'(\mathbf{1}, \boldsymbol{X})\boldsymbol{B}. \qquad (6.11.12$$

We have only to premultiply (6.11.12) by $[(\mathbf{1}, \boldsymbol{X})'(\mathbf{1}, \boldsymbol{X})]^{-1}$ and interchange the two sides of the equation, to obtain the desired relation (6.11.11).

In particular, the regression of a measurement matrix $\boldsymbol{Y}$ on a measurement vector $\boldsymbol{x}$ is given by

$$\boldsymbol{Y} \perp (\mathbf{1}, \boldsymbol{x}) = (\mathbf{1}, \boldsymbol{x})\boldsymbol{B},$$

where

$$\boldsymbol{B} = [(\mathbf{1}, \boldsymbol{x})'(\mathbf{1}, \boldsymbol{x})]^{-1}(\mathbf{1}, \boldsymbol{x})'\boldsymbol{Y}, \tag{6.11.13}$$

and the regression of a measurement vector $\boldsymbol{y}$ on a measurement matrix $\boldsymbol{X}$ is given by

$$\boldsymbol{y} \perp (\mathbf{1}, \boldsymbol{X}) = (\mathbf{1}, \boldsymbol{X})\boldsymbol{b},$$

where

$$\boldsymbol{b} = [(\mathbf{1}, \boldsymbol{X})'(\mathbf{1}, \boldsymbol{X})]^{-1}(\mathbf{1}, \boldsymbol{X})'\boldsymbol{y}. \tag{6.11.14}$$

EXAMPLE 6.11.3. Find the regression of $\boldsymbol{Y}$ on $\boldsymbol{x}$, where

$$\boldsymbol{x} = \begin{matrix} 1 \\ 2 \\ 3 \\ 4 \\ 5 \end{matrix} \qquad \mathrm{Y} = \begin{matrix} 1 & 4 \\ 1 & 6 \\ 3 & 9 \\ 4 & 10 \\ 6 & 16 \end{matrix}$$

Solution. $\quad \boldsymbol{Y} \perp (\mathbf{1}, \boldsymbol{X}) = (\mathbf{1}, \boldsymbol{x})\boldsymbol{B}$

where $\quad \boldsymbol{B} = [(\mathbf{1}, \boldsymbol{x})'(\mathbf{1}, \boldsymbol{x})]^{-1}(\mathbf{1}, \boldsymbol{x})'\boldsymbol{Y}$

$$(\mathbf{1}, \boldsymbol{x})'(\mathbf{1}, \boldsymbol{x}) = \begin{matrix} 5 & 15 \\ 15 & 55 \end{matrix} \qquad (\mathbf{1}, \boldsymbol{x})'\boldsymbol{Y} = \begin{matrix} 15 & 45 \\ 57 & 164 \end{matrix}$$

$$[(\mathbf{1}, \boldsymbol{x})'(\mathbf{1}, \boldsymbol{x})]^{-1} = \begin{matrix} 1{\cdot}1 & -0{\cdot}3 \\ -0{\cdot}3 & 0{\cdot}1 \end{matrix}$$

whence $\quad \boldsymbol{B} = \begin{matrix} -0{\cdot}6 & 0{\cdot}3 \\ 1{\cdot}2 & 2{\cdot}9 \end{matrix}$

Thus the regression of $\boldsymbol{Y}$ on $\boldsymbol{x}$ may be illustrated by the cluster of the two sets of five points each $(x_i\ y_{1i})$, $(x_i\ y_{2i})$ about the two regression lines

$$y = -0{\cdot}6+1{\cdot}2x \quad \text{and} \quad y = 0{\cdot}3+2{\cdot}9x.$$

EXAMPLE 6.11.4. Find the regression of $\boldsymbol{y}$ on $\boldsymbol{X}$ where

$$\boldsymbol{X} = \begin{matrix} 2 & 4 \\ 4 & 16 \\ 1 & 1 \\ 5 & 25 \\ 3 & 9 \end{matrix} \quad \text{and} \quad \boldsymbol{y} = \begin{matrix} 0 \\ 8 \\ 1 \\ 27 \\ 1 \end{matrix}$$

Solution. $\quad \boldsymbol{y} \perp (\mathbf{1}, \boldsymbol{X}) = (\mathbf{1}.\ \boldsymbol{X})\boldsymbol{b},$

where $\quad \boldsymbol{b} = [(\mathbf{1}, \boldsymbol{X})'(\mathbf{1}, \boldsymbol{X})]^{-1}(\mathbf{1}, \boldsymbol{X})'\boldsymbol{y}$

$$(\mathbf{1}, \boldsymbol{X})'(\mathbf{1}, \boldsymbol{X}) = \begin{pmatrix} 5 & 15 & 55 \\ 15 & 55 & 225 \\ 55 & 2225 & 979 \end{pmatrix} \quad \text{and} \quad (\mathbf{1}, \boldsymbol{X})'\boldsymbol{y} = \begin{pmatrix} 37 \\ 171 \\ 813 \end{pmatrix}$$

$$700[(\mathbf{1}, \boldsymbol{X})'(\mathbf{1}, \boldsymbol{X})]^{-1} = \begin{pmatrix} 3220 & -2310 & 350 \\ -2310 & 1870 & -300 \\ 350 & -300 & 50 \end{pmatrix}$$

$$\therefore \quad 700\ \boldsymbol{b} = \begin{pmatrix} 8680 \\ -9600 \\ 2300 \end{pmatrix} \quad \text{and} \quad \boldsymbol{b} = \begin{pmatrix} 12{\cdot}4000 \\ -13{\cdot}7143 \\ 3{\cdot}2857 \end{pmatrix}$$

In this example the regression equation is

$$y = 12{\cdot}4000-13{\cdot}7143x_1+3{\cdot}2857x_2, \tag{6.11.15}$$

and this is illustrated by Fig. 9.

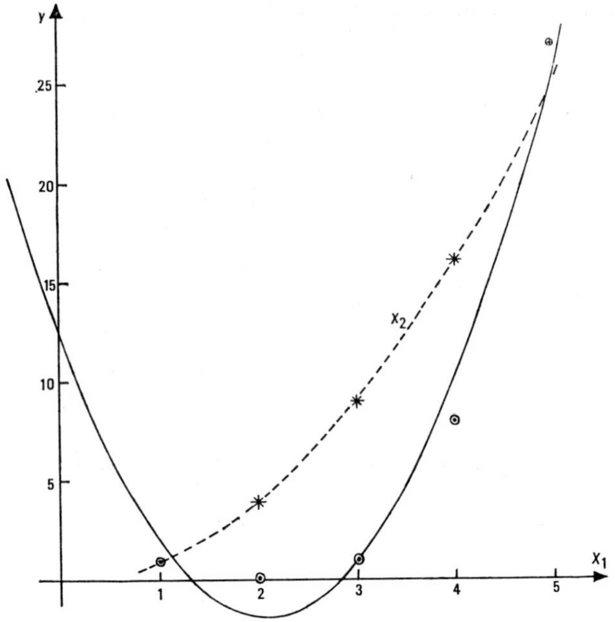

Fig. 9. Graph of a regression equation
⊙ ⊙ ⊙ ⊙ ⊙ observations of y
——— regression $(12{\cdot}4-13{\cdot}71x_1+3{\cdot}29x_2)$

6.12. The Least-Squares Property of $\boldsymbol{y}\cdot(\mathbf{1},\boldsymbol{x})$

If $\boldsymbol{y}$ is any measurement vector and $\boldsymbol{X}$ is any measurement matrix, then the vector

$$\boldsymbol{y}\cdot(\mathbf{1},\boldsymbol{X}) = \boldsymbol{y}-\boldsymbol{y}\perp(\mathbf{1},\boldsymbol{X}) \tag{6.12.1}$$

is the deviation of $\boldsymbol{y}$ from its regression on $\boldsymbol{X}$. This vector $\boldsymbol{y}\cdot(\mathbf{1},\boldsymbol{X})$ has the further property that, if $\boldsymbol{z}$ is any other vector of the form

$$\boldsymbol{z} = \boldsymbol{y}+(\mathbf{1},\boldsymbol{X})\boldsymbol{b} \qquad (\boldsymbol{b}\text{ any vector}), \tag{6.12.2}$$

then

$$\boldsymbol{z}'\boldsymbol{z} \geqslant \{\boldsymbol{y}\cdot(\mathbf{1},\boldsymbol{X})\}'\{\boldsymbol{y}\cdot(\mathbf{1},\boldsymbol{X})\}. \tag{6.12.3}$$

In other words, the sum of the squares of the deviations of the elements of $\boldsymbol{y}$ from the corresponding elements of $\boldsymbol{y}\perp(\mathbf{1},\boldsymbol{X})$, is the least possible for the sum of the squares of the deviations of the elements of $\boldsymbol{y}$ from the corresponding elements of any linear function of $(\mathbf{1},\boldsymbol{X})$. On this account, the regression of $\boldsymbol{y}$ on $\boldsymbol{X}$, namely $\boldsymbol{y}\perp(\mathbf{1},\boldsymbol{X})$, is known also as the least-squares regression of $\boldsymbol{y}$ on $\boldsymbol{X}$.

It will be found that in Example 6.11.4 the vector of deviations

$\boldsymbol{y}-(\mathbf{1}, \boldsymbol{X})\boldsymbol{b}$ consists of elements which are fairly small in comparison with those of $\boldsymbol{y}$. The sum of the squares of the elements is the smallest possible for any vector of the form

$$\boldsymbol{z} = \boldsymbol{y}-(\mathbf{1}, \boldsymbol{X})\boldsymbol{b}, \quad (\boldsymbol{b} \text{ being any 3-element vector}). \tag{6.12.4}$$

EXAMPLE 6.12.1. Establish the least-squares property of $\boldsymbol{y} \cdot (\mathbf{1}, \boldsymbol{X})$ when $\boldsymbol{X}$ is any $(n \times 2)$ matrix such that $(\mathbf{1}, \boldsymbol{X})'(\mathbf{1}, \boldsymbol{X})$ is non-singular.

Solution. Let $\boldsymbol{b}$ be any vector of three elements, and let

$$\phi(\boldsymbol{b}) = \{\boldsymbol{y}-(\mathbf{1}, \boldsymbol{X})\boldsymbol{b}\}'\{\boldsymbol{y}-(\mathbf{1}, \boldsymbol{X})\boldsymbol{b}\}.$$

We are required to establish that $\phi(\boldsymbol{b})$ will be at its minimum when

$$\boldsymbol{y}-(\mathbf{1}, \boldsymbol{X})\boldsymbol{b} = \boldsymbol{y} \cdot (\mathbf{1}, \boldsymbol{X}),$$

namely when

$$(\mathbf{1}, \boldsymbol{X})'\{\boldsymbol{y}-(\mathbf{1}, \boldsymbol{X})\boldsymbol{b}\} = \mathbf{0}. \tag{A}$$

We may expand the expression for $\phi(\boldsymbol{b})$ as

$$\phi(\boldsymbol{b}) = \boldsymbol{y}'\boldsymbol{y}-2\boldsymbol{b}'(\mathbf{1}, \boldsymbol{X})'\boldsymbol{y}+\boldsymbol{b}'(\mathbf{1}, \boldsymbol{X})'(\mathbf{1}, \boldsymbol{X})\boldsymbol{b}.$$

This is quadratic in $\boldsymbol{b}$ and can never be negative; hence it must be at a minimum at its stationary point. A sufficient condition for a minimum is then that $d\phi/d\boldsymbol{b} = \mathbf{0}$. This condition is

$$-2(\mathbf{1}, \boldsymbol{X})'\boldsymbol{y}+2(\mathbf{1}, \boldsymbol{X})'(\mathbf{1}, \boldsymbol{X})\boldsymbol{b} = \mathbf{0}.$$

But this condition is the same as (A). Hence $\phi(\boldsymbol{b})$ will be at its minimum when (A) is satisfied. This establishes the least-squares property.

EXAMPLE 6.12.2. Give a geometrical interpretation to the least-squares property of $\boldsymbol{y} \cdot (\mathbf{1}, \boldsymbol{x})$ when $\boldsymbol{x}, \boldsymbol{y}$ are any two measurement vectors.

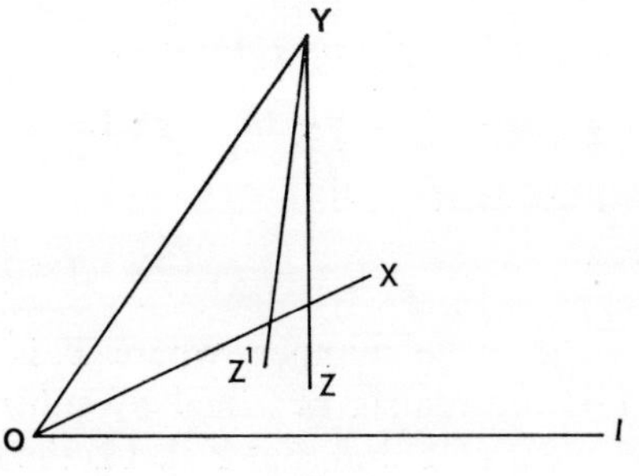

Fig. 10

Solution. Let OI, OX and OY represent the vectors $\mathbf{1}$, $\boldsymbol{x}$ and $\boldsymbol{y}$ in n-dimensional space. Then let Z be the foot of the perpendicular from Y onto the plane XOI. As in the illustration of § **10**, the vector $\boldsymbol{y} \cdot (\mathbf{1}, \boldsymbol{x})$ will be represented by the ray ZY. The sum of squares

$$\{\boldsymbol{y} \cdot (\mathbf{1}, \boldsymbol{x})\}'\{\boldsymbol{y} \cdot (\mathbf{1}, \boldsymbol{x})\}$$

will equal the square of the distance ZY.

Consider any vector $(\mathbf{1}b_0 + \boldsymbol{x}b_1)$. This will be represented by any point in the plane XOI. The least squares property of $\boldsymbol{y} \cdot (\mathbf{1}, \boldsymbol{x})$ is that $\{\boldsymbol{y} \cdot (\mathbf{1}, \boldsymbol{x})\}'\{\boldsymbol{y} \cdot (\mathbf{1}, \boldsymbol{x})\}$, is the least possible value of

$$\{\boldsymbol{y} - (\mathbf{1}b_0 + \boldsymbol{x}b_1)\}'\{\boldsymbol{y} - (\mathbf{1}b_0 + \boldsymbol{x}b)\}.$$

The geometrical representation of this property is that the square, YZ^2, of the distance of Y from Z, the foot of the perpendicular from Y onto the plane XOI, is the least possible for the square of the distance from Y to *any* point in the plane XOI. This result can be established by Pythagoras' Theorem.

6.13. Deviations from the Mean

Where $\boldsymbol{y}$ is any measurement vector, let

$$\bar{y} = \boldsymbol{f}'\boldsymbol{y}. \tag{6.13.1}$$

be the arithmetic mean of the elements of $\boldsymbol{y}$.

One often considers the vector obtained by subtracting $\bar{y}$ from each element of $\boldsymbol{y}$. This vector can be written as $(\boldsymbol{y} - \mathbf{1}\bar{y})$, but it may be written also as $\boldsymbol{y} \cdot \mathbf{1}$, since, where

$$\boldsymbol{y} \cdot \mathbf{1} = \boldsymbol{y} - \mathbf{1}b \tag{6.13.2}$$

and

$$\mathbf{1}'(\boldsymbol{y} \cdot \mathbf{1}) = 0, \tag{6.13.3}$$

we find, since

$$fn = 1, \tag{6.13.4}$$

that

$$\bar{y} = \boldsymbol{f}'\boldsymbol{y} = \boldsymbol{f}'\mathbf{1}b = b. \tag{6.13.5}$$

Hence

$$\boldsymbol{y} - \mathbf{1}\bar{y} = \boldsymbol{y} - \mathbf{1}b = \boldsymbol{y} \cdot \mathbf{1}. \tag{6.13.6}$$

It follows from (6.13.6) that

$$\boldsymbol{y} \perp \mathbf{1} = \mathbf{1}\bar{y}. \tag{6.13.7}$$

It is clear that in the same manner, where $\boldsymbol{Y}$ is any measurement matrix, then $\boldsymbol{Y} \cdot \mathbf{1}$ is the matrix obtained by subtracting the column mean from each element of $\boldsymbol{Y}$ and that $\boldsymbol{Y} \perp \mathbf{1}$ is the matrix obtained by substituting the column mean for each element of $\boldsymbol{Y}$.

Furthermore, if

$$\bar{y}' = f'Y \tag{6.13.8}$$

is the row vector of the column means of Y, then we may write

$$Y \perp 1 = 1\bar{y}'. \tag{6.13.9}$$

6.14. Covariance and Variance

If x and y are any two measurement vectors, the mean of the products of the deviations of corresponding elements from the means $\bar{x}$ and $\bar{y}$ is called the *covariance* of x and y. Thus

D. 43 The covariance of any two measurement vectors x and y is given by

$$n \operatorname{cov}(x, y) = (x \cdot 1)'(y \cdot 1).$$

EXAMPLE 6.14.1. Find the covariance of x and y where

$$x' = (1, 2, 3, 4, 5) \quad \text{and} \quad y' = (1, 4, 9, 16, 25).$$

Solution. $\bar{x} = 3$, $\bar{y} = 11$, and hence

$$(x \cdot 1)' = (-2, -1, 0, 1, 2) \quad (y \cdot 1)' = (-10, -7, -2, 5, 14)$$

$$(x \cdot 1)'(y \cdot 1) = 60 \qquad n = 5.$$

$$\therefore \quad 5 \operatorname{cov}(x, y) = 60 \qquad \therefore \quad \operatorname{cov}(x, y) = 12.$$

EXAMPLE 6.14.2. Show that

$$\operatorname{cov}(x, y) = \{(x'y)/n\} - \bar{x}\bar{y} \text{ where } \bar{x} = f'x \text{ and } \bar{y} = f'y.$$

Solution

$$\operatorname{cov}(x, y) = (x \cdot 1)'(y \cdot 1)/n = \{x - 1\bar{x}\}'\{y - 1\bar{y}\}/n.$$

$$\therefore \quad \operatorname{cov}(x, y) = (x'y/n) - (\bar{x}1'y)/n - (\bar{x}1'1\bar{y})/n + (\bar{x}1'1\bar{y})/n$$

$$\therefore \quad \operatorname{cov}(x, y) = (x'y/n) - \bar{x}\bar{y}.$$

Where y is any measurement vector, the mean of the squares of the deviations of its elements from their mean $\bar{y}$ is called the *variance* of y. The positive square root of the variance is called the *standard deviation.*

D. 44 The variance of any measurement vector y is given by

$$n \operatorname{var}(y) = (y \cdot 1)'(y \cdot 1).$$

The standard deviation of y is given by

$$s_y = \sqrt{\{\operatorname{var}(y)\}} \geqslant 0.$$

The standard deviation of any measurement vector $\mathbf{y}$ is a convenient measure of the extent to which the elements are spread out from the mean $\bar{y}$.

EXAMPLE 6.14.3. Calculate the variance and standard deviation for the two vectors $\mathbf{x}$ and $\mathbf{y}$, where

$$\mathbf{x}' = (1, 2, 3, 4, 5); \quad \mathbf{y}' = (1, 4, 9, 16, 25).$$

Solution. $n = 5, \quad \bar{x} = 3, \quad \bar{y} = 11.$

$$(\mathbf{x} \cdot \mathbf{1})' = (-2, -1, 0, 1, 2) \qquad (x \cdot 1)'(x \cdot 1) = 10$$
$$\text{var } \mathbf{x} = 2 \qquad s_x = \sqrt{2}$$

$$(\mathbf{y} \cdot \mathbf{1})' = (-10, -7, -2, 5, 14) \qquad (y \cdot 1)'(y \cdot 1) = 374$$
$$\text{var } \mathbf{y} = 74{\cdot}8 \qquad s_y = 8{\cdot}65$$

EXAMPLE 6.14.4. Show that var $\mathbf{y} = (\mathbf{y}'\mathbf{y}/n) - \bar{y}^2$.

Solution. In the second example following D. 43 we showed that

$$\text{cov}\,(\mathbf{x}, \mathbf{y}) = (\mathbf{x}'\mathbf{y}/n) - \bar{x}\bar{y}$$

$$\therefore \quad \text{var}\,(\mathbf{y}) = \text{cov}\,(\mathbf{y}, \mathbf{y}) = (\mathbf{y}'\mathbf{y}/n) - \bar{y}^2.$$

For instance, in the last example

$$\text{var}\,(\mathbf{y}) = (979/5) - 11^2 = 195{\cdot}8 - 121{\cdot}0 = 74{\cdot}8.$$

6.15. Correlation Coefficient

The correlation coefficient between two measurement vectors is defined as follows:

D. 45 The correlation coefficient between any two measurement vectors $\mathbf{x}$ and $\mathbf{y}$ is given by

$$r(\mathbf{x}, \mathbf{y}) = \text{cov}\,(\mathbf{x}, \mathbf{y})\ /\ s_x s_y.$$

EXAMPLE 6.15.1. Find the correlation coefficient between the measurement vectors $\mathbf{x}$ and $\mathbf{y}$, where

$$\mathbf{x}' = (1, 2, 3, 4, 5) \qquad \mathbf{y}' = (1, 4, 9, 16, 25).$$

Solution. We have found in examples in § **6.14** that

$$\text{cov}\,(\mathbf{x}, \mathbf{y}) = 12 \qquad s_x = \sqrt{2} \qquad s_y = \sqrt{74{\cdot}8}$$

$$\therefore \quad r(\mathbf{x}, \mathbf{y}) = 12/\sqrt{149{\cdot}6} = 0{\cdot}981.$$

EXAMPLE 6.15.2. Show that

$$r(\boldsymbol{x}, \boldsymbol{y}) = \frac{n\boldsymbol{x}'\boldsymbol{y} - (\boldsymbol{1}'\boldsymbol{x})(\boldsymbol{1}'\boldsymbol{y})}{\{n\boldsymbol{x}'\boldsymbol{x} - (\boldsymbol{1}'\boldsymbol{x})^2\}^{1/2}\{n\boldsymbol{y}'\boldsymbol{y} - (\boldsymbol{1}'\boldsymbol{y})^2\}^{1/2}}.$$

Proof. We have shown in the examples of § **6.14** that

$$n \operatorname{cov}(\boldsymbol{x}, \boldsymbol{y}) = \boldsymbol{x}'\boldsymbol{y} - n\bar{x}\bar{y}$$

$$\therefore \quad n^2 \operatorname{cov}(\boldsymbol{x}, \boldsymbol{y}) = n\boldsymbol{x}'\boldsymbol{y}' - (\boldsymbol{1}'\boldsymbol{x})(\boldsymbol{1}'\boldsymbol{y})$$

$$\therefore \quad n^2 \operatorname{var} \boldsymbol{x} = n\boldsymbol{x}'\boldsymbol{x} - (\boldsymbol{1}'\boldsymbol{x})^2$$

$$\therefore \quad ns_x = \{n\boldsymbol{x}'\boldsymbol{x} - (\boldsymbol{1}'\boldsymbol{x})^2\}^{1/2} \quad \text{and} \quad n^2 \operatorname{var} \boldsymbol{y} = n\boldsymbol{y}'\boldsymbol{y} - (\boldsymbol{1}'\boldsymbol{y})^2$$

$$\therefore \quad r(\boldsymbol{x}, \boldsymbol{y}) = \operatorname{cov}(\boldsymbol{x}, \boldsymbol{y})/s_x s_y = \frac{n\boldsymbol{x}'\boldsymbol{y} - (\boldsymbol{1}'\boldsymbol{x})(\boldsymbol{1}'\boldsymbol{y})}{\{n\boldsymbol{y}'\boldsymbol{y} - (\boldsymbol{1}'\boldsymbol{y})^2\}^{1/2}\{n\boldsymbol{x}'\boldsymbol{x} - (\boldsymbol{1}'\boldsymbol{x})^2\}^{1/2}}$$

6.16. Variance-Covariance Matrix

The concept of covariance can be extended to apply to more than two measurement vectors. Thus if a measurement matrix $\boldsymbol{X}$ is composed of measurement column vectors, then

D. 46 The variance-covariance matrix for the measurement matrix $\boldsymbol{X}$, is $\boldsymbol{W}(\boldsymbol{X})$, given by

$$n\,\boldsymbol{W}(\boldsymbol{X}) = (\boldsymbol{X} \cdot \boldsymbol{1})'(\boldsymbol{X} \cdot \boldsymbol{1}).$$

The off-diagonal elements of $W(X)$ are the covariances between the various pairs of the vectors of which $\boldsymbol{X}$ is composed, whereas the diagonal elements are the variances of these measurement vectors.

EXAMPLE 6.16.1. Find the variance-covariance matrix for the measurement matrix $\boldsymbol{X} = (\boldsymbol{x}, \boldsymbol{y})$ where

$$\boldsymbol{x}' = (1, 2, 3, 4, 5), \quad \boldsymbol{y}' = (1, 4, 9, 16, 25).$$

Solution

$$\boldsymbol{X} = \begin{pmatrix} 1 & 1 \\ 2 & 4 \\ 3 & 9 \\ 4 & 16 \\ 5 & 25 \end{pmatrix} \quad f'\boldsymbol{X} = (3, 11) \quad \therefore \quad (\boldsymbol{X} \cdot \boldsymbol{1}) = \begin{pmatrix} -2 & -10 \\ -1 & -7 \\ 0 & -2 \\ 1 & 5 \\ 2 & 14 \end{pmatrix}$$

$$(\boldsymbol{X} \cdot \boldsymbol{1})'(\boldsymbol{X} \cdot \boldsymbol{1}) = \begin{pmatrix} 10 & 60 \\ 60 & 374 \end{pmatrix} \quad n = 5 \quad \therefore \quad \boldsymbol{W}(\boldsymbol{X}) = \begin{pmatrix} 2 & 12 \\ 12 & 74{\cdot}8 \end{pmatrix}$$

Alternatively, we could use our earlier results to examples in § **6.14** and write

$$W(X) = \begin{pmatrix} \text{var } \boldsymbol{x} & \text{cov }(\boldsymbol{x}, \boldsymbol{y}) \\ \text{cov }(\boldsymbol{y}, \boldsymbol{x}) & \text{var } \boldsymbol{y} \end{pmatrix} = \begin{pmatrix} 2 & 12 \\ 12 & 74{\cdot}8 \end{pmatrix}$$

6.17. The Additive Property of a Projection $Y \perp X$

Suppose that a measurement matrix $\boldsymbol{X}$ consists of an orthogonal set of submatrices $\boldsymbol{X}_1, \boldsymbol{X}_2, \ldots, \boldsymbol{X}_k$. Then it is a convenient fact that the projection of any measurement vector $\boldsymbol{y}$ onto $\boldsymbol{X}$ may be decomposed without residue into the k separate projections of $\boldsymbol{y}$ onto $\boldsymbol{X}_1$, onto $\boldsymbol{X}_2, \ldots$, and onto $\boldsymbol{X}_k$. More generally the following theorem holds.

THEOREM 6.17.1. If $\boldsymbol{Y}$ is any measurement matrix and $\boldsymbol{X}$ is any orthogonal set of matrices $\{\boldsymbol{X}_1, \boldsymbol{X}_2, \ldots, \boldsymbol{X}_k\}$, then

$$\boldsymbol{Y} \perp \boldsymbol{X} = \sum_{h=1}^{k} (\boldsymbol{Y} \perp \boldsymbol{X}_h).$$

EXAMPLE 6.17.1. Establish the theorem in the case where $\boldsymbol{Y}$ is a vector and when the $\boldsymbol{X}_h$ are orthogonal vectors.

Solution. Let $\boldsymbol{X}$ be composed of the k orthogonal measurement vectors $\boldsymbol{x}_1, \boldsymbol{x}_2 \ldots \boldsymbol{x}_k$. Let $\boldsymbol{Y}$ be the measurement vector $\boldsymbol{y}$.

Let $\boldsymbol{y} \perp \boldsymbol{x}_h = \boldsymbol{x}_h b_h; \qquad h = 1, 2, \ldots, k;$

then $\boldsymbol{x}_h'\boldsymbol{y} = (\boldsymbol{x}_h'\boldsymbol{x}_h)b_h; \qquad h = 1, 2, \ldots, k.$

Again, let $\boldsymbol{y} \perp \boldsymbol{X} = \boldsymbol{X\beta}.$

then $\boldsymbol{X}'\boldsymbol{y} = \boldsymbol{X}'\boldsymbol{X\beta} = \hat{\boldsymbol{c}}\boldsymbol{\beta},$

where $\boldsymbol{c}' = (\boldsymbol{x}_1'\boldsymbol{x}_1, \boldsymbol{x}_2'\boldsymbol{x}_2, \ldots, \boldsymbol{x}_k'\boldsymbol{x}_k),$

since $\boldsymbol{x}_h'\boldsymbol{x}_j = 0$, when $h \neq j$.

Now $(\boldsymbol{X}'\boldsymbol{y})' = (\boldsymbol{x}_1'\boldsymbol{y}, \boldsymbol{x}_2'\boldsymbol{y}, \ldots, \boldsymbol{x}_k'\boldsymbol{y});$

$\therefore \quad \boldsymbol{\beta}' = \{\hat{\boldsymbol{c}}^{-1}\boldsymbol{X}'\boldsymbol{y}\}' = (b_1, b_2, \ldots, b_k)$

since $b_h = (\boldsymbol{x}_h'\boldsymbol{x}_h)^{-1}(\boldsymbol{x}_h'\boldsymbol{y}).$

Finally, $\boldsymbol{y} \perp \boldsymbol{X} = \boldsymbol{X\beta} = \sum_{h=1}^{k} \boldsymbol{x}_h b_h = \sum_{h=1}^{k} (\boldsymbol{y} \perp \boldsymbol{x}_h).$

EXAMPLE 6.17.2. Establish the theorem in the case where Y is a vector $\boldsymbol{y}$ and $k = 2$.

Solution. Let $\boldsymbol{y} \perp \boldsymbol{X}_1 = \boldsymbol{X}_1\boldsymbol{b}, \boldsymbol{y} \perp \boldsymbol{X}_2 = \boldsymbol{X}_2\boldsymbol{c}$

and $$\boldsymbol{y} \perp (\boldsymbol{X}_1, \boldsymbol{X}_2) = (\boldsymbol{X}_1, \boldsymbol{X}_2)\begin{pmatrix} g \\ h \end{pmatrix}.$$

Then $$\boldsymbol{X}_1'\boldsymbol{y} = \boldsymbol{X}_1'\boldsymbol{X}_1\boldsymbol{b},$$
$$\boldsymbol{X}_2'\boldsymbol{y} = \boldsymbol{X}_2'\boldsymbol{X}_2\boldsymbol{c}$$

and $$(\boldsymbol{X}_1, \boldsymbol{X}_2)'\boldsymbol{y} = (\boldsymbol{X}_1', \boldsymbol{X}_2)'(\boldsymbol{X}_1, \boldsymbol{X}_2)g\begin{pmatrix} g \\ h \end{pmatrix}.$$

$$\therefore \quad \begin{pmatrix} \boldsymbol{X}_1'\boldsymbol{y} \\ \boldsymbol{X}_2'\boldsymbol{y} \end{pmatrix} = \begin{pmatrix} \boldsymbol{X}_1'\boldsymbol{X}_1 & 0 \\ 0 & \boldsymbol{X}_2'\boldsymbol{X}_2 \end{pmatrix}\begin{pmatrix} g \\ h \end{pmatrix}$$

since $\boldsymbol{X}_1$ and $\boldsymbol{X}_2$ are orthogonal

$$\therefore \quad \boldsymbol{X}_1'\boldsymbol{y} = \boldsymbol{X}_1'\boldsymbol{X}\boldsymbol{g}$$

$$\boldsymbol{X}_2'\boldsymbol{y} = \boldsymbol{X}_2'\boldsymbol{X}\boldsymbol{h}$$

$$\boldsymbol{g} = (\boldsymbol{X}_1{}'\boldsymbol{X}_1)^{-1}\boldsymbol{X}_1'\boldsymbol{y} = \boldsymbol{b}$$

and $$\boldsymbol{h} = (\boldsymbol{X}_2\boldsymbol{X}_2)^{-1}\boldsymbol{X}_2'\boldsymbol{y} = \boldsymbol{c}$$

and $$\boldsymbol{y} \perp (\boldsymbol{X}_1, \boldsymbol{X}_2) = (\boldsymbol{X}_1 g + \boldsymbol{X}_2 h) = \boldsymbol{X}_1\boldsymbol{b} + \boldsymbol{X}_2\boldsymbol{c}$$

$$\therefore \quad \boldsymbol{y} \perp (\boldsymbol{X}_1, \boldsymbol{X}_2) = \boldsymbol{y} \perp \boldsymbol{X}_1 + \boldsymbol{y} \perp \boldsymbol{X}_2.$$

This proof can be modified to prove the general theorem.

6.18. A Further Theorem on Projections

It is often useful to be able to quote the following theorem.

THEOREM 6.18.1. If $\boldsymbol{X}$, $\boldsymbol{Y}$ are measurement matrices and $\boldsymbol{C}$ is any non-singular matrix postmultiplying $\boldsymbol{X}$, then

$$\boldsymbol{Y} \perp \boldsymbol{X}\boldsymbol{C} = \boldsymbol{Y} \perp \boldsymbol{X}. \tag{6.18.1}$$

The proof is as follows:

Let $$\boldsymbol{Y} \perp \boldsymbol{XC} = \boldsymbol{XCH}$$

then $$\boldsymbol{C}'\boldsymbol{X}'\boldsymbol{Y} = \boldsymbol{C}'\boldsymbol{X}'\{\boldsymbol{Y} \perp \boldsymbol{XC}\};$$

$$\therefore \quad \boldsymbol{X}'\boldsymbol{Y} = \boldsymbol{X}'\{\boldsymbol{Y} \perp \boldsymbol{XC}\}, \text{ since } \boldsymbol{C} \text{ is non-singular.}$$

Hence $\boldsymbol{Y} \perp \boldsymbol{XC}$ satisfies both the conditions defining $\boldsymbol{Y} \perp \boldsymbol{X}$.

Hence, since $\boldsymbol{Y} \perp \boldsymbol{X}$ is unique,

$$\boldsymbol{Y} \perp \boldsymbol{XC} = \boldsymbol{Y} \perp \boldsymbol{X}.$$

EXAMPLE 6.18.1. Show that the regression of $\boldsymbol{Y}$ on $\boldsymbol{X}$ is

$$\boldsymbol{Y} \perp (\boldsymbol{1}, \boldsymbol{X} \cdot \boldsymbol{1}).$$

Solution. By definition the regression of $\boldsymbol{Y}$ on $\boldsymbol{X}$ is $\boldsymbol{Y} \perp (\mathbf{1}, \boldsymbol{X})$.

Now
$$(\mathbf{1}, \boldsymbol{X} \cdot \mathbf{1}) = (\mathbf{1}, \boldsymbol{X})\boldsymbol{C}^*, \text{ where}$$

$$\boldsymbol{C}^* = \begin{pmatrix} 1-\bar{x}_1 & -\bar{x}_2 \dots \bar{x}_k \\ 1 & 0 \\ & 1 \\ 0 & & \ddots \\ & & & 1 \end{pmatrix}$$

and $\bar{x}_1, \bar{x}_2, \dots, \bar{x}_k$ denote the means of the columns of $\boldsymbol{X}$.

Hence $\boldsymbol{Y} \perp (\mathbf{1},\boldsymbol{X} . \mathbf{1}) = \boldsymbol{Y} \perp (\mathbf{1},\boldsymbol{X})$ by the above Theorem and thus gives the regression of $\boldsymbol{Y}$ on $\boldsymbol{X}$.

EXAMPLE 6.18.2. Verify that where

$$\boldsymbol{Y} = \begin{pmatrix} 1 & 1 \\ 1 & 4 \\ 0 & 6 \\ 1 & 4 \\ 2 & 1 \end{pmatrix} \quad \boldsymbol{Z}_1 = \begin{pmatrix} 1 \\ 1 \\ 1 \\ 1 \\ 1 \end{pmatrix} \text{ and } \boldsymbol{Z}_2 = \begin{pmatrix} -2 & -10 \\ -1 & -7 \\ 0 & -2 \\ 1 & 5 \\ 2 & 14 \end{pmatrix},$$

$$\boldsymbol{Y} \perp (\boldsymbol{Z}_1, \boldsymbol{Z}_2) = \boldsymbol{Y} \perp \boldsymbol{Z}_1 + \boldsymbol{Y} \perp \boldsymbol{Z}_2.$$

Solution. We find by the methods of §**6.11** that, where

$$\boldsymbol{X} = \begin{pmatrix} 1 & 1 \\ 2 & 4 \\ 3 & 9 \\ 4 & 16 \\ 5 & 25 \end{pmatrix}, \quad \boldsymbol{Y} \perp (\mathbf{1}, \boldsymbol{X}) = (1/70) \begin{pmatrix} 82 & 64 \\ 36 & 304 \\ 30 & 384 \\ 64 & 304 \\ 138 & 64 \end{pmatrix}.$$

Now
$$(\boldsymbol{Z}_1, \boldsymbol{Z}_2) = (\mathbf{1}, \boldsymbol{X}) \begin{pmatrix} 1 & -3 & -11 \\ 0 & 1 & 0 \\ 0 & 0 & 1 \end{pmatrix}, \tag{6.18.2}$$

so that we may apply (6.18.1) to establish

$$\boldsymbol{Y} \perp (\boldsymbol{Z}_1, \boldsymbol{Z}_2) = (1/70) \begin{pmatrix} 82 & 64 \\ 36 & 304 \\ 30 & 384 \\ 64 & 304 \\ 138 & 64 \end{pmatrix}. \tag{6.18.3}$$

Now let $Y \perp Z_1 = Z_1 b$; then

$$b = (Z_1' Z_1)^{-1} Z_1' Y = (1, 3{\cdot}2)$$

so that
$$Y \perp Z_1 = \begin{pmatrix} 1 & 3{\cdot}2 \\ 1 & 3{\cdot}2 \\ 1 & 3{\cdot}2 \\ 1 & 3{\cdot}2 \\ 1 & 3{\cdot}2 \end{pmatrix} = (1/70)\begin{pmatrix} 70 & 224 \\ 70 & 224 \\ 70 & 224 \\ 70 & 224 \\ 70 & 224 \end{pmatrix}. \quad (6.18.4)$$

Again, let $Y \perp Z_2 = Z_2 B$;

then
$$B = (Z_2' Z_2)^{-1}(Z_2' Y) = \begin{pmatrix} 10 & 60 \\ 60 & 374 \end{pmatrix}^{-1} \begin{pmatrix} 2 & 0 \\ 16 & -16 \end{pmatrix}$$

$$\therefore \quad B = (1/70)\begin{pmatrix} 187 & -30 \\ -30 & 5 \end{pmatrix}\begin{pmatrix} 2 & 0 \\ 16 & -16 \end{pmatrix} = (1/70)\begin{pmatrix} -106 & 480 \\ 20 & -80 \end{pmatrix}$$

so that
$$Y \perp Z_2 = (1/70)\begin{pmatrix} 12 & -160 \\ -34 & 80 \\ -40 & 160 \\ -6 & 80 \\ 68 & -160 \end{pmatrix}. \quad (6.18.5)$$

Comparing (6.18.3), (6.18.4) and (6.18.5), we see that

$$Y \perp (Z_1, Z_2) = Y \perp Z_1 + Y \perp Z_2. \quad (6.18.6)$$

6.19. The Deviation of the Regression of Y on X

The regression of Y on X is $Y \perp (\mathbf{1}, X)$. We have seen in the example of § **6.18** that

$$Y \perp (\mathbf{1}, X) = Y \perp (\mathbf{1}, X \cdot \mathbf{1}). \quad (6.19.1)$$

It follows from Theorem 6.17.1 that the regression of Y on X is

$$Y \perp (\mathbf{1}, X) = Y \perp \mathbf{1} + Y \perp (X \cdot \mathbf{1}). \quad (6.19.2)$$

The deviation of the regression of Y on X is given by

$$\{Y \perp (\mathbf{1}, X)\} \cdot \mathbf{1} = (Y \perp \mathbf{1}) \cdot \mathbf{1} + \{Y \perp (X \cdot \mathbf{1})\} \cdot \mathbf{1}$$
$$= \{Y \perp (X \cdot \mathbf{1})\} \cdot \mathbf{1}. \quad (6.19.3)$$

Suppose

$$\{Y \perp (\mathbf{1}, X)\} \cdot \mathbf{1} = Y \perp (X \cdot \mathbf{1}) - \mathbf{1}b$$
$$= (X \cdot \mathbf{1})B - \mathbf{1}b, \text{ say.} \quad (6.19.4)$$

Then

$$\mathbf{1}'\mathbf{1}b = \mathbf{1}'(X \cdot \mathbf{1})B = \mathbf{0}. \quad (6.19.5)$$

$\therefore \quad \boldsymbol{b} = \mathbf{0}$, and the regression of $\boldsymbol{Y}$ on $\boldsymbol{X}$ is given by

$$\{\boldsymbol{Y} \perp (\mathbf{1}, \boldsymbol{X})\} \cdot \mathbf{1} = \boldsymbol{Y} \perp (\boldsymbol{X} \cdot \mathbf{1}). \tag{6.19.6}$$

EXAMPLE 6.19.1. Give the geometrical interpretation of (6.19.6) when $\boldsymbol{Y}$ and $\boldsymbol{X}$ are measurement *vectors*.

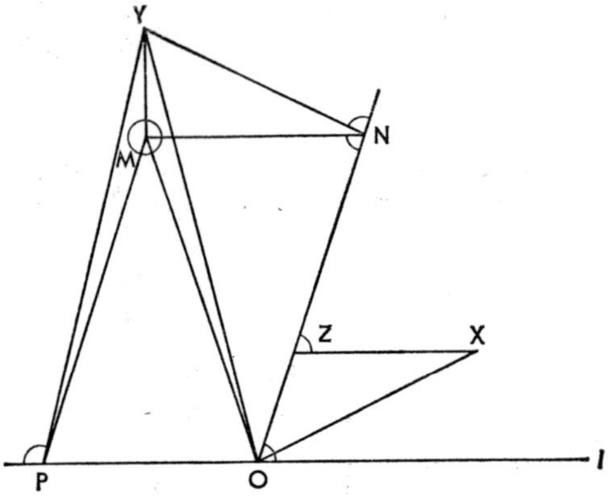

Fig. 11

Solution. Let the vectors $\mathbf{1}$, $\boldsymbol{x}$ and $\boldsymbol{y}$ be represented by the rays OI, OX and OY.

In the diagram O, X, I and all the labelled points except Y lie in the same horizontal plane and YM is the vertical perpendicular from Y onto the plane. ON is drawn perpendicular to OI in the horizontal plane and N is the foot of the perpendicular from Y onto ON. Z is the foot of the perpendicular from X onto ON. P is the foot of the perpendicular from Y onto OI. Then writing $*$ for 'represents' we find

$$\text{OZ} * \boldsymbol{x} \cdot \mathbf{1} \qquad \text{ON} * \boldsymbol{y} \perp (\boldsymbol{x} \cdot \mathbf{1})$$

$$\text{OM} * \boldsymbol{y} \perp (\mathbf{1}, \boldsymbol{X}) \qquad \text{ON} * \{\boldsymbol{y} \perp (\mathbf{1}, \boldsymbol{x})\} \cdot \mathbf{1}.$$

The equation

$$\{\boldsymbol{y} \perp (\mathbf{1}, \boldsymbol{x})\} \cdot \mathbf{1} = \boldsymbol{y} \perp (\boldsymbol{x} \cdot \mathbf{1})$$

corresponds to the fact that N is the foot both of the perpendicular from Y onto OZ and of the perpendicular from M onto OZ.

6.20. Analysis of the Regression of Y on X

Suppose that $\boldsymbol{X}$ is a measurement matrix composed of an *uncorrelated* set of matrices $\boldsymbol{X}_h$ ($h = 1, 2, \ldots, k$). Then the deviation of the regression of $\boldsymbol{Y}$ (any other measurement matrix) on $\boldsymbol{X}$ may be analysed into the separate deviations of the regressions of Y onto the X_h. Thus

$$Y \perp (X \cdot \mathbf{1}) = \sum_{h=1}^{k} \{Y \perp (X_h \cdot \mathbf{1})\}. \tag{6.20.1}$$

This follows immediately from Theorem 6.17.1, since if X_h are an uncorrelated set, $(X_h \cdot \mathbf{1})$ $(h = 1, 2, \ldots, k)$ are an orthogonal set of measurement matrices.

EXAMPLE 6.20.1. The measurement matrix X is composed of the uncorrelated matrices $\begin{pmatrix} 1 & 0 \\ 1 & 0 \\ 0 & 1 \\ 0 & 1 \end{pmatrix}$ and $\begin{pmatrix} 1 & 0 \\ 0 & 1 \\ 1 & 0 \\ 0 & 1 \end{pmatrix}$

If y is the measurement vector $\begin{pmatrix} 1 \\ 2 \\ 5 \\ 8 \end{pmatrix}$,

analyse $y \perp (X \cdot \mathbf{1})$ into the two components associated with the submatrices.

Solution

$$y \perp (X \cdot \mathbf{1}) = \begin{pmatrix} -3{\cdot}5 \\ -1{\cdot}5 \\ 1{\cdot}5 \\ 3{\cdot}5 \end{pmatrix} = \begin{pmatrix} -2{\cdot}5 \\ -2{\cdot}5 \\ 2{\cdot}5 \\ 2{\cdot}5 \end{pmatrix} + \begin{pmatrix} -1 \\ 1 \\ -1 \\ 1 \end{pmatrix}.$$

In the particular case where X consists of uncorrelated measurement *vectors* x_h $(h = 1, 2, \ldots, k)$, we shall have

$$Y \perp (X \cdot \mathbf{1}) = \sum_{h=1}^{k} \{Y \perp (x_h \cdot \mathbf{1})\}. \tag{6.20.2}$$

If the vectors of which X is composed are not an uncorrelated set, we can still form an uncorrelated set z_h $(h = 1, 2, \ldots, k)$ by taking

$$\begin{aligned}
z_1 &= x_1 \cdot \mathbf{1} \\
z_2 &= x_2 \cdot (\mathbf{1}, x_1) \\
z_3 &= x_3 \cdot (\mathbf{1}, x_1, x_2) \\
&\vdots \\
z_h &= x_h \cdot (\mathbf{1}, x_1, \ldots, x_{h-1}) \\
&\vdots \\
z_k &= x_k \cdot (\mathbf{1}, x_1, \ldots, x_{k-1}).
\end{aligned} \tag{6.20.3}$$

Then

$$Y \perp (X \cdot 1) = \sum_{h=1}^{k} \{Y \perp (z \cdot 1)\} = \sum_{h=1}^{k} \{Y \perp z_k\}. \tag{6.20.4}$$

The analysis (6.20.4) is not on the face of it as convenient as the more straightforward analysis,

$$\{Y \perp (X \cdot 1)\} = (X \cdot 1)B \text{ where } (X \cdot 1)'(X \cdot 1)B = (X \cdot 1)'Y,$$

whence

$$\{Y \perp (X \cdot 1)\} = \sum_{h=1}^{k} (x_h \cdot 1)B_h' \text{ (say)}, \tag{6.20.5}$$

where $x_h \cdot 1$ and B_h' are the h-th column of $X_h \cdot 1$ and the h-th row of B. (6.20.6)

But (6.20.4) has the advantage over (6.20.5) that the *variance* of $\{Y \perp (X \cdot 1)\}$ can be conveniently analysed into terms associated with the separate terms of the R.H.S. of (6.20.4), whereas it cannot so easily be analysed into terms associated with the separate terms of the R.H.S. of (6.20.5).

EXAMPLE 6.20.2. Give a geometrical illustration of the two breakdowns given by equations (6.20.4) and (6.20.5) when Y is a vector, y, and X has two columns only.

Solution. In this case the two equations may be written as

$$y \perp X \cdot 1 = y \perp x_1 \cdot 1 + y \perp x_2 \cdot (1_1 x_1)$$

and

$$y \perp X \cdot 1 = (x_1 \cdot 1)b_1 + (x_2 \cdot 1)b_2.$$

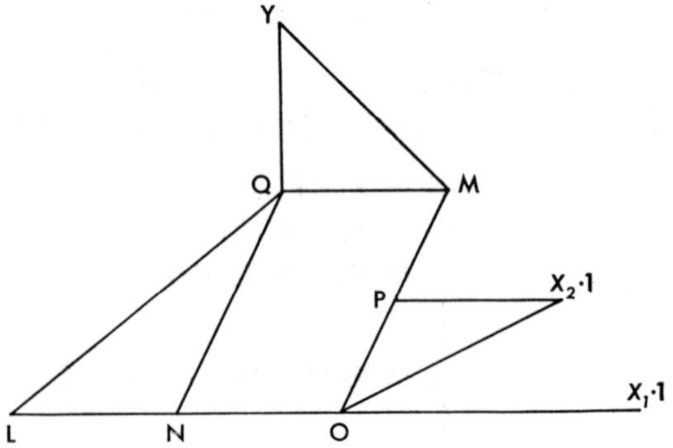

Fig. 12

In the diagram, all the labelled points lie in the same horizontal plane except Y, and YQ is the perpendicular from Y onto the horizontal plane. $OX_1 \cdot \mathbf{1}$, $OX_2 \cdot \mathbf{1}$ and OY represent $\boldsymbol{x}_1 \cdot \mathbf{1}$, $\boldsymbol{x}_2 \cdot \mathbf{1}$ and $\boldsymbol{y}$. QN is the perpendicular from Q onto $OX_1 \cdot \mathbf{1}$, OMQN is a rectangle, and YM is the perpendicular from Y onto OM and $X_2 \cdot \mathbf{1}$, P is perpendicular fron $X_2 \cdot \mathbf{1}$ onto OM. QL is drawn parallel to $OX_2 \cdot \mathbf{1}$ to cut $OX_1 \cdot \mathbf{1}$ in L. The two equations are represented by

$$OQ = ON+OM \quad \text{and}$$

$$OQ = OL+LQ = (OX_1 \cdot \mathbf{1})b_1(OX_2 \cdot \mathbf{1})b_2 \text{ (say).}$$

6.21. Analysis of Sums of Squares

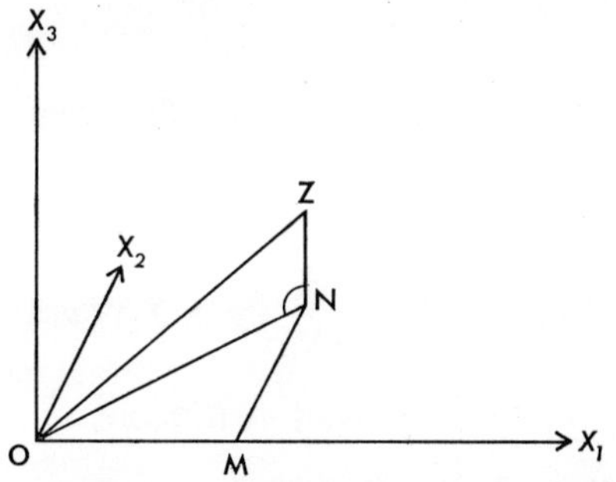

Fig. 13

The sum of the squares of the elements of any vector $\boldsymbol{z}$ is given by $\boldsymbol{z}'\boldsymbol{z}$. If $\boldsymbol{z}$ is represented geometrically by the ray OZ, where Z has for coordinates the elements of $\boldsymbol{z}$, then by the repeated application of Pythagoras' Theorem

$$OZ^2 = \boldsymbol{z}'\boldsymbol{z} \tag{6.21.1}$$

where OZ stands for the scalar length of the ray OZ.

EXAMPLE 6.21.1. Illustrate and establish (6.21.1) when $\boldsymbol{z}$ has three elements.

Solution. Let OX_1 OX_2 OX_3 be three mutually perpendicular axes. Let Z be any point in the space generated by OX_1 OX_2 OX_3.

Let $(Z_1\ Z_2\ Z_3)$ be the coordinates of Z.

If $\boldsymbol{z}$ denotes the vector $\begin{pmatrix} Z_1 \\ Z_2 \\ Z_3 \end{pmatrix}$ then $\boldsymbol{z}'\boldsymbol{z} = Z_1^2+Z_2^2+Z_3^2$.

Let ZN be the perpendicular from Z onto the plane OX_1X_2 and let M be the foot of the perpendicular from N onto OX_1. Then

$$Z_1 = OM,\ Z_2 = MN \text{ and } Z_3 = NZ$$

$$\therefore\quad \mathbf{z}'\mathbf{z} = (OM^2+MN^2)+NZ^2 = ON^2+NZ^2 = OZ^2$$

by two applications of Pythagoras.

$\therefore$ (6.21.1) is established.

The above geometrical example also illustrates a special case of the following result.

If $\boldsymbol{y}$ is a measurement vector and $\boldsymbol{X}$ is a measurement matrix composed of k orthogonal sub matrices $\boldsymbol{X}_h$ $(h = 1, 2, \ldots, k)$ then corresponding to the breakdown of $\boldsymbol{y} \perp \boldsymbol{X}$ according to

$$\boldsymbol{y} \perp \boldsymbol{X} = \sum_{h=1}^{k} (\boldsymbol{y} \perp \boldsymbol{X}_h) \quad \text{(by Theorem 6.17.1)} \qquad (6.21.2)$$

we may analyse the sum of squares $(\boldsymbol{y} \perp \boldsymbol{X})'(\boldsymbol{y} \perp \boldsymbol{X})$ as

$$(\boldsymbol{y} \perp \boldsymbol{X})'(\boldsymbol{y} \perp \boldsymbol{X}) = \sum_{h=1}^{k} (\boldsymbol{y} \perp \boldsymbol{X}_h)'(\boldsymbol{y} \perp \boldsymbol{X}_h). \qquad (6.21.3)$$

The result (6.21.3) may be proved as follows.

Let $$\boldsymbol{y} \perp \boldsymbol{X}_h = \boldsymbol{X}_h\boldsymbol{B}_h \quad (h = 1, 2, \ldots, k); \qquad (6.21.4)$$

then $$\boldsymbol{y} \perp \boldsymbol{X} = \sum_{h=1}^{k} (\boldsymbol{y}_h \perp \boldsymbol{X}_h) \qquad (6.21.2)$$

$$\therefore\quad (\boldsymbol{y} \perp \boldsymbol{X}')'(\boldsymbol{y} \perp \boldsymbol{X}) = \sum_{h=1}^{k} (\boldsymbol{X}_h\boldsymbol{B}_h)' \sum_{i=1}^{k} \boldsymbol{X}_i\boldsymbol{B}_i$$

$$= \sum_{h=1}^{k} \boldsymbol{B}_h' \sum_{i=1}^{k} \boldsymbol{X}_h'\boldsymbol{X}_i\boldsymbol{B}_i = \sum_{h=1}^{k} (\boldsymbol{B}_h'\boldsymbol{X}_h'\boldsymbol{X}_h\boldsymbol{B}_h) \qquad (6.21.5)$$

since for $h \neq i$ $\boldsymbol{X}_h'\boldsymbol{X}_i = \mathbf{0}$, because the $\boldsymbol{X}_h$ are orthogonal,

$$\therefore\quad (\boldsymbol{y} \perp \boldsymbol{X}')'(\boldsymbol{y} \perp \boldsymbol{X}) = \sum_{h=1}^{k} (\boldsymbol{X}_h\boldsymbol{B}_h)'(\boldsymbol{X}_h\boldsymbol{B}_h)$$

$$= \sum_{h=1}^{k} (\boldsymbol{y} \perp \boldsymbol{X}_h)'(\boldsymbol{y} \perp \boldsymbol{X}_h). \qquad (6.21.6)$$

EXAMPLE 6.21.2. Explain how the previous geometrical example illustrates the result (6.21.3).

Solution. Let $\boldsymbol{X}$ be composed of the three orthogonal vectors

$\boldsymbol{x}_1$, $\boldsymbol{x}_2$, $\boldsymbol{x}_3$, and let $\boldsymbol{y}$ be any vector. Let rays OY, OX_1, OX_2, OX_3 represent the vectors $\boldsymbol{y}$, $\boldsymbol{x}_1$, $\boldsymbol{x}_2$ and $\boldsymbol{x}_3$. Then in the diagram of the last example, $\boldsymbol{y} \perp \boldsymbol{x}_1$, $\boldsymbol{y} \perp \boldsymbol{x}_2$, $\boldsymbol{y} \perp \boldsymbol{x}_3$ and $\boldsymbol{y} \perp \boldsymbol{X}$ are represented by the feet of the perpendiculars from Y onto the space $OX_1X_2X_3$, onto OX_1, onto OX_2 and onto OX_3. Y itself lies outside $OX_1X_2X_3$ and is therefore not shown in the diagram.

Let Z be the foot of the perpendicular from Y onto $OX_1X_2X_3$.

Then the feet of the perpendiculars from Y onto the three rays OX_1 OX_2 OX_3 are also the feet of the perpendiculars from Z onto OX_1 OX_2 and OX_3. Hence $\boldsymbol{y} \perp \boldsymbol{x}_1$, $\boldsymbol{y} \perp \boldsymbol{x}_2$ and $\boldsymbol{y} \perp \boldsymbol{x}_3$ are represented by OM, MN and NZ.

[We use the fact that the opposite sides of any rectangle are equal and parallel.]

Hence $$(\boldsymbol{y} \perp \boldsymbol{X})'(\boldsymbol{y} \perp \boldsymbol{X}) = \mathrm{OZ}^2,$$

and

$$\sum_{h=1}^{3} (\boldsymbol{y} \perp \boldsymbol{x}_h)'(\boldsymbol{y} \perp \boldsymbol{x}_h) = \mathrm{OM}^2+\mathrm{MN}^2+\mathrm{NZ}^2$$

and the equation

$$(\boldsymbol{y} \perp \boldsymbol{X})'(\boldsymbol{y} \perp \boldsymbol{X}) = \sum_{h=1}^{3} (\boldsymbol{y} \perp \boldsymbol{X}_h)'(\boldsymbol{y} \perp \boldsymbol{X})$$

is represented by the geometrical relation

$$\mathrm{OZ}^2 = \mathrm{OM}^2+\mathrm{MN}^2+\mathrm{NZ}^2.$$

EXAMPLE 6.21.3. Illustrate the result (6.21.3) with the vector $\boldsymbol{y}$ and matrix $\boldsymbol{X}$.

$$\boldsymbol{y} = \begin{pmatrix} 1 \\ 2 \\ 5 \\ 8 \end{pmatrix} \quad \boldsymbol{X}_1 = \begin{pmatrix} 1 & 0 \\ 1 & 0 \\ 0 & 1 \\ 0 & 1 \end{pmatrix} \quad \boldsymbol{X}_2 = \begin{pmatrix} 1 & 1 \\ -1 & -1 \\ 1 & -1 \\ -1 & 1 \end{pmatrix} \quad \boldsymbol{X} = (\boldsymbol{X}_1, \boldsymbol{X}_2)$$

Solution

$$\boldsymbol{y} \perp \boldsymbol{X}_1 = \begin{pmatrix} 1\frac{1}{2} \\ 1\frac{1}{2} \\ 6\frac{1}{2} \\ 6\frac{1}{2} \end{pmatrix} \quad \boldsymbol{y} \perp \boldsymbol{X}_2 = \begin{pmatrix} -\frac{1}{2} \\ \frac{1}{2} \\ -1\frac{1}{2} \\ 1\frac{1}{2} \end{pmatrix} \quad \boldsymbol{y} \perp \boldsymbol{X} = \begin{pmatrix} 1 \\ 2 \\ 5 \\ 8 \end{pmatrix}$$

$$(\boldsymbol{y} \perp \boldsymbol{X}_1)'(\boldsymbol{y} \perp \boldsymbol{X}_1) = 89 \qquad (\boldsymbol{y} \perp \boldsymbol{X}_2)'(\boldsymbol{y} \perp \boldsymbol{X}_2) = 5$$

$$(\boldsymbol{y} \perp \boldsymbol{X})'(\boldsymbol{y} \perp \boldsymbol{X}) = 94 = \sum_{h=1}^{2} (\boldsymbol{y} \perp \boldsymbol{X}_h)'(\boldsymbol{y} \perp \boldsymbol{X}_h)$$

6.22. Analysis of Variance

Now suppose that the measurement matrix $\boldsymbol{X}$ is composed of the *uncorrelated* measurement matrices $\boldsymbol{X}_h$ $(h = 1, 2, \ldots, k)$. Then as a special case of (6.21.3) we have

$$(\boldsymbol{y} \perp \boldsymbol{X} \cdot \mathbf{1})'(\boldsymbol{y} \perp \boldsymbol{X} \cdot \mathbf{1}) = \sum_{h=1}^{k} (\boldsymbol{y} \perp \boldsymbol{X}_h \cdot \mathbf{1})'(\boldsymbol{y} \perp \boldsymbol{X}_h \cdot \mathbf{1}). \quad (6.22.1)$$

But, by (6.19.6), $(\boldsymbol{y} \perp \boldsymbol{X} \cdot \mathbf{1})'(\boldsymbol{y} \perp \boldsymbol{X} \cdot \mathbf{1})$ is the sum of squares of deviations of the regression of $\boldsymbol{y}$ on $\boldsymbol{X}$, and thus

$$(\boldsymbol{y} \perp \boldsymbol{X} \cdot \mathbf{1})'(\boldsymbol{y} \perp \boldsymbol{X} \cdot \mathbf{1}) = n \operatorname{var} \{\boldsymbol{y} \perp (\boldsymbol{X}, \mathbf{1})\}. \quad (6.22.2)$$

Similarly,

$$(\boldsymbol{y} \perp \boldsymbol{X}_h \cdot 1)'(\boldsymbol{y} \perp \boldsymbol{X}_h \cdot \mathbf{1}) = n \operatorname{var} \{\boldsymbol{y} \perp (\boldsymbol{X}_h, \mathbf{1})\}.$$

Hence

$$\operatorname{var} \{\boldsymbol{y} \perp (\mathbf{1}, \boldsymbol{X})\} = \sum_{h=1}^{k} \operatorname{var} \{\boldsymbol{y} \perp (\mathbf{1}, \boldsymbol{X}_h)\}, \quad (6.22.3)$$

when $\boldsymbol{X}$ consists of the uncorrelated matrices $\boldsymbol{X}_h$.

Thus the variance of the regression of $\boldsymbol{y}$ on $\boldsymbol{X}$ decomposes into the separate variances of the regressions of $\boldsymbol{y}$ on the uncorrelated matrices $\boldsymbol{X}_h$ of which $\boldsymbol{X}$ is composed.

EXAMPLE 6.22.1. Analyse the variance of the regression of $\boldsymbol{y}$ on $\boldsymbol{X}$ in the example of § **6.20**.

Solution $\quad \boldsymbol{y} = \begin{pmatrix} 1 \\ 2 \\ 5 \\ 8 \end{pmatrix} \quad \boldsymbol{X}_1 = \begin{pmatrix} 1 & 0 \\ 1 & 0 \\ 0 & 1 \\ 0 & 1 \end{pmatrix} \quad \boldsymbol{X}_2 = \begin{pmatrix} 1 & 0 \\ 0 & 1 \\ 1 & 0 \\ 0 & 1 \end{pmatrix}$

$$\boldsymbol{y} \perp (\mathbf{1}, \boldsymbol{X}_1) = \begin{pmatrix} 1\frac{1}{2} \\ 1\frac{1}{2} \\ 6\frac{1}{2} \\ 6\frac{1}{2} \end{pmatrix} \quad \boldsymbol{y} \perp (\mathbf{1}, \boldsymbol{X}_2) = \begin{pmatrix} 3 \\ 5 \\ 3 \\ 5 \end{pmatrix} \quad \boldsymbol{y} \perp (\mathbf{1}, \boldsymbol{X}) = \begin{pmatrix} \frac{1}{2} \\ 2\frac{1}{2} \\ 5\frac{1}{2} \\ 7\frac{1}{2} \end{pmatrix}$$

$$\operatorname{var} \{\boldsymbol{y} \perp (\mathbf{1}, \boldsymbol{X})\} = 6\tfrac{1}{4}, \quad \operatorname{var} \{\boldsymbol{y} \perp (\mathbf{1}, \boldsymbol{X}_2)\} = 1,$$

$$\operatorname{var} \{\boldsymbol{y} \perp (\mathbf{1}, \boldsymbol{X})\} = 7\tfrac{1}{4} = \sum_{h=1}^{2} \operatorname{var} \{\boldsymbol{y} \perp (\mathbf{1}, \boldsymbol{X}_h)\}.$$

The result (6.22.3) generalises immediately to the case where the measurement vector $\boldsymbol{y}$ is replaced by a measurement matrix $\boldsymbol{Y}$.

Thus if $\boldsymbol{W}[\boldsymbol{Y} \perp (\mathbf{1}, \boldsymbol{X})]$ denotes the variance-covariance matrix of the regression of the matrix $\boldsymbol{Y}$ on the matrix $\boldsymbol{X}$, and if $\boldsymbol{W}[\boldsymbol{Y} \perp (\mathbf{1}, \boldsymbol{X}_h)]$ denotes the same for $\boldsymbol{X}_h$ $(h = 1, 2, \ldots, k)$, then

$$W[Y \perp (1, X)] = \sum_{h=1}^{k} W[Y \perp (1, X_h)]. \tag{6.22.4}$$

To prove (6.22.4) it is sufficient to prove that for any two measurement vectors (y_1, y_2)

$$\text{cov}\,[y_1 \perp (1, X), y_2 \perp (1, X)] = \sum_{h=1}^{k} \text{cov}\,[y_1 \perp (1, X_h), y_2 \perp (1, X_h)]. \tag{6.22.5}$$

Now by (6.22.3)

$$\text{var}\,[(y_1+y_2) \perp (1, X)] = \sum_{h=1}^{k} \text{var}\,[(y_1+y_2) \perp (1, X_h)], \tag{6.22.6}$$

and

$$\text{var}\,[(y_1-y_2) \perp (1, X)] = \sum_{h=1}^{k} \text{var}\,[(y_1-y_2) \perp (1, X_h)]. \tag{6.22.7}$$

Subtracting (6.22.7) from (6.22.6) we find

$$4\,\text{cov}\,[y_1 \perp (1, X), y_2 \perp (1, X)] = 4 \sum_{h=1}^{k} \text{cov}\,[y_1 \perp (1, X_h), y_2 \perp (1, X_h)] \tag{6.22.8}$$

and (6.22.5) and (6.22.4) follow.

6.23. Formal Analysis of Sums of Squares

Let y be a measurement vector and let X be a measurement matrix composed of an uncorrelated set of matrices $X_1, X_2, \ldots, X_k$.

Then

$$y'y = (y \perp 1)'(y \perp 1) + (y \cdot 1)'(y \cdot 1) \tag{6.23.1}$$

and

$$y' \cdot y = (y \perp 1)'(y \perp 1) + [y \perp (X \cdot 1)]'[y \perp (X \cdot 1)] + [y \,.\, (1, X)]'[y \cdot (1, X)], \tag{6.23.2}$$

since 1, $X \cdot 1$ and $y \cdot (1, X)$ are orthogonal.

Again

$$[y \perp (X \cdot 1)]'[y \perp (X \cdot 1)] = \sum_{h=1}^{k} [y \perp (X_h \cdot 1)]'[y \perp (X_h \cdot 1)], \tag{6.23.3}$$

since the X_h are uncorrelated.

Hence, by (6.23.1), (6.23.2) and (6.23.3),

$$(y \cdot 1)'(y \cdot 1) = \sum_{h=1}^{k} [y \perp (X_h \,.\, 1)]'[y \perp (X_h \cdot 1)] + [y \cdot (1, X)]'[y \cdot (1, X)]. \tag{6.23.4}$$

If we adopt the notation

$$S.S.(z) = z'z \tag{6.23.5}$$

to indicate the sum of squares of the elements of any measurement vector z, equation (6.23.4) may be written as

$$S.S.(y \cdot 1) = \sum_{h=1}^{k} [S.S.\{y \perp (X_h \cdot 1)\}] + S.S.[y \cdot (1, X)]. \tag{6.23.6}$$

The sums of squares of the residuals of y may accordingly be broken down into the sums of squares of the residuals of the regressions of y onto the various uncorrelated sets X_h and the sum of squares of the residuals of y from the regression onto the whole set X.

For purposes of computation, a rearrangement of (6.23.6) will often prove convenient. By (6.20.1)

$$S.S.(y \cdot 1) = S.S.(y) - n\bar{y}^2 \tag{6.23.7}$$

and similarly

$$S.S.[y \perp (X_h \cdot 1)] = S.S.[y \perp (1, X_h)] - n\bar{y}^2. \tag{6.23.8}$$

Hence (6.23.6) leads to

$$S.S.(y) - n\bar{y}^2 = \sum_{h=1}^{k} [S.S.[y \perp (1, X_h)] - n\bar{y}^2] + S.S.[y \cdot (1, X)]. \tag{6.23.9}$$

$S.S.[y \cdot (1, X)]$, the sum of squares of the residual of y from its regression on X, is usually called the *residual sum of squares.* Equation (6.23.9) can be rearranged so as to yield a convenient formula for its calculation, namely,

$$S.S.[y \cdot (1, X)] = S.S.[y] - \sum_{h=1}^{k} [S.S.\{y \perp (1, X_h)\}] + (k-1)n\bar{y}^2. \tag{6.23.10}$$

To calculate $n\bar{y}^2$ in formula (6.23.10) one would use the equation

$$n\bar{y}^2 = (1'y)^2/n, \tag{6.23.11}$$

where $1'y$ is simply the sum of the elements of the measurement vector y.

6.24. Table for Analysis of the Sums of Squares

A formal analysis of the sum of squares $(y \cdot 1)'(y \cdot 1)$ of the deviation of a measurement vector y may be based on equation (6.23.10). It is usually set out as follows.

TABLE 6.24.1

Factor	Sum of squares	Degrees of freedom	Variance ratio	F-ratio
$\mathbf{X}_1$	$S.S.[\mathbf{y}\perp(\mathbf{X}_1\cdot\mathbf{1})]$	m_1	v_1	v_1/w
$\mathbf{X}_2$	$S.S.[\mathbf{y}\perp(\mathbf{X}_2\cdot\mathbf{1})]$	m_2	v_2	v_2/w
.				
.				
.				
$\mathbf{X}_k$	$S.S.[\mathbf{y}\perp(\mathbf{X}_k\cdot\mathbf{1})]$	m_k	v_k	v_k/w
Residual	$S.S.[\mathbf{y}\cdot(\mathbf{1},\mathbf{X})]$	$n-m-1$	w	. . .
Total	$S.S.[\mathbf{y}\cdot\mathbf{1}]$	$n-1$	v	. . .

In this chapter we are mainly concerned with the first two columns of the table, but a very brief explanation will be given of the method of calculation and interpretation of the other three columns. For a more thorough explanation, the reader is referred to the Appendix to Chapter 11 in Volume 2.

The number of degrees of freedom associated with the sum of squares $S.S.[\mathbf{y}\perp(\mathbf{X}_h\cdot\mathbf{1})]$ is the number of linearly independent columns of the matrix $\mathbf{X}_h\cdot\mathbf{1}$. Often this will be the same as the number of columns in the matrix $\mathbf{X}_h$, but in the important case where the columns of $\mathbf{X}_h$ are a disjoint exhaustive set of attribute-vectors, the number of degrees of freedom will be one less than the number of columns in $\mathbf{X}_h$, since in this case $(\mathbf{X}_h\cdot\mathbf{1})'\mathbf{1}=\mathbf{0}$. (In this equation the numbers of elements in the two vectors $\mathbf{1}$ must equal the numbers of rows and of columns in the matrix $\mathbf{X}_h$.)

The variance-ratio shown in column 4 is obtained by dividing the sum of squares shown in column 2, by the number of degrees of freedom shown in column 3. The *F*-ratio given in column 5 is to be obtained by dividing the variance ratio shown in column 4 by the residual variance-ratio, w, given in the penultimate row of column 4.

The value of the *F*-ratio is useful as giving an indication of the importance of the 'effect' of each group of factors on the measurements $\mathbf{y}$. Where n is the number of degrees of freedom, *F*-ratios of less than $1+5/2\sqrt{n}$ can quite well arise by chance (with normal variation $P>0{\cdot}05$) and should not be taken to indicate a serious effect. On the other hand, provided the number of degrees of freedom in the residual row is substantial, values of the *F*-ratio as large as $[1+3{\cdot}6(n-\sqrt{n})^{-1/2}]$ will be decidedly rare in the absence of a real effect. (With normal variation $P<0{\cdot}01$.)

6.25. Moments

In § **5.26** we explained how moments may be used to summarise the properties of the probability function of a variate. In exactly the same way, they may be used to summarise the properties of the frequency function of a measurement vector or matrix.

Corresponding to D. 36 in § **5.26** we may write:

D. 47 For any measurement vector $\boldsymbol{x}$, the moments about the origin are given by

$$m_r' = \boldsymbol{f}'(\boldsymbol{x}^r) \quad (r = 0, 1, 2, \ldots)$$

and the moments about the mean by

$$m_r = \boldsymbol{f}'(\boldsymbol{x} \cdot \boldsymbol{1})^r \quad (r = 0, 1, 2, \ldots)$$

where $\boldsymbol{f}'$ is the scaling vector.

A more convenient way of writing D. 47 is as follows:

D. 47*a* For any measurement vector $\boldsymbol{x}$ the moments about the origin are given by

$$m_r' = (1/n) \sum_{t=1}^{n} x_t^r \quad (r = 0, 1, 2, \ldots)$$

and the moments about the mean by

$$m_r = (1/n) \sum_{t=1}^{n} (x_t - \bar{x})^r.$$

The moments about the origin are the arithmetic means of the various powers of the n measurements and the moments about the mean are the arithmetic means of the various powers of the deviations of the n measurements from their arithmetic mean.

Similarly, the definition D. 37 for bivariate moments for variates may be matched by the following definition for measurements.

D. 48 For any pair of measurement vectors $\boldsymbol{x}$ and $\boldsymbol{y}$, the bivariate moments about the origin are given by

$$m_{rs}' = \boldsymbol{f}'\hat{\boldsymbol{x}}^r\boldsymbol{y}^s = (1/n) \sum_{t=1}^{n} x_t^r y_t^s$$

and the bivariate moments about the mean by

$$m_{rs} = \boldsymbol{f}'\widehat{(\boldsymbol{x}_r \cdot \boldsymbol{1})}^r(\boldsymbol{y} \cdot \boldsymbol{1})^s = (1/n) \sum_{t=1}^{n} (x_t - \bar{x})^r (y_t - \bar{y})^s.$$

The skewness and kurtosis may be defined by the same formulae in terms of moments for a measurements vector as for a variate. Thus

$$\gamma_1 = m_3/m^{3/2} \qquad \text{(skewness)}$$

and

$$\gamma_2 = (m_4/m_2^2) - 3 \qquad \text{(kurtosis)}.$$

In § **5.27** it was pointed out that the sums representing the moments of a variate will not always converge. In the case of measurement matrices there is never any such difficulty, since the sums giving the moments consist of a finite number of terms only.

6.26. Cumulants

As an alternative to moments, one may use a set of measures called *cumulants*. We shall see in later chapters that these measures have useful properties where one is concerned with the frequency distributions of the means of several samples each of size n. At the present stage, we will merely give their definition and explain how the first few cumulants are related to the first few moments.

First we may define the Moment-generating-function $M'(\alpha)$ by

D. 49 The moment-generating function for any measurement vector $\boldsymbol{x}$ is given by

$$M'(\alpha) = (1/n) \sum_{t=1}^{n} e^{x_t \alpha} = \sum_{r=0}^{\infty} m_r' \alpha^r / r!$$

and the cumulant generating function $K(\alpha)$ by

D. 50 The cumulant generating function for any measurement vector $\boldsymbol{x}$ is given by

$$K(\alpha) = \log M'(\alpha) = \sum_{r=1}^{\infty} K_r \alpha^r / r!. \tag{6.26.1}$$

The coefficients K_r are called the cumulants for $\boldsymbol{x}$.

To obtain formulae for the cumulants in terms of the moments it is convenient to define the moment generating function about the mean as

$$M(\alpha) = e^{-\bar{x}\alpha} M'(\alpha) = (1/n) \sum_{t=1}^{n} e^{(x_t - \bar{x})\alpha} = \sum_{r=0}^{\infty} m_r \alpha^r / r! \tag{6.26.2}$$

and to note that by D. 50

$$K(\alpha) = \bar{x}\alpha + \log M(\alpha). \tag{6.26.3}$$

Provided α is taken sufficiently small, we may expand $\log M(\alpha)$ by using the formula

$$\log (1+y) = y - (y^2/2) + (y^3/3) - (y^4/4) \ldots \tag{6.26.4}$$

putting

$$y = (m_2\alpha^2/2) + (m_3\alpha^3/6) + (m_4\alpha^4/24) + \ldots \tag{6.26.5}$$

We thus find that

$$K(\alpha) = \bar{x}\alpha + m_2\alpha^2/2 + m_3\alpha^3/6 + (m_4 - 3m_2^2)\alpha^4/24 + \dots \quad (6.26.6)$$

By equating coefficients in (6.26.1) and (6.26.5) we obtain

$$K_1 = \bar{x} = m_1' \quad K_2 = m_2 \quad K_3 = m_3 \quad K_4 = m_4 - 3m_2^2. \quad (6.26.7)$$

EXAMPLE 6.26.1. The following table relates to the households co-operating in the Ministry of Labour's enquiry into household expenditure. Calculate the first four cumulants and the coefficients γ_1, γ_2, of skewness and kurtosis.

TABLE 6.26.1

Analysis of Co-operating Households by Size

Number of persons x	1	2	3	4	5	6	7	8	9	10	11	*Total*
Number of households per 1000 $f(x)$	101	283	249	196	98	40	19	8	4	1	1	1000

Solution. First we calculate

$$1000\, m_1' = \sum_{x=1}^{11} f(x)x = 3\,182$$

$$1000\, m_2' = \sum_{x=1}^{11} f(x)x^2 = 12\,488$$

$$1000\, m_3' = \sum_{x=1}^{11} f(x)x^3 = 58\,382$$

$$1000\, m_4' = \sum_{x=1}^{11} f(x)x^4 = 317\,336$$

Then we may use the formulae (calculated in question 2 at the end of this chapter) to obtain

$$m_2 = m_2' - m_1'^2 = 2{\cdot}362\,876$$

$$m_3 = m_3' - 3m_2'm_1' + 2m_1'^3 = 3{\cdot}607\,841$$

$$m_4 = m_4' - 4m_3'm_1' + 6m_2'm_1'^2 - 3m_1'^4 = 25{\cdot}350\,787$$

We now obtain the first four cumulants as

$$K_1 = m_1' = 3{\cdot}182 \qquad K_2 = m_2 = 2{\cdot}362\,876$$

$$K_3 = m_3 = 3{\cdot}607\,841 \qquad K_4 = m_4 - 3m_2^2 = 6{\cdot}601\,238 \quad (6.26.10)$$

and the coefficients of skewness and kurtosis as

$$\gamma_1 = K_3/K_2^{3/2} = 0{\cdot}993\,1$$

$$\gamma_2 = K_4/K_2^2 \quad = 1{\cdot}182\,3 \qquad (6.26.11)$$

Instead of using the formulae (6.26.9) and (6.26.10) to obtain the cumulants we could have equated coefficients of the first four powers of α in (6.16.1) after expanding

$$\log_e (1+m_1'\alpha+m_2'\alpha^2/2+m_3'\alpha^3/6+m_4'\alpha^4/24 \ldots)$$

n powers of α to obtain

$$K_1 = m_1', \quad K_2 = m_2'-m_1'^2, \quad K_3 = m_3'-3m_2'm_1'+2m_1'^3, \quad \text{and}$$

$$K_4 = m_4'-4m_3'm_1'+12m_2'm_1'^2-3m_2'^2-6m_1'^4. \qquad (6.26.12)$$

6.27. Histograms and Grouped Frequency Distributions

Fig. 14 shows a histogram for the grouped frequency distribution of years, according to the proportion of the strikes in each year which lasted for one week or for longer.

In the histogram the number of years in each range is measured by the area of the pillar within that range of the horizontal scale. In this particular example all the ranges are of equal width, so it is possible to measure the areas of the pillars along the single vertical scale. But

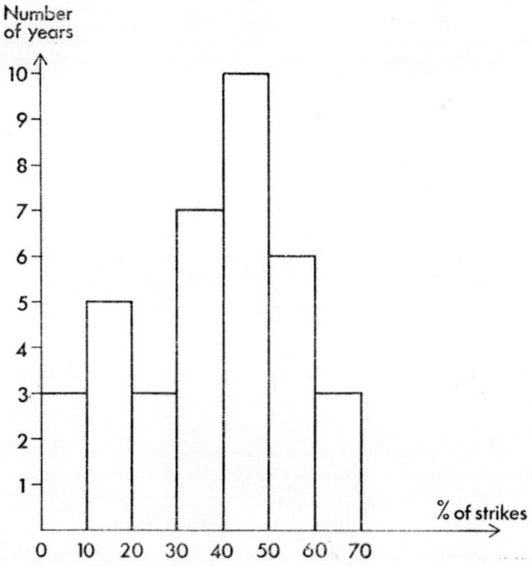

Fig. 14. Frequency distribution of years with percentages of strikes lasting one week or longer

TABLE 6.27.1

Frequency distribution of strikes lasting over one week as per cent of all strikes. U.K., 1911-47

Percentage	0–9·9	10–19·9	20–29·9	30–39·9	40–49·9	50–59·9	60–69·9	70–	*Total*
Number of years	3	5	3	7	10	6	3	0	37
Number of years per 100	8·1	13·5	8·1	18·9	27·0	16·2	8·1	0	100·0

in cases where the widths are unequal, it will not be possible to read off the numbers portrayed by a scale giving the heights of the pillars.

The histogram and the grouped frequency distribution only present part of the information contained in the complete set of observations (in this case the actual percentages in each of the 37 years), so that one cannot obtain from them exact estimates of such measures as the

TABLE 6.27.2

Cumulative frequencies calculated from Table 6.27.1

Percentage	x	0	10·0	20·0	30·0	40·0	50·0	60·0	70·0	∞
Cumulative frequency	$F(x)$	0	3	8	11	18	28	34	37	37

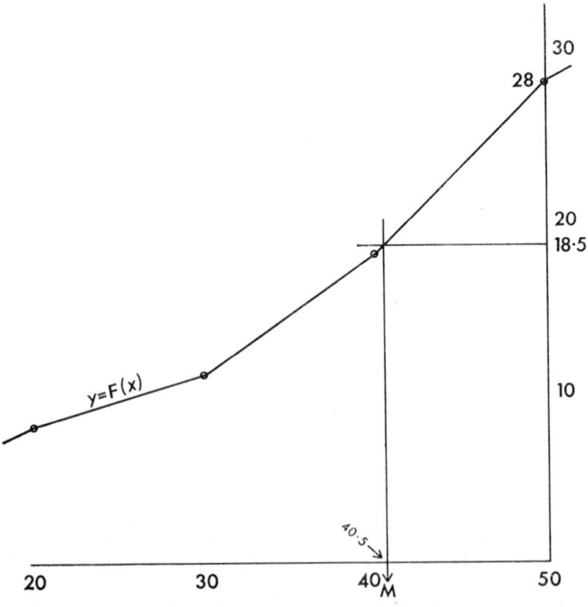

Fig. 15

moments, the median or the mode (most frequent value) of the distribution.

There are however standard formulae for obtaining estimates of these statistics in the case where the intervals of the grouped frequency distribution are all of the same width h.

From the grouped frequency distribution it is possible to calculate the cumulative frequencies as in Table 6.27.2.
Let X denote the value such that $F(X) < N/2$ and $F(X+h) > N/2$ (where N is the total number of observations). Then the median may be estimated as

$$M = X + \frac{(N/2) - F(X)}{F(X+h) - F(X)}\, h. \qquad (6.27.1)$$

Again let $g_i = F(x_i+h) - F(x_i) \quad (i = 1, 2, \ldots, k) \qquad (6.27.2)$
and let g_I be the greatest g_i. Then the mode may be estimated as

$$m = x_I + \frac{[g_I - g_{I-1}]h}{2g_I - g_{I+1} - g_{I-1}}. \qquad (6.27.3)$$

Figs. 15 and 16 illustrate the corresponding geometrical constructions.

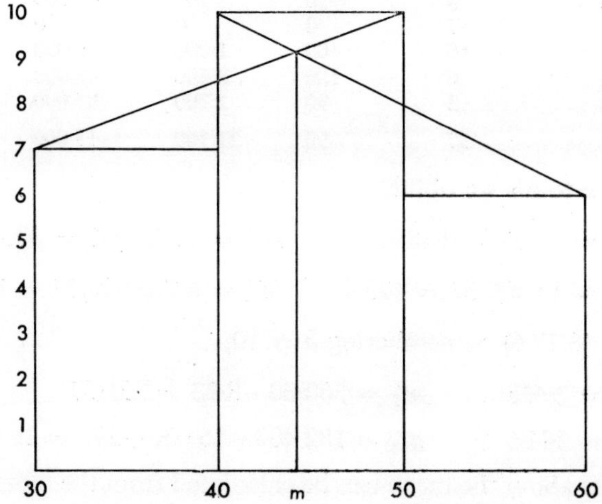

Fig. 16

In our example $N = 37$, $X = 40{\cdot}0$, $h = 10{\cdot}0$, $F(X) = 18$ and $F(X+h) = 28$, so that, by (6.27.1),

$$M = 40{\cdot}0 + \frac{0{\cdot}5 \times 10}{28 - 18} = 40{\cdot}5. \qquad (6.27.4)$$

Again $g_I = 10$ and $x_I = 40$, $g_{I-1} = 7$ and $g_{I+1} = 6$, so that by (6.27.2)

$$m = 40+\tfrac{30}{7} = 44{\cdot}3. \qquad (6.27.5)$$

Again, let M_1' M_2' M_3' and M_4' be the first four moments about the origin calculated on the assumption that all the observations in any one interval are concentrated at the centre of the interval; then the true moments about the origin may be estimated as

$$m_1' = M_1',\ m_2' = M_2'-\frac{h^2}{12},\ m_3' = M_3'-M_1'\frac{h^2}{4},\ m_4' = M_4'-M_2'\frac{h^2}{2}+\frac{7h^4}{240} \qquad (6.27.6)$$

Thus, in our example we may calculate M_i' and m_i' $(i = 1, 2, 3, 4)$ as follows, where, to lighten the arithmetic, we have shifted the origin of measurement of x from 0 to 35.

TABLE 6.27.3

Centre of interval measured from 35 x	*Number of years* g	gx	gx^2	gx^3	gx^4
-30	3	-90	+2 700	-81 000	2 430 000
-20	5	-100	+2 000	-40 000	800 000
-10	3	-30	+300	-3 000	30 000
0	7	0	0	0	0
10	10	100	1 000	10 000	100 000
20	6	120	2 400	48 000	960 000
30	3	90	2 700	81 000	2 430 000
Total	37	90	11 100	15 000	6 750 000

From the totals we obtain

$$M_1' = 90/37 = 2{\cdot}428 \qquad M_2' = 11\,100/37 = 300$$

$$M_3' = 15\,000/37 = 405{\cdot}4 \qquad M_4' = 6\,750\,000/37 = 182\,400$$

so that by (6.27.6), remembering $h = 10$,

$$m_1' = 2{\cdot}428 \qquad m_2' = 300{\cdot}00-8{\cdot}33 = 291{\cdot}67$$

$$m_3' = 344{\cdot}6 \qquad m_4' = 182\,400-15\,000+292 = 167\,692.$$

Moments about the mean can be calculated from the moments about the origin by the formulae

$$m_1 = 0, \quad m_2 = m_2'-m_1'^2, \quad m_3 = m_3'-3m_2'm_1'+2m_1'^3,$$

$$m_4 = m_4'-4m_3'm_1'+6m_2'm_1'^2-3m_1'^4. \qquad (6.27.7)$$

(See Exercise 2 at end of this Chapter.)

The methods of this section are not appropriate if the observations refer to a discrete variate (such as the number of persons in household) which is constrained to be one of a small number of possible values. The methods are to be used when the variate concerned (e.g. ages or wages) is free to vary over a scale which is effectively continuous.

Exercises 6

6.1. $\boldsymbol{w}$ denotes the measurement-vector of 700 elements giving the number of w's in each of 700 lines of a certain passage of prose. It is found by counting that the frequency function $g(w) = \boldsymbol{f'r}[\boldsymbol{w}=w]$ for $\boldsymbol{w}$ satisfies

w	0	1	2	3	4	5	>5
$700\,g(w)$	276	269	109	38	5	3	0

Where $\boldsymbol{x}$ denotes the variate $\boldsymbol{w}-\mathbf{1}$, obtain the first four moments m_s' for $\boldsymbol{x}$ by using the formula

$$m_s' = \sum_{x=-1}^{\infty} \boldsymbol{f'r}[\boldsymbol{x}=x]x^s$$

6.2. Use the binomial expansion $(p+q)^n = \sum_{r=0}^{n} \binom{n}{r} p^r q^{n-r}$ to obtain the following formulae giving moments about the mean in terms moments about the origin,

$$m_1 = 0, \quad m_2 = m_2' - m_1'^2, \quad m_3 = m_3' - 3m_2'm_1' + 2m_1'^3,$$
$$m_4 = m_4' - 4m_3'm_1' + 6m_2'm_1'^2 - 3m_1'^4.$$

6.3. Use the solutions of questions 1 and 2 to obtain the first four moments about the mean for $\boldsymbol{x}$, and hence also for $\boldsymbol{w}$, to three places of decimals.

6.4. Use the results of questions 1 and 3, and the formulae

$$K_1 = m_1', \ K_2 = m_2, \ K_3 = m_3 \text{ and } K_4 = m_4 - 3m_2^2$$

to obtain the first four cumulants for $\boldsymbol{w}$.

6.5. Confirm the formulae given in question 4 by equating coefficients in

$$\sum_{r=1}^{\infty} [K_r\alpha^r/r!] = K(\alpha) = \log_e [M'(\alpha)],$$

where

$$1 + \sum_{r=1}^{\infty} [m_r'\alpha^r/r!] = M'(\alpha) = e^{m_1'\alpha}M(\alpha)$$

$$1 + \sum_{r=1}^{\infty} [m_r\alpha^r/r!] = M(\alpha)$$

and $\log_e (1+x) = x - (x^2/2) + (x^3/3) \ldots \quad (-1 < x < 1)$.

6.6. $y' = (1, 3, 5, 5, 7)$ and $X = \begin{pmatrix} 1 & -2 & 2 \\ 1 & -1 & -1 \\ 1 & 0 & -2 \\ 1 & 1 & -1 \\ 1 & 2 & 2 \end{pmatrix}$

Use the formulae $y \cdot x = y - Xb$ and $b = (X'X)^{-1}X'y$ to obtain the corrected measurement vector $y \cdot X$. Hence obtain the projection $y \perp X$.

6.7. Use the results of question 6 to find the regression of y on Z where

$$y' = (1, 3, 5, 5, 7) \text{ and } Z' = \begin{pmatrix} -2, & -1, & 0, & 1, & 2 \\ 2, & -1, & -2, & -1, & 2 \end{pmatrix}.$$

6.8. Verify that, where

$$Y' = \begin{pmatrix} 1, & 1, & 0, & -1, & -1 \\ 4, & 1, & 0, & 1, & 4 \end{pmatrix} \text{ and } X_1' = (1, 1, 1, 1, 1)$$

$$\text{and } X_2' = \begin{pmatrix} -3, & -2, & -1, & 2, & 4 \\ -1, & -1, & -1, & 0, & 3 \end{pmatrix},$$

$$Y \perp (X_1, X_2) = Y \perp X_1 + Y \perp X_2.$$

6.9. The measurement matrix X' is composed of the uncorrelated matrices $\begin{pmatrix} 1, & 1, & 0, & 0 \\ 0, & 0, & 1, & 1 \end{pmatrix}$ and $\begin{pmatrix} 1, & 0, & 1, & 0 \\ 0, & 1, & 0, & 1 \end{pmatrix}$. If y' is the measurement vector (1, 1, 3, 6), analyse the regression of y on X into the two components associated with the submatrices X_1 and X_2.

Verify that the variance of the regression of y on X can be analysed as the sum of the variances of the regressions of y on X_1 and on X_2.

6.10. Carry out an analysis of the sum of squares for the observations in the following table.

x_1	1	2	3	4	5
			Values of y		
x_2					
17	120	122	147	130	152
12	97	103	84	73	90
9	73	87	71	63	56
7	48	58	46	36	33
5	20	33	33	22	7

y = Average weekly household payments (shillings) on alcoholic drink
x_1 = Number of persons in household
x_2 = Weekly income of household (pounds)
$x_3 = (x_1 - 3)^2$
$x_4 = (x_1 - 3)(x_2 - 10)$

6.11. The following data relate to indicators of the productivity of individual workers in random samples drawn from two different establishments in the same industry:

Sample from Establishment A		*Sample from Establishment B*	
Productivity indicator	*Number of workers*	*Productivity indicator*	*Number of workers*
4 or less	1	34 or less	1
4·1–6	7	34·1–36	6
6·1–7	15	36·1–38	28
7·1–8	26	38·1–40	33
8·1–9	36	40·1–42	31
9·1–10	44	42·1–44	23
10·1–11	34	44·1–46	17
11·1–12	22	46·1–50	20
12·1–13	11	50·1–54	16
13·1 or more	4	54·1–60	11
		60·1–66	8
		66·1 or more	6
Total	200	Total	200

Compare the statistical characteristics of the two distributions and briefly consider possible explanations of their differences.
[Cambridge Economics Tripos Part 2, 1962]

6.12.

Numbers of Females and Age at Marriage (numbers in thousands)

Age	*England and Wales* 1951	1961	*Scotland* 1951	1961
Under 21	92	130	10·1	14·7
21–24	133	120	15·5	15·0
25–29	62	40	8·6	5·5
30–34	25	16	3·1	1·9
35–44	26	18	2·6	1·9
45–54	13	12	1·0	1·0
55 and over	8	10	0·4	0·6

[Source: *Annual Abstract of Statistics*]

Make such quick calculations of appropriate summary measures as enable you to comment on the main features disclosed by these data.
[Cambridge Economics Tripos Part 2, 1964]

6.13. What is skewness? Give examples of economic frequency distributions with high skewness. Find the mean and median of the sample described in the following table.

Sample of twenty estates in Great Britain on which duty was paid in 1958–9 (in order of magnitude)

Net value in £'000

26·2	21·9	17·4	12·5	9·9	8·4	7·1	6·7	6·2	5·5
5·5	5·1	5·0	4·7	4·3	3·9	3·5	3·2	3·0	3·0

[Cambridge Economics Tripos Part 2, 1962]

6.14. X_1 is consumption of milk per head, X_2 the mean price of milk, and X_3 the national income per head. Time series of the three variables are rendered trend-free and the standard deviations and correlation coefficients calculated:

$$\sigma_1 = 7{\cdot}22, \quad \sigma_2 = 5{\cdot}47, \quad \sigma_3 = 6{\cdot}87$$

$$r_{12} = -0{\cdot}83, \; r_{13} = 0{\cdot}92, \; r_{23} = -0{\cdot}61.$$

Calculate the regression of X_1 on X_2 and X_3. Interpret the regression as a demand equation.

[Cambridge Economic Tripos Part 2, 1964]

NOTE. In Exercises 6.11 to 6.13 only parts of the questions originally set in the examinations have been retained.

Chapter 7

RANDOM SAMPLES

7.1. Populations and Samples

In order to study the properties of some body of material or group of persons, institutions, or other entities, it is often convenient to select a part of the group for intensive study. When this is done, the whole body or group is regarded as the *population* and the selected part is regarded as the *sample*.

Provided the sample has been selected in such a way that it is likely to be representative, as regards the properties studied, of the population, we can then use descriptive measures of the sample as estimates of the corresponding descriptive measures of the population.

However they will only be estimates, and we should in general expect some divergence between the descriptive measures of the sample and the corresponding, often unknown, descriptive measures of the population. The nature of this divergence will depend on the method of selection of the sample; if chance enters into the method of selection of the sample, then chance will partly determine the degree of divergence between the measures of the sample and the corresponding measures of the population.

In such cases, it is appropriate to regard the descriptive measures of the sample as themselves being variates. In cases where the procedure for selecting the sample has been very carefully stated with precise specification of the way in which chance enters into that procedure, it may be possible to obtain the probability functions, or some simple properties of the probability functions, of the descriptive measures (or statistics) of the sample, in terms of the (often unknown) properties of the population. In later chapters we shall see how such knowledge may be used in estimating the properties of the population from those of the sample.

7.2. Simple Random Samples with Replacement

The simplest kind of sample, although it is one which is seldom employed in its pure form, is the *random sample with replacement*. For such a sample, each member $x(j)$ $(j = 1, 2, \ldots, N)$ of the population, has at each stage the same probability, $1/N$, of being drawn as the next member of the sample, whether or not it has already been drawn.

Thus if $z(h)$ $(h = 1, 2, \ldots, n)$ denote the n members of the sample, each $z(h)$ may be regarded as an independent variate whose probability function is the same as the frequency function for the population $x(j)$ $(j = 1, 2, \ldots, N)$.

Thus
$$p'r\{z(h) = x\} = f'r\{x = x\} \tag{7.2.1}$$

where x denotes the measurement vector for the population; and the multivariate probability distribution of the variates $z(h)$ $(h = 1, 2, \ldots, n)$ is
$$p' \bigcap_{h=1}^{n} r\{z(h) = x(h)\} = \prod_{h=1}^{n} f'r\{x = x(h)\}. \tag{7.2.2}$$

Thus the probability function for any particular sequence of sample values $x(h)$ $(h = 1, 2, \ldots, n)$ is the product of the values of the population frequency function at these values $x(h)$.

7.3. Samples and their Descriptive Measures regarded as a Set of Variates

In the case of the simple sample with replacement, the n sample values $z(h)$ may be regarded, as we have seen, as n independent variates, each with an identical probability function which coincides with the frequency function of the population.

This view of the individual sample-values as being a set of variates with a probability function depending on the frequency function of the population may be adopted, whenever the method of selection of the sample is sufficiently well specified. Only in the case of the simple random sample with replacement is the relation between the probability function of the sample values and the frequency function of the population so easy and direct; but there are many other cases where the relation is sufficiently clear for it to be useful in estimation.

Since the sample values are variates, so also are any descriptive measures of the sample which may be calculated on the basis of the sample values. These descriptive measures will have their own probability functions, and these probability functions and the associated expected values, variances etc. will be related to the frequency function of the population, in a manner dependent upon the method of selection of the sample.

In this chapter we shall investigate such relations in the cases of some of the more elementary methods of selecting samples.

7.4. The Expected Value of the Mean of a Simple Random Sample with Replacement

Let $z(h)$ $(h = 1, 2, \ldots, n)$ be a simple random sample with replacement from a population of values with frequency function $f(x)$.

Then each $z(h)$ may be regarded as an independent variate with probability function

$$p'r\{z(h) = x\} = f(x). \tag{7.4.1}$$

Hence the expected value of each $z(h)$ $(h = 1, 2, \ldots, n)$ is

$$p'z(h) = \sum_x f(x)x = \bar{x},$$

where $\bar{x}$ is the arithmetic mean of the population.

Now let $$\bar{z} = n^{-1} \sum_{h=1}^{n} z(h)$$

be the arithmetic mean of the sample-values then z is a variate with expected value

$$p'\bar{z} = n^{-1} \sum_{h=1}^{n} p'z(h) = \bar{x}. \tag{7.4.2}$$

Hence the expected value of the arithmetic mean of a random sample with replacement is equal to the arithmetic mean of the population from which the sample is drawn.

7.5. The Variance of the Arithmetic Mean of a Random Sample with Replacement

Let $z(h)$ $(h = 1, 2, \ldots, n)$ be a random sample with replacement from a population with frequency function $f(x)$.

In considering the variance of the sample there will be no loss of generality in supposing that the origin of measurement has been so chosen that the arithmetic mean $\bar{x}$ of the population values is zero. Thus

$$\bar{x} = \sum_x f(x)x = 0. \tag{7.5.1}$$

In this case, the $z(h)$ are n independent variates, each with expected value $$p'z(h) = 0$$

and each with variance

$$\text{var}\,\{z(h)\} = p'\{z(h)\}^2 = \text{var}\,x, \tag{7.5.2}$$

where $$\text{var}\,x = \sum_x f(x)x^2$$

is the variance of the population.

If $z(h)$, $z(h')$ are any two distinct members of the sample, then, since they are independent,

$$p'\hat{z}(h)\,z(h') = p'z(h)\,p'z(h') = 0. \tag{7.5.3}$$

Let
$$\bar{z} = n^{-1} \sum_{h=1}^{n} z(h) \tag{7.5.4}$$
be the arithmetic mean of the sample; then
$$\text{var } \bar{z} = \boldsymbol{p}'\bar{z}^2, \quad \text{since} \quad \boldsymbol{p}'\bar{z} = 0.$$
$$\therefore \quad \text{var } \bar{z} = n^{-2}\boldsymbol{p}' \sum_{h=1}^{n} \sum_{h'=1}^{n} \hat{z}(h)\, z(h') = n^{-2} \sum_{h=1}^{n} \boldsymbol{p}'\{z(h)\}^2$$
using (7.5.3), so that by (7.5.2)
$$\text{var } \bar{z} = n^{-1} \text{ var } \boldsymbol{x}. \tag{7.5.5}$$
We see then that the variance of the arithmetic mean of a simple random standard deviation of the population by the square root of the number in the sample.

Taking square roots in (7.5.5), we further obtain
$$\sigma_{\bar{z}} = n^{-1/2}\sigma_{\boldsymbol{x}}, \tag{7.5.6}$$
showing that the standard error of the arithmetic mean of a simple random sample with replacement may be obtained by dividing the standard deviation of the population by the square root of the number in the sample.

7.6. The Expected Value of the Variance of a Random Sample with Replacement

The variance of a sample of n values $z(h)$ $(h = 1, 2, \ldots, n)$ taken from a population with frequency function $f(x)$ is
$$\boldsymbol{v} = \text{var } z = n^{-1} \sum_{h=1}^{n} \{z(h) - \bar{z}\}^2 = n^{-1} \sum_{h=1}^{n} \{z(h)\}^2 - n^{-2} \left\{ \sum_{h=1}^{n} z(h) \right\}^2, \tag{7.6.1}$$
where
$$\bar{z} = n^{-1} \sum_{h=1}^{n} z(h). \tag{7.6.2}$$

In finding the expected value $\boldsymbol{p}'\boldsymbol{v}$, there will be no loss of generality if we suppose the origin of measurement to have been so chosen that the population mean $\bar{x}$, and hence also the expected value $\boldsymbol{p}'\bar{z}$ is zero.

Then
$$\boldsymbol{p}'\boldsymbol{v} = n^{-1} \sum_{h=1}^{n} \boldsymbol{p}'\{z(h)\}^2 - \boldsymbol{p}'\bar{z}^2. \tag{7.6.3}$$
Now we have seen in (7.5.5) that $\boldsymbol{p}'\bar{z}^2 = n^{-1} \text{ var } \boldsymbol{x}$

and in (7.5.2) that $\boldsymbol{p}'\{z(h)\}^2 = \text{var } \boldsymbol{x}$; (7.6.4) hence

$$p'\boldsymbol{v} = \{(n-1)/n\} \text{ var } \boldsymbol{x} \tag{7.6.5}$$

Thus the expected value of the sample variance is only $\{(n-1)/n\}$ times the population variance.

7.7. Simple Random Samples without Replacement

If a simple random sample is taken without replacement from a population of N members, it means that no member of the population may be drawn more than once, but that subject to this limitation, all the (remaining) members of the population have the same probability $(N+1-r)^{-1}$ of being selected at the next, the r-th draw.

Suppose that $z(h)$ $(h = 1, 2, \ldots, n)$ are the members of a sample drawn at random without replacement from the population $x(j)$ $(j = 1, 2, \ldots, N)$. Suppose that the origin has been so chosen that the population mean $\bar{x}$ is zero.

Although the $z(h)$ are not independent, it is clear from symmetry that for each h the unconditional probability function of $z(h)$ is the same as that for $z(1)$, namely

$$\boldsymbol{p'r}\{z(h) = x\} = \boldsymbol{p'r}\{z(1) = x\} = f(x), \tag{7.7.1}$$

where $f(x)$ is the frequency function of the population: for, as in the case of a simple random sample with replacement, the probability function for $z(1)$ is given by (7.2.1).

Hence $$\boldsymbol{p'}z(h) = \sum_x f(x)x = \bar{x} = 0. \tag{7.7.2}$$

as in the case of each member of a sample, with replacement from a population with zero mean.

Hence also $\boldsymbol{p'}\bar{z} = 0$, where $\bar{z}$ is the sample mean.

Hence, in general, the expected value of the arithmetic mean of a simple random sample without replacement equals the population arithmetic mean; for any shift of the origin of measurement will equally affect the population arithmetic mean and the expected value of the sample arithmetic mean.

Thus, in general, $\boldsymbol{p'}\bar{z} = \bar{x}$. (7.7.3)

7.8. Variance of Arithmetic Mean of Simple Random Sample without Replacement

Let us retain the assumptions of the last section; in particular let us suppose that the population mean $\bar{x}$ is zero.

The variance of the sample arithmetic mean $\bar{z}$ will then be given by

$$\text{var } \bar{z} = p'\bar{z}^2 = n^{-2} \sum_{h=1}^{n} \sum_{h'=1}^{n} p'\hat{z}(h)z(h'). \tag{7.8.1}$$

It is clear from symmetry that, when $h \neq h'$,

$$p'\hat{z}(h)z(h') = c_N \tag{7.8.2}$$

where c_N is a constant whose value depends only on N, the number in the population and not on n.

Furthermore, for each value of h ($h = 1, 2, \ldots, n$) the unconditional variance

$$p'\{z(h)\}^2 = p'\{z(1)\}^2 = \text{var } x, \tag{7.8.3}$$

as given by (7.5.2).

Hence, from (7.8.1), (7.8.2), (7.8.3)

$$\text{var } \bar{z} = n^{-1}\{\text{var } x + (n-1)c_N\}. \tag{7.8.4}$$

To evaluate c_N, we note that, if we put $n = N$, the sample will be identical with the population, so that $\bar{z} = \bar{x} = 0$ and

$$\text{var } \bar{z} = 0 \quad (\text{when } n = N). \tag{7.8.5}$$

Combining (7.8.4), (7.8.5)

$$\text{var } x + (N-1)c_N = 0;$$

$$\therefore \quad c_N = -\text{var } x \,/\, (N-1). \tag{7.8.6}$$

Finally, from (7.8.4), (7.8.6) we have for all n ($n = 1, 2, \ldots, N$)

$$\text{var } \bar{z} = \{(N-n)\,/\,(N-1)\}n^{-1} \text{ var } x. \tag{7.8.7}$$

This shows that the formula (7.5.5) for the variance of the arithmetic mean of a random sample *with* replacement, must, in the case of a random sample *without* replacement, be modified by multiplication by the factor $\{(N-n)\,/\,(N-1)\}$, where n and N are the numbers in the sample and in the population, respectively.

7.9. The Sampling of Attributes

One often requires to estimate from a sample the proportion of a population which possesses a certain attribute.

Suppose that M out of a total of N members of a population possess a certain attribute, and that in a simple random sample of n members, m of these members are found to possess that attribute. Let

$$y = m/n; \tag{7.9.1}$$

then we may use the formulae established in earlier sections to discover the expected value and variance of the sample proportion $\mathbf{y}$.

To do this, we define $x(j)$ $(j = 1, 2, \ldots, N)$ to take the value 1 for those members of the population which possess the attribute and 0 for those members which do not. Then their arithmetic mean is

$$\bar{x} = N^{-1} \sum_{j=1}^{N} x(j) = M/N. \tag{7.9.2}$$

Let $\mathbf{z}(h)$ $(h = 1, 2, \ldots, n)$ be the values of $x(j)$ for the n members of the sample, then

$$\sum_{h=1}^{n} \mathbf{z}(h) = \mathbf{m}, \tag{7.9.3}$$

since $\mathbf{z}(h) = \mathbf{1}$ for those members of the sample which possess the attribute, whereas $\mathbf{z}(h) = 0$ for those which do not.

From (7.9.1), (7.9.3) the arithmetic mean of the sample is

$$\bar{\mathbf{z}} = n^{-1} \sum_{h=1}^{n} \mathbf{z}(h) = \mathbf{y}. \tag{7.9.4}$$

The expected value and variance of the sample proportion $\mathbf{y}$ may therefore be directly obtained from our previous formulae relating to $\bar{\mathbf{z}}$.

In the case of simple random samples both with and without replacement, we obtain from (7.4.2) and (7.7.3) and (7.9.2) and (7.9.4)

$$E(\mathbf{y}) = \mathbf{p}'\mathbf{y} = \mathbf{p}'\bar{\mathbf{z}} = \bar{x} = M/N = Y, \quad \text{(say)} \tag{7.9.5}$$

that is to say the expected value of the sample proportion (possessing the attribute) is equal to the corresponding proportion in the population.

We may next note that the variance of the $x(j)$ in the population is

$$\text{var}\, \mathbf{x} = N^{-1} \sum_{j=1}^{N} \{x(j) - \bar{x}\}^2 = N^{-1} \left[\sum_{j=1}^{N} \{x(j)\}^2 - N\bar{x}^2 \right] = Y(1-Y). \tag{7.9.6}$$

In the case of a simple random sample with replacement, it follows from formula (7.5.5) and (7.9.6) that the variance of $\mathbf{y}$, the sample proportion is

$$\text{var}\, \mathbf{y} = n^{-1} Y(1-Y), \tag{7.9.7}$$

where Y is the population proportion.

In the case of a simple random sample *without* replacement, it follows from formula (7.8.7) and from (7.9.6) that the variance of the sample proportion is

$$\text{var}\, \mathbf{y} = \{(N-n)/(N-1)\} n^{-1} Y(1-Y), \tag{7.9.8}$$

where Y is the population proportion.

7.10. Stratified Samples

Suppose that N measurements $x(t)$ are classified into m subsets containing N_j ($j = 1, 2, \ldots, m$) measurements, $x_j(1), x_j(2), \ldots, x_j(N_j)$ where $\sum_{j=1}^{m} N_j = N$.

Let
$$\bar{x}_j = N_j^{-1} \sum_{t=1}^{N_j} x_j(t), \tag{7.10.1}$$

and let
$$S_j^2 = N_j^{-1} \sum_{t=1}^{N_j} \{x_j(t) - \bar{x}_j\}^2. \tag{7.10.2}$$

Then
$$\bar{x} = N^{-1} \sum_{j=1}^{m} N_j \bar{x}_j. \tag{7.10.3}$$

Now suppose that samples of size n_j are taken from each subset ($j = 1, 2, \ldots, m$) and let the arithmetic means of these samples by $\bar{z}_j$ ($j = 1, 2, \ldots, m$). Then, if they are simple random samples with replacement, the $\bar{z}_j$ may be regarded as m independent variates with expected values

$$E(\bar{z}_j) = \mathbf{p}'\bar{z}_j = \bar{x}_j \quad (j = 1, 2, \ldots, m) \tag{7.10.4}$$

and with variances

$$\operatorname{var} \bar{z}_j = n_j^{-1} S_j^2. \tag{7.10.5}$$

Now consider the variate

$$\mathbf{z} = N^{-1} \sum_{j=1}^{m} N_j \bar{z}_j. \tag{7.10.6}$$

Because of (7.10.4),

$$E(\mathbf{z}) = \mathbf{p}'\mathbf{z} = N^{-1} \sum_{j=1}^{m} N_j \mathbf{p}'\bar{z}_j = N^{-1} \sum_{j=1}^{m} N_j \bar{x}_j = \bar{x}, \tag{7.10.7}$$

so that the expected value of $\mathbf{z}$ is $\bar{x}$, the population mean.

Similarly we may find the variance of the variate $\mathbf{z}$. Since the $\bar{z}_j$ are independent variates.

$$\operatorname{var} \mathbf{z} = \operatorname{var} \left\{ \sum_{j=1}^{m} N^{-1} N_j \bar{z}_j \right\} = \sum_{j=1}^{m} N^{-2} N_j^2 \operatorname{var} \bar{z}_j;$$

Hence, by (7.10.5),

$$\operatorname{var} \mathbf{z} = N^{-2} \sum_{j=1}^{m} N_j^2 n_j^{-1} S_j^2. \tag{7.10.8}$$

This variance may, under suitable circumstances, to be illustrated in the next two sections, be much less than the variance of the mean of a simple random sample of the same total size from the whole set of observations.

We shall find that it is therefore often in the interests of precise estimation to stratify a population, before taking samples from it and obtaining measurements of the sample. For this may enable a more precise estimate of the arithmetic mean of measurements for the whole population to be obtained at no greater cost.

7.11. Example Illustrating the Gain in Precision to be found by Stratifying a Population

Suppose that a town can be divided into three districts containing $N_1 = 10\,000$, $N_2 = 30\,000$ and $N_3 = 60\,000$ households, respectively.

It is required to estimate the arithmetic mean incomes of these 100 000 households.

Suppose that the unknown mean incomes are in fact

$$\bar{x}_1 = £10\,000 \quad \bar{x}_2 = £3000 \quad \bar{x}_3 = £1000,$$

so that the mean for all 100 000 households is $\bar{x} = £2500$.

Suppose that the unknown standard deviations of income in the three districts are

$$s_1 = £5000 \qquad s_2 = £2000 \qquad s_3 = £500.$$

The standard deviation for the whole town can then be found as s where

$$Ns^2 = \sum_{j=1}^{3} N_j\{s_j^2 + \bar{x}_j^2\} - N\bar{x}^2, \quad \text{where} \quad N = \sum_{j=1}^{3} N_j;$$

whence $$s = £3300 \text{ approx.}$$

Suppose we took a random sample of 100 households (with replacement) and found their arithmetic mean income, it would have a standard error of $\sigma = s/10 = £330$.

But now, suppose, instead, that we took separate random samples (with replacement) from each district in proportion to their populations, i.e. of sizes 10, 30 and 60.

Let their arithmetic means be $\bar{z}_1$ $\bar{z}_2$ and $\bar{z}_3$; the standard error of the arithmetic mean

$$z = (1/10)(\bar{z}_1 + 3\bar{z}_2 + 6\bar{z})$$

is (by equation (7.10.8)) given by

$$\sigma_{\bar{z}}^2 = 10^{-3}\{s_1^2+3s_2^2+6s_3^2\}$$

$$\therefore \quad \sigma_{\bar{z}}^2 = 10^{+3}\{25+12+1{\cdot}5\} \quad (£^2)$$

$$\therefore \quad \sigma_{\bar{z}}^2 = 38\,500 \quad (£^2)$$

Thus $\qquad \sigma_{\bar{z}} = £196 \quad \text{approx.}$

This is much less that the figure $\sigma = £330$ for the simple random sample.

Thus the stratified sample gives a more precise estimate of the arithmetic mean.

7.12. Example of Stratified Sampling with Unequal Sampling Fractions

In the example just considered we took the same sampling fraction (1 in 1000) in each of the three districts.

But now suppose, instead, that we take samples of sizes

40, 40 and 20

from the three districts, giving the same sample size of 100 in total, as before.

If we now work out

$$z = (1/10)(\bar{z}_1+3\bar{z}_2+6\bar{z}_3),$$

as before it will still be true that

$$E(z) = £2500;$$

but the standard error $\sigma_{\bar{z}}$ of z in £ will now be given by

$$\sigma_{\bar{z}}^2 = 10^{-3}\{0{\cdot}25s_1^2+2{\cdot}25s_2^2+18s_3^2\}$$

$$= 10^3\{6{\cdot}25+9{\cdot}00+4{\cdot}50\}$$

$$= 19\,750.$$

Hence $\qquad \sigma_{\bar{z}} = £140 \text{ approx.}$

This is considerably less than the value $\sigma_{\bar{z}} = £196$ obtained by the uniform 1 in 1000 stratified sample, and it illustrates the gain in precision which may sometimes be obtained by sampling more intensively from those strata of a population which have higher variance than others. This topic will be discussed in greater detail in Chapter 14 (Volume 2) on the Design of Social Surveys.

7.13. The Expected Value of a Sample Regression Coefficient

It cannot, in general, be taken for granted that, even in a simple sample with replacement, the expected value of a regression coefficient will agree with the corresponding coefficient in the population.

However, in certain special cases, a regression coefficient calculated from the sample values can be given this property.

The most important case is where the population can be stratified into sections, within each of which the value of x is constant, and where the sampling fraction is made the same in each stratum.

Suppose that there are m strata, and that in stratum i, $x = X_i$ and there are N_i population members and that the sum of the y-values of these N_i members is Y_i $(i = 1, 2, \ldots, m)$.

Then in the population, the regression coefficient is

$$\beta = \operatorname{cov}(\boldsymbol{x}, \boldsymbol{y}) \,/\, \operatorname{var} \boldsymbol{x} \tag{7.13.1}$$

where

$$\operatorname{var} \boldsymbol{x} = N^{-1} \sum_{i=1}^{m} N_i X_i^2 - N^{-2} \left[\sum_{i=1}^{m} N_i X_i \right]^2 \quad \left(N = \sum_{i=1}^{m} N_i \right)$$

$$\operatorname{cov}(\boldsymbol{x}, \boldsymbol{y}) = N^{-1} \sum_{i=1}^{m} X_i Y_i - N^{-2} \left[\sum_{i=1}^{m} N_i X_i \right] \sum_{i=1}^{m} Y_i. \tag{7.13.2}$$

Let the same sampling fraction ϕ be used for each stratum, so that the number of members of the sample is $n = \phi N$, and the number in each stratum is

$$n_i = \phi N_i \quad (i = 1, 2, \ldots, m). \tag{7.13.3}$$

The regression coefficient for the sample members $(\boldsymbol{z}, \boldsymbol{w})$ is

$$\boldsymbol{b} = \operatorname{cov}(\boldsymbol{z}, \boldsymbol{w}) \,/\, \operatorname{var} \boldsymbol{z}, \tag{7.13.4}$$

where

$$\operatorname{var} \boldsymbol{z} = n^{-1} \sum_{i=1}^{m} n_i X_i^2 - n^{-2} \left[\sum_{i=1}^{m} n_i X_i \right]^2 = \operatorname{var} \boldsymbol{x}.$$

$$E[\operatorname{cov}(\boldsymbol{z}, \boldsymbol{w}] = n^{-1} \sum_{i=1}^{m} X_i E(W_i) - n^{-2} \left[\sum_{i=1}^{m} n_i X_i \right] \sum_{i=1}^{m} E(W_i), \tag{7.13.5}$$

where W_i is the sum of the y-values of the sample members in the i-th stratum $(i = 1, 2, \ldots, m)$.

$$E(W_i) = \phi Y_i \tag{7.13.6}$$

$$\therefore\; E[\operatorname{cov}(\boldsymbol{z}, \boldsymbol{w})] = n^{-1} \phi \sum_{i=1}^{m} X_i Y_i - n^{-2} \left[\phi \sum_{i=1}^{m} X_i \right] \phi \sum_{i=1}^{m} Y_i$$

$$= \operatorname{cov}(\boldsymbol{x}, \boldsymbol{y}). \tag{7.13.7}$$

By (7.13.1), (7.13.4), (7.13.5) and (7.13.7)

$$E[\boldsymbol{b}] = \beta. \tag{7.13.8}$$

7.14. The Probability Function of the Arithmetic Mean of a Simple Random Sample with Replacement

The probability function for the arithmetic mean of a simple random sample with replacement depends solely on the sample-size and on the frequency function of the population from which the sample is drawn. However, the relation between the sample-mean probability function and the population frequency function is extremely involved.

It is shown in § **8.24** that a far simpler relation exists between the cumulants for the probability function of the sample mean and the cumulants for the frequency function for the population.

Indeed, if the cumulants for the population frequency function are K_r $(r = 1, 2, 3, \ldots)$, then those for the probability function of the arithmetic mean of a simple random sample of size n, with replacement are

$$\bar{K}_r = n^{(1-r)}K_r \quad (r = 1, 2, 3, \ldots). \tag{7.14.1}$$

We have already met the two particular cases of this relation where $r = 1$ and $r = 2$.

Thus when $r = 1$, we find

$$\bar{K}_1 = K_1$$

whence

$$\bar{\mu}'_1 = m'_1 \tag{7.14.2}$$

where μ'_1 is the expected value of the sample mean and m'_1 is the arithmetic mean of the population: this is the result expressed above in equation (7.5.2).

Again when $r = 2$, we find

$$\bar{K}_2 = n^{-1}K_2;$$

whence

$$\bar{\mu}_2 = n^{-1}m_2, \tag{7.14.3}$$

where $\bar{\mu}_2$ is the variance of the sample mean and m_2 is the variance of the population from which the sample is drawn.

This is the result expressed above in equation (7.5.5).

7.15. Simple Random Samples from a Bivariate Population

Let $[z(h), w(h)]$ $(h = 1, 2, \ldots, n)$ be a simple random sample from a bivariate population of N members with arithmetic mean $(\bar{x}, \bar{y})$, variances σ_x^2, σ_y^2 and covariance cov (x, y).

Let $\boldsymbol{u}(h) = z(h)+w(h)$ and $v(h) = z(h)-w(h)$ $(h = 1, 2, \ldots, n)$,
(7.15.1)

so that by Exercise 5.5 $\boldsymbol{u}(h)$ is a variate with expected value $(\bar{x}+\bar{y})$ and variance $\sigma_x^2+\sigma_y^2+2\operatorname{cov}(x, y)$ and similarly $\boldsymbol{v}(h)$ is a variate with expected value $(\bar{x}-\bar{y})$ and with variance $\sigma_x^2+\sigma_y^2-2\operatorname{cov}(x, y)$.

We may apply the results to the two sets $\boldsymbol{u}(h)$ and $\boldsymbol{v}(h)$ $(h = 1, 2,\ldots, n)$ regarded as simple random samples. If their arithmetic means are $\bar{\boldsymbol{u}}$ and $\bar{\boldsymbol{v}}$, then $\bar{\boldsymbol{u}}$ and $\bar{\boldsymbol{v}}$ must be variates with expected values

$$E(\bar{\boldsymbol{u}}) = \bar{x}+\bar{y};\ E(\bar{\boldsymbol{v}}) = \bar{x}-\bar{y}, \tag{7.15.2}$$

and with variances

$$\begin{aligned} \operatorname{var}(\bar{\boldsymbol{u}}) &= (1/n)[\sigma_x^2+\sigma_y^2+2\operatorname{cov}(\boldsymbol{x}, \boldsymbol{y})] \\ \operatorname{var}(\bar{\boldsymbol{v}}) &= (1/n)[\sigma_x^2+\sigma_y^2-2\operatorname{cov}(\boldsymbol{x}, \boldsymbol{y})] \end{aligned} \tag{7.15.3}$$

if sampling is with replacement, and with variances

$$\begin{aligned} \operatorname{var}(\bar{\boldsymbol{u}}) &= (1-f)(1/n)[\sigma_x^2+\sigma_y^2+2\operatorname{cov}(x, y)] \\ \operatorname{var}(\bar{\boldsymbol{v}}) &= (1-f)(1/n)[\sigma_x^2+\sigma_y^2-2\operatorname{cov}(x, y)], \\ f &= (n-1)\,/\,(N-1). \end{aligned} \tag{7.15.4}$$

Let $\bar{z}$ and $\bar{\boldsymbol{w}}$ denote the arithmetic means of the $z(h)$ and of the $w(h)$, then

$$\bar{\boldsymbol{u}} = \bar{z}+\bar{\boldsymbol{w}} \quad \text{and} \quad \bar{\boldsymbol{v}} = \bar{z}-\bar{\boldsymbol{w}}; \tag{7.15.5}$$

hence, whether sampling is with or without replacement,

$$\begin{aligned} \operatorname{var}(\bar{\boldsymbol{u}}) &= \operatorname{var}(\bar{z})+\operatorname{var}(\bar{\boldsymbol{w}})+2\operatorname{cov}(\bar{z}, \bar{\boldsymbol{w}}) \\ \operatorname{var}(\bar{\boldsymbol{v}}) &= \operatorname{var}(\bar{z})+\operatorname{var}(\bar{\boldsymbol{w}})-2\operatorname{cov}(\bar{z}, \bar{\boldsymbol{w}}) \end{aligned}$$

and

$$\operatorname{var}(\bar{\boldsymbol{u}})-\operatorname{var}(\bar{\boldsymbol{v}}) = 4\operatorname{cov}(\bar{z}, \bar{\boldsymbol{w}}). \tag{7.15.6}$$

Substituting from (7.15.3) and (7.15.4), we obtain expressions for the covariance of the arithmetic means $\bar{z}$ and $\bar{\boldsymbol{w}}$;

$$\operatorname{cov}(\bar{z}, \bar{\boldsymbol{w}}) = (1/n)\operatorname{cov}(\boldsymbol{x}, \boldsymbol{y}),\ \operatorname{cov}(\bar{z}, \bar{\boldsymbol{w}}) = [(1-f)/n]\operatorname{cov}(\boldsymbol{x}, \boldsymbol{y}) \tag{7.15.7}$$

in the cases of sampling with and without replacement.

It follows that the covariance between the sample totals $\boldsymbol{z} = \sum z(h)$ and $\boldsymbol{w} = \sum w(h)$ is given by

$$\operatorname{cov}(\boldsymbol{z}, \boldsymbol{w}) = n\operatorname{cov}(\boldsymbol{x}, \boldsymbol{y}) \text{ and by } \operatorname{cov}(\boldsymbol{z}, \boldsymbol{w}) = (1-f)n\operatorname{cov}(\boldsymbol{x}, \boldsymbol{y}). \tag{7.15.8}$$

in the cases of sampling with and without replacement.

7.16. Simple Random Samples of Clusters of Elements

It is often inconvenient or expensive to obtain simple random samples of elements from a population, because the elements are grouped into clusters, and the cost of taking a whole cluster into the sample may be not much greater than taking a single element from the cluster. In this case one may take a simple random sample of clusters and apply the theory of simple random sampling to the totals for the clusters.

Suppose that a population consists of K clusters of observations C_i $(i = 1, 2, \ldots, K)$ where the i-th cluster, C_i, consists of N_i elements x_{ij} $(j = 1, 2, \ldots, N_i)$.

Let
$$N = \sum_{i=1}^{K} N_i, \tag{7.16.1}$$

and let
$$x_i = \sum_{j=1}^{N_i} x_{ij}, \tag{7.16.2}$$

be the totals for the clusters and let
$$x = \sum_{i=1}^{K} x_i \tag{7.16.3}$$

be the total for the population, and let
$$\bar{x} = x/N \tag{7.16.4}$$

be the arithmetic mean for the population.

If a simple random sample of k clusters is taken, and the totals for these k clusters are $z(h)$ $(h = 1, 2, \ldots, k)$, then the $z(h)$ are a simple random sample of k from the K totals x_i.

Consider the estimate
$$\mathbf{x}^* = Kz/k$$

where
$$\mathbf{z} = \sum_{h=1}^{k} z(h) \tag{7.16.5}$$

is the sample total. Evidently, whether sampling is with or without replacement,
$$E(\mathbf{x}^*) = (K/k)E[\textstyle\sum z(h)] = (K/k)(kx/K) = x, \tag{7.16.6}$$

and
$$\text{var}(\mathbf{x}^*) = (K/k)^2 \text{var}(\mathbf{z}) = K^2 \text{var}(\bar{\mathbf{z}}), \tag{7.16.7}$$

where $\bar{\mathbf{z}}$ is the arithmetic mean of the sample cluster totals.

It follows from (7.5.5) that if sampling is with replacement,

$$\text{var}\,(\boldsymbol{x}^*) = (K^2/k)\sigma_x^2, \tag{7.16.8}$$

and from (7.8.7) that, if sampling is without replacement,

$$\text{var}\,(x^*) = (1-f)(K^2/k)\sigma_x^2 \quad (f = (k-1)\,/\,(K-1)), \tag{7.16.9}$$

where σ_x^2 is the variance of the population of cluster totals x_i.

As an estimator of σ_x^2 we should take

$$\sigma_x^{*2} = [k/(k-1)]s_z^2 \tag{7.16.10}$$

if sampling is with replacement, and

$$\sigma_x^{*2} = [k(K-1)\,/\,(k-1)K]s_z^2 \tag{7.16.11}$$

if sampling is without replacement, where s_z^2 is the variance of the k cluster-totals $z(h)$.

Substituting these values of σ_x^{*2} for σ_x^2 in (7.16.8) and (7.16.9), we obtain estimates of var (x^*) as

$$\sigma_{x^*}^{*2} = \frac{K^2}{k-1}\,s_z^2 \tag{7.16.12}$$

if sampling is with replacement, and as

$$\sigma_{x^*}^{*2} = \frac{K(K-k)}{k-1}\,s_z^2 \tag{7.16.13}$$

if sampling is without replacement.

In particular, N, the total number of elements in the population, may be estimated as

$$N^* = \frac{K}{k}\,n, \tag{7.16.14}$$

where n is the total number of elements in the sample of clusters, namely

$$n = n(h), \tag{7.16.15}$$

where $n(h)$ denotes the number of elements in cluster h of the sample.

Similarly, we may estimate the variance of N^* as

$$\sigma_{N^*}^{*2} = \frac{K^2}{k-1}\,s_n^2 \tag{7.16.16}$$

if sampling is with replacement, and as

$$\sigma_{N^*}^{*2} = \frac{K(K-k)}{k-1}\,s_n^2 \tag{7.16.17}$$

if sampling is without replacement, where s_n^2 is the variance of the sample frequency distribution of cluster-sizes $n(h)$.

Having estimated the total for the population and the size of the population as x^* and N^* as given by (7.16.5) and (7.16.14), we may estimate the population arithmetic mean per element as

$$\bar{x}^* = \frac{x^*}{N^*} = \frac{z}{n} = \bar{\bar{z}} \qquad (7.16.18)$$

where z is the arithmetic mean per element of the sample.

However it cannot be assumed that the expected value of our estimate $\bar{x}^*$ will equal the true value $\bar{\bar{x}}$. The relation of $E(\bar{x}^*)$ to $\bar{x}$ will be discussed in Volume 2 in Chapter 14, on sample surveys.

EXAMPLE 7.16.1. A simple sample, without replacement, of ten clusters of households is taken from a population of two hundred clusters of households, for each of which wealth (y) and income (x) can be estimated. In each cluster in the sample, the number $n(h)$ and the total income ($n(h)$) are obtained only for those households which are recorded as having zero wealth. A summary of the information so obtained is set out in Table 7.16.1.

Estimate the number N of households recorded as having zero wealth and the total income x and mean income $\bar{x}$ of these households, in the population of 200 clusters. Indicate which of your estimates is unbiased (in the sense that the expected value of the estimator is the true value) and which is biased. Estimate the standard errors of your estimates of N and of x.

TABLE 7.16.1

Incomes of households with zero wealth

Cluster number h	*Number* $n(h)$ *of households in cluster*	*Total income* $z(h)$ *of households in cluster* '000
1	4	25·82
2	6	20·84
3	2	17·94
4	4	26·28
5	7	26·57
6	2	13·44
7	1	7·21
8	1	8·07
9	0	0·00
10	1	8·68
Total	28 n	154·85 z

$$\sum_{h=1}^{10} n(h)^2 = 128$$

$$\sum_h z(h)^2 = 3\,176\,000\,000$$

$$K = 200 \qquad k = 10$$

Solution

$$N^* = \frac{K}{k}n = \frac{200}{10}28 = 560 \quad \text{(unbiased)}$$

$$x^* = \frac{K}{k}z = \frac{200}{10}154\,850 = 3\,097\,000 \quad \text{(unbiased)}$$

$$\bar{x}^* = \frac{z}{n} = \frac{154\,850}{28} = 5530 \quad \text{(probably biased)}$$

$$s_n^2 = \frac{128}{10} - \left(\frac{28}{10}\right)^2 = 4{\cdot}96$$

$$s_z^2 = 317\,600\,000 - (15\,485)^2 = 77\,815\,000$$

$$\sigma_{N^*}^{*2} = \frac{200 \times 190}{9} s_n^2 = 21\,000 \qquad \sigma_{N^*}^* = 145$$

$$\sigma_{x^*}^{*2} = \frac{200 \times 190}{9} s_z^2 = 328\,550\,000\,000 \qquad \sigma_{**}^* = 573\,000$$

SUMMARY:

Number of households in population with zero wealth = 560 (s.e. 145)

Total income of these households = 3 100 000 dollars (s.e. \$573 000)

Mean income of these households = 5530 dollars (s.e. not obtained) (may be biased).

7.17. Sets of Observations Regarded as Random Samples

Suppose that $z(h)$ ($h = 1, 2, \ldots, n$) are a set of independent observations of the same variate $\boldsymbol{x}$ with probability function

$$f(x) = \boldsymbol{p}'\boldsymbol{r}(\boldsymbol{x} = x) \qquad \text{(rational).} \tag{7.17.1}$$

Then the multivariate probability function of the set of variates $\boldsymbol{z}(h)$ will be exactly what it would have been if they had been instead a simple random sample of size n drawn from a population $\boldsymbol{x}$, with frequency function

$$f(x) = \boldsymbol{f}'\boldsymbol{r}(\boldsymbol{x} = x) \tag{7.17.2}$$

It is thus common to regard a sequence of independent observations of a variate as though they were a simple random sample with replacement from a population.

It should be noted however that this representation is strictly legitimate only if the probabilities $f(x)$ are all *rational*, since a frequency function is necessarily limited to rational numbers.

A further difficulty arises if the number of possible values of the variate is unlimited.

Consider, for example, the experiment of repeatedly spinning an unbiased coin until heads is obtained; and let the variate $\boldsymbol{x}$ denote the total number of spins. Then

$$p'\boldsymbol{r}(\boldsymbol{x} = x) = 2^{-x} \quad (x = 1, 2, 3, \ldots). \tag{7.17.3}$$

But it is clearly not possible to regard a sequence of n observations of $\boldsymbol{x}$ as being a simple random sample size n from any population of finite size N, since we can always find a value x so large that

$$p'\boldsymbol{r}(\boldsymbol{x} = x) < N^{-1}. \tag{7.17.4}$$

7.18. Sample Size Regarded as a Variate

For certain purposes it is appropriate to regard the number of observations in the sample as being itself a variate.

For example, in what are known as 'sequential' sampling schemes (see Chapter 17 in Vol. 2), the choice of the size of sample depends on the contents of the sample so far selected, so that it is essential to regard the size of a sample so obtained as a variate.

Moreover we shall sometimes wish to contemplate the sample size as tending to infinity when in Volume 2 we come to discuss the properties of 'estimators'. In such cases very great care will be needed in handling the notation.

Let $\boldsymbol{n}$ denote the number of observations contained in a sample, and let n denote any particular value which the variate $\boldsymbol{n}$ may take. Then when discussing the probability distribution of some measure $\boldsymbol{v}$ (such as the arithmetic mean or standard error etc.) based on a sample of given size n, we must consider probabilities, given the event $\boldsymbol{r}(\boldsymbol{n} = n)$ and not unrestricted probabilities.

For example, if $\boldsymbol{v}$ denotes the variance of the sample, we should denote the expected value of $\boldsymbol{v}$, when the sample is of size n, not as

$$\boldsymbol{p}'\boldsymbol{v}, \quad \text{but as} \quad \boldsymbol{p}'_{|\,\boldsymbol{r}(\boldsymbol{n} = n)}\boldsymbol{v}. \tag{7.18.1}$$

We have seen (7.6.5) that in the case of simple random samples with replacement

$$\boldsymbol{p}'_{|\,\boldsymbol{r}(\boldsymbol{n} = n)}\boldsymbol{v} = \{(n-1)/n\} \text{ var } \boldsymbol{x}, \tag{7.18.2}$$

where var $\boldsymbol{x}$ is the population variance.

If we wish to consider the value which is approached by $\boldsymbol{p}'_{|\,\boldsymbol{r}(\boldsymbol{n} = n)}\boldsymbol{v}$

as the sample size is indefinitely increased, we may employ the notation

$$\lim_{n \to \infty} \boldsymbol{p}'_{| \boldsymbol{r}(\boldsymbol{n} = n)} \boldsymbol{v}. \tag{7.18.3}$$

Thus if $\boldsymbol{v}$ has the interpretation given above,

$$\lim_{n \to \infty} \boldsymbol{p}'_{| \boldsymbol{r}(\boldsymbol{n} = n)} \boldsymbol{v} = \text{var } \boldsymbol{x}. \tag{7.18.4}$$

7.19. Limiting Form of Distribution Functions in Large Samples

Suppose that $\boldsymbol{z}(h)$ $(h = 1, 2, \ldots, n)$ is a simple random sample with replacement from a population of only two members, namely 0 and 1. Let

$$\boldsymbol{t}(n) = \sum_{h=1}^{n} 2^{-h} \boldsymbol{z}(h) \tag{7.19.1}$$

be calculated for the sample. Then $\boldsymbol{t}(n)$ is a rectangular variate whose probability function is

$$f_n(t) = \boldsymbol{p}'\boldsymbol{r}\{\boldsymbol{t}(n) = t\} = 2^{-n} \text{ for } t = 0, 2^{-n}, 2.2^{-n}, 3.2^{-n}, \ldots, (2^n - 1)\, 2^{-n}$$

namely for

$$t = 0{\cdot}00 \ldots 00, 0{\cdot}00 \ldots 01, 0{\cdot}00 \ldots 10, 0{\cdot}00 \ldots 11, \ldots, 0{\cdot}11 \ldots 11$$

$$(n \text{ places}) \quad \text{when expressed in the scale of 2.} \tag{7.19.2}$$

The distribution function for $\boldsymbol{t}(n)$ in samples of size n is

$$F(n, t) = \boldsymbol{p}'_{| \boldsymbol{r}(\boldsymbol{n} = n)} \boldsymbol{r}\{\boldsymbol{t}(\boldsymbol{n}) \leqslant t\} = 2^{-n}[2^n t] \text{ for } 0 \leqslant t \leqslant 1, \tag{7.19.3}$$

where [ ] denotes 'the integral part of'.

Thus the distribution function is a step function proceeding from (0, 0) to (1, 1) by 2^n equal steps of height 2^{-n} each.

We may consider the limiting case as $n \to \infty$, and find

$$\lim_{n \to \infty} F(n, t) = \lim_{n \to \infty} \boldsymbol{p}'_{| \boldsymbol{r}(\boldsymbol{n} = n)} \boldsymbol{r}\{\boldsymbol{t}(\boldsymbol{n}) \leqslant t\} = t \text{ for } 0 \leqslant t \leqslant 1. \tag{7.19.4}$$

This limiting form of the distribution function in large samples is no longer a step function, but the simple continuous function t. This cannot itself represent the distribution function of any variate in the probability model we have so far considered, since all those distribution functions must be step functions.

Yet it might be very convenient to approximate the rather elaborate distribution function (7.19.3) for large samples by the almost equivalent, but far simpler, continuous distribution function (7.19.4).

This is but one extreme example of many, where a highly elaborate form of distribution function of a statistic based on a large sample can be closely approximated by a far simpler distribution function, which is, however, continuous and therefore cannot be the distribution function of any variate in the probability model so far considered.

In order to take advantage of the great simplifications obtainable by approximating the distribution functions of such large-sample statistics by continuous functions, it will prove well worthwhile to enlarge the scope of our probability model to allow a new kind of variate $\boldsymbol{x}$, whose distribution function

$$F(x) = \boldsymbol{p'r}(\boldsymbol{x} \leqslant x)$$

may be a continuous monotonic increasing function of x. This new kind of variate is known as a *continuous variate*, and the revised form of probability model required to accommodate it forms the subject of the next chapter.

Exercises 7

7.1. Explain what is meant by a simple random sample (*i*) with replacement, (*ii*) without replacement.

7.2. A simple random sample of size n is taken (with replacement) from a list of firms, and the following statistics are worked out for the firms in the sample:

X_1 = the arithmetic mean of gross profits
X_2 = $[n/(n-1)]$ times the sample variance of gross profits
$X_3 = \sqrt{X_2}$ $\quad X_4 = X_2/n$ $\quad X_5 = \sqrt{X_4}$.

Which of the following assertions are true?

(*i*) $E[X_1]$ = the arithmetic mean of gross profits for all the firms in the list.
(*ii*) $E[X_2]$ = the variance of gross profits for all the firms in the list.
(*iii*) $E[X_3]$ = the standard deviation of gross profits for all the firms in the list.
(*iv*) $E[X_4]$ = the variance of X_1, regarded as a variate.
(*v*) $E[X_5]$ = the standard error of X_1, regarded as a variate.

7.3. Establish a formula giving the variance of the proportion of red balls in a sample of n balls taken at random, (a) with (b) without replacement, from a bag containing M red and $(N-M)$ other balls. [*Hint*: Count a red ball as 1 and any other ball as 0 and consider the mean score of the balls selected.]

7.4. A population of size 100 is subdivided into groups of sizes 20, 40 and 40. Random samples with replacement of sizes 4, 16 and 5 have means X_1, X_2, X_3 and variances V_1, V_2, V_3. How would you choose w_1, w_2, w_3 in order that, where $Y = w_1X_1+w_2X_2+w_3X_3$, $E(Y)$ should equal the arithmetic mean of the population?

How would you choose u_1, u_2, u_3 in order that, where $V = u_1V_1+u_2V_2+u_3V_3$, $E(V)$ should equal the variance of the population? How would you choose λ_1, λ_2, λ_3 in order that where $v_Y = \lambda_1V_1+\lambda_2V_2+\lambda_3V_3$, $E(v_Y)$ should equal the variance of Y where Y is regarded as a variate?

7.5. Give an account of the advantages which may be derived from using stratified samples.

7.6. Why is it particularly difficult to estimate the standard error of the mean of a sample with high skewness?

Find the mean and the median of the sample of the Table given in Exercise 6.13.

[Cambridge Economics Tripos Part 2, 1962]

7.7. Define the terms: (a) Distribution function, (b) Covariance, (c) Correlation coefficient, (d) Mode.

A pack of 52 cards contains four suits, S, H, D and C, of 13 cards each; s and h denote the numbers of cards of suits S and H in a random sample of 8 cards taken without replacement from the pack; x denotes the least number of cards in any suit in the sample.

Obtain the variance-covariance matrix for s and h and the correlation coefficient for s and h.

Calculate numerically the distribution function for x and sketch it. What is the mode of the distribution of x?

[Cambridge Economics Tripos Part 2, 1963]

7.8. A frequency distribution is divided into k groups with proportions of items ρ_i $(i = 1, 2, \ldots, k)$. A sample of N produces numbers N_i $(i = 1, 2, \ldots, k)$ in the groups. Show that the correlation coefficient between N_i and N_j $(i \neq j)$ in repeated sampling, with N constant, is $-[\rho_i\rho_j / (1-\rho_i)(1-\rho_j)]$.

[Cambridge Economics Tripos Part 2, 1964]

7.9. A sample survey is to be carried out to estimate the total value of consumers' durables in British households. It is known that the average value per household varies considerably between income groups; show that the precision of the estimate obtained from a sample of given size can be increased by the use of stratified sampling. If it is also known that the variance of the value of durables per household varies between income groups, show how this knowledge can be used to increase further the precision of the estimate.

[Cambridge Economics Tripos Part 2, 1965]

Chapter 8

A MODEL OF DISCRETE AND CONTINUOUS VARIATES

8.1. Alternative Form for the Model for Discrete Variates

The use of vectors to represent events, variates, and relations between variates, enabled us very economically to represent simple probability situations in a compact model which could be handled by means of the elementary operations of matrix algebra.

Alternatively, we could have introduced discrete variates not as vectors, but as fundamental concepts whose properties are summed up by their probability functions. To any discrete variate, $\boldsymbol{x}$, there then corresponds a function of the real variable x:

$$f(x; \boldsymbol{x}) = \boldsymbol{p}'\boldsymbol{r}(\boldsymbol{x} = x), \qquad (8.1.1)$$

which gives the probability that $\boldsymbol{x} = x$.

In this alternative form of introduction, $\boldsymbol{p}'$, $\boldsymbol{r}$ and $\boldsymbol{x}$ are no longer to be regarded as vectors. Where $\boldsymbol{r}(\boldsymbol{x} = x)$ may denote the relation or event that the variate $\boldsymbol{x}$ takes the value x, and $\boldsymbol{p}'\boldsymbol{r}(\boldsymbol{x} = x)$ denotes the probability of this event, $\boldsymbol{p}'\boldsymbol{r}$ may simply be interpreted to mean 'the probability that'.

In order to comply with the convention that the probability of the union (or disjunction) of two or more disjoint events should equal the sum of the probabilities of those events, we should then define the distribution function for the discrete variate $\boldsymbol{x}$ to be

$$F(x; \boldsymbol{x}) = \boldsymbol{p}'\boldsymbol{r}(\boldsymbol{x} \leqslant x) = \sum_{X \leqslant x} \boldsymbol{p}'\boldsymbol{r}(\boldsymbol{x} = X). \qquad (8.1.2)$$

Here again $\boldsymbol{p}'\boldsymbol{r}$ is to be interpreted simply to mean 'the probability that': it must not now be regarded as the scalar product of two vectors $\boldsymbol{p}'$ and $\boldsymbol{r}$.

8.2. Models Based on Variates and their Distribution Functions

Instead of taking as fundamental concepts the variate $\boldsymbol{x}$ and its probability function $f(x; \boldsymbol{x}) = \boldsymbol{p}'\boldsymbol{r}(\boldsymbol{x} = x)$ and then deriving the distribution function, we might, instead, have taken as fundamental the variate $\boldsymbol{x}$

and its distribution function, $F(x; \boldsymbol{x}) = \boldsymbol{p'r}(\boldsymbol{x} \leqslant x)$, and then derived the probability function in terms of the distribution function.

We would have stipulated

$$f(x; \boldsymbol{x}) = \boldsymbol{p'r}(\boldsymbol{x} = x) = \boldsymbol{p'r}(\boldsymbol{x} \leqslant x) - \boldsymbol{p'r}(\boldsymbol{x} < x)$$
$$= F(x; \boldsymbol{x}) - \lim_{\varepsilon \to 0} F(\boldsymbol{x} - \varepsilon^2; \boldsymbol{x}). \qquad (8.2.1)$$

There are, as a matter of fact, certain technical advantages in proceeding in this way, taking it as axiomatic that every variate $\boldsymbol{x}$ has its distribution function $F(x; \boldsymbol{x})$. This procedure will enable us to enlarge the scope of our probability model so as to accommodate a new type of variate to be called a *continuous* variate in addition to the type of discrete variates which were covered by the model of Chapter 5. A continuous variate may take any value along a continuous scale, whereas a discrete variate was limited to an enumerable sequence of discrete values.

The basic modification which is now to be made to our earlier probability model (of Chapters 4 and 5), is that variates and events are no longer to be represented by vectors. But it will still be convenient to use the same notation as before, with a very few exceptions to be distinguished later: moreover we shall interpret the symbols in the same manner as earlier, whenever we relate the model to any actual situation. We shall, indeed, suppose that all the results obtained for our previous model carry over to the new model, so long, of course, as we are still dealing with events and relations and *discrete* variates, which were the subject matter of our earlier model. Analogous results for continuous variates will require some separate justification.

As in Chapter 5, if R denotes any logical relation between variates, the event that R is true will be denoted by $\boldsymbol{r}(R)$ and the probability that R is true will be denoted by $\boldsymbol{p'r}(R)$. Again, as in Chapter 5, if $\boldsymbol{x}$ is any variate and x any real number, then the probability that $\boldsymbol{x} \leqslant x$ will be denoted by $\boldsymbol{p'r}(\boldsymbol{x} \leqslant x)$, and this function of x may alternatively be denoted by $F(x; \boldsymbol{x})$ and called the distribution function of $\boldsymbol{x}$. Usually, provided it is clear from the context that $\boldsymbol{x}$ is the variate concerned, this distribution function may be written more shortly as $F(x)$.

Whenever $\boldsymbol{A}$ and $\boldsymbol{B}$ are any two events or logical relations, we shall continue to denote their intersections by $\boldsymbol{A} \cap \boldsymbol{B}$ or else by $\widehat{\boldsymbol{A}\boldsymbol{B}}$. For example

$$\{\boldsymbol{r}(\boldsymbol{x} \leqslant x) \cap \boldsymbol{r}(\boldsymbol{y} \leqslant y)\} \text{ and } \hat{\boldsymbol{r}}(\boldsymbol{x} \leqslant x)\boldsymbol{r}(\boldsymbol{y} \leqslant y)$$

are each valid ways of denoting the intersection of the events that $\boldsymbol{x} \leqslant x$ and that $\boldsymbol{y} \leqslant y$.

Whenever $\boldsymbol{x}$ and $\boldsymbol{y}$ are two discrete variates, we shall continue to denote their product by $\hat{\boldsymbol{x}}\boldsymbol{y}$ or by $\hat{\boldsymbol{y}}\boldsymbol{x}$.

An exception will now be indicated to our normal procedure of carrying on with the earlier notation. We shall now avoid the use of the notation $\boldsymbol{p}'\boldsymbol{x}$ to denote the expected value of a variate $\boldsymbol{x}$. We shall prefer to denote the expected value of any variate $\boldsymbol{x}$ by $E(\boldsymbol{x})$ and in the case of a discrete variate $\boldsymbol{x}$, we may regard its expected value as being *defined* by equation (5.8.6) where $f(x)$ denotes the probability function for $\boldsymbol{x}$.

Although we shall no longer denote $E(\boldsymbol{x})$ by $\boldsymbol{p}'\boldsymbol{x}$, we shall continue to use the notation $\boldsymbol{p}'\boldsymbol{x}$ to denote the probability of any event or logical relation $\boldsymbol{x}$. But the symbols $\boldsymbol{p}'$ and $\boldsymbol{x}$ must no longer be regarded as vectors.

In general, the expected value of any function $\phi(\boldsymbol{x})$ of a discrete variate must now be defined as

$$E\{\phi\} = \sum_{x} f(x)\phi(x), \tag{8.2.2}$$

as in (5.8.6).

Just as we have assumed that to each variate $\boldsymbol{x}$ there belongs a distribution function $F(x)$, we shall assume also that to every group of variates $\boldsymbol{x}_h$ ($h = 1, 2, \ldots, m$ there) belongs a multivariate distribution function.

$$F(x_1, x_2, \ldots, x_m) = \boldsymbol{p}' \bigcap_{h=1}^{m} \boldsymbol{r}(\boldsymbol{x}_h \leqslant x_h). \tag{8.2.3}$$

In (8.2.3), $\boldsymbol{p}'$ stands as usual for 'the probability of' and $\bigcap_{h=1}^{m} \boldsymbol{r}(\boldsymbol{x}_h \leqslant x_h)$ denotes, as before, the intersection of the m events $\boldsymbol{r}(\boldsymbol{x}_h \leqslant x_h)$ ($h = 1, 2, \ldots, m$).

We shall assume that as in the earlier model, whenever $\boldsymbol{r}(R)$ is logically false then $\boldsymbol{p}'\boldsymbol{r}(R) = 0$.

We shall take over from the earlier model the result that

$$\boldsymbol{p}'\{\boldsymbol{j}(1) \cup \boldsymbol{j}(2)\} + \boldsymbol{p}'\{\boldsymbol{j}(1) \cap \boldsymbol{j}(2)\} = \boldsymbol{p}'\boldsymbol{j}(1) + \boldsymbol{p}'\boldsymbol{j}(2). \tag{8.2.4}$$

It follows easily from these assumptions that for $x_2 \geqslant x_1$ and where $F(x)$ denotes the distribution function for $\boldsymbol{x}$,

$$\boldsymbol{p}'\boldsymbol{r}(x_1 < \boldsymbol{x} \leqslant x_2) = F(x_2) - F(x_1). \tag{8.2.5}$$

Again, where $F(x, y)$ denotes the distribution function for the variate pair $(\boldsymbol{x}, \boldsymbol{y})$,

$$\boldsymbol{p}\{\boldsymbol{r}(x_1 < \boldsymbol{x} \leqslant x_2) \cap \boldsymbol{r}(y_1 < \boldsymbol{y} \leqslant y_2)\} \\ = F(x_2, y_2) - F(x_2, y_1) - F(x_1, y_2) + F(x_1, y_1). \tag{8.2.6}$$

8.3. Further Properties of the Distribution Function of a Discrete Variate

In the model of Chapter 5, for any given probability vector $\boldsymbol{p}'$ and any given variate $\boldsymbol{x}$, one could prove that

$$\lim_{x \to \infty} \boldsymbol{p}'\boldsymbol{r}(\boldsymbol{x} \leqslant x) = 1, \tag{8.3.1}$$

and that

$$\lim_{x \to -\infty} \boldsymbol{p}'\boldsymbol{r}(\boldsymbol{x} \leqslant x) = 0. \tag{8.3.2}$$

Hence the distribution function for any discrete variate $\boldsymbol{x}$ must in that model satisfy

$$\lim_{x \to \infty} F(x) = 1; \quad \lim_{x \to -\infty} F(x) = 0. \tag{8.3.3}$$

We shall assume that in the new model, every distribution function retains the property (8.3.3).

It is also obvious that every distribution function of a *discrete* variate $\boldsymbol{x}$ must, in virtue of equation (8.2.1), be a step function; whenever it is differentiable, its derivative must be zero. (8.3.4)

8.4. Continuous Variates

For many purposes it will be convenient to consider a new kind of variate $\boldsymbol{x}$ with a distribution function $F(x)$ which still satisfies conditions (8.3.3) but, instead of obeying (8.3.4), satisfies $F(x)$ is continuous for all x, and is differentiable with

$$\frac{dF}{dx} \geqslant 0 \text{ for all } x \text{ for which } 0 < F(x) < 1. \tag{8.4.1}$$

Such a variate will be called a *continuous variate.*

The meaning of the distribution function $F(x)$ for a continuous variate $\boldsymbol{x}$, is the probability that $\boldsymbol{x} \leqslant x$; but the variate $\boldsymbol{x}$ and its differentiable distribution function $F(x)$ are quite foreign to the model of Chapter 5, because no continuous variate $\boldsymbol{x}$ can possibly be represented by a vector in that model so as to enable the scalar product $\boldsymbol{p}'\boldsymbol{r}(\boldsymbol{x} \leqslant x)$ to represent, for all real x, the probability that $\boldsymbol{x} \leqslant x$.

However, it is shown, in Appendix 8, that, given any function $F(x)$ which satisfies (8.3.1), (8.3.2), (8.3.3) and (8.4.1), it is possible to define a pair of discrete variates $\boldsymbol{n}$ and $\boldsymbol{x}$; where $\boldsymbol{n}$ may take the values 1, 2, 3, . . . , in such a way that

$$\lim_{n \to \infty} \boldsymbol{p}'_{|\boldsymbol{r}(\boldsymbol{n}=n)}\boldsymbol{r}(\boldsymbol{x} \leqslant x) = F(x). \tag{8.4.2}$$

It is therefore proper to regard the distribution function of any continuous variate as the limit towards which the distribution function of a discrete variate may tend as some condition n tends to infinity.

It is indeed in order to obtain good simple approximations for the distribution functions of discrete variates in cases where some n is large, that one most often requires to discuss continuous variates.

8.5. Probability Density Functions

In virtue of property (8.4.1), the derivative $f(x)$ of the distribution function $F(x)$ of any continuous variate will exist at all values x (except possibly at the end points where $F(x) = 0$ or $F(x) = 1$), and will be non-negative: furthermore, by (8.3.2) and (8.3.3) it must be such that

$$\int_{-\infty}^{\infty} f(x)\,dx = 1. \tag{8.5.1}$$

The function $f(x)$ is known as the probability density function for x.

8.6. Illustrative Examples of Probability Density Functions

Certain families of probability density functions arise so often in statistical models that they have been given special names. A few examples of these functions are given in Table 8.6.1, together with their expected values and variances. Figs. 17 and 18 illustrate some of them.

Not all the names in Table 8.6.1 are already in current usage; in particular, the names 'pseudo-chi-squared variate' and 'dry-ice variate'

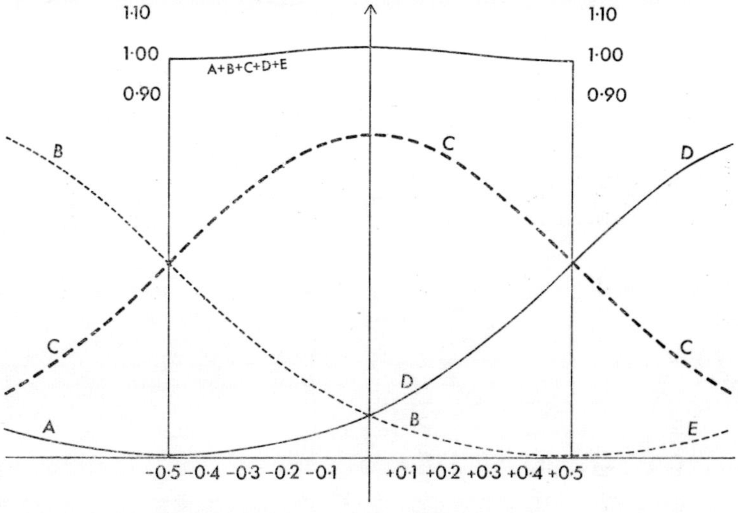

Fig. 17. Density function for rounding-off errors of $N\,(0, 0{\cdot}25)$ variate

TABLE 8.6.1

Name of variate	*Ref. symbol*	*Para-meters*	μ_1'	μ_2	*Probability density function*	
Rectangular	x_1	a, b	$\frac{a+b}{2}$	$\frac{(b-a)^2}{12}$	$f(x) = (b-a)^{-1} \quad a<x<b$	(8.6.1)
Normal	x_2	a, σ^2	a	σ^2	$f(x) = \frac{1}{\sqrt{(2\pi)}\sigma} e^{-(x-a)^2/2\sigma^2}$	(8.6.2)
Student's t	x_3	λ	0	$1+\frac{2}{\lambda-2}$	$f(x) = \frac{1}{B\left(\frac{1}{2},\frac{\lambda}{2}\right)} \frac{\lambda^{-\frac{1}{2}}}{\left(1+\frac{x^2}{\lambda}\right)^{\frac{\lambda+1}{2}}}$	(8.6.3)
Chi-squared (χ^2)	x_4	γ	γ	2γ	$f(x) = \frac{x^{\frac{1}{2}\gamma-1} e^{-\frac{1}{2}x}}{2^{\frac{1}{2}\gamma}\Gamma(\frac{1}{2}\gamma)} \quad : x>0$	(8.6.4)
Pseudo-chi-squared	x_5	γ, λ	$\frac{\gamma}{1-\frac{2}{\lambda}}$	$\frac{\left(1+\frac{\gamma}{\lambda-2}\right)}{\left(1-\frac{2}{\lambda}\right)\left(1-\frac{4}{\lambda}\right)}2\gamma$	$f(x) = \frac{\Gamma\left(\frac{\gamma+\lambda}{2}\right)}{\left(\frac{\lambda}{2}\right)^{\gamma/2}\Gamma\left(\frac{\lambda}{2}\right)} \frac{x^{\frac{1}{2}\gamma-1}\left(1+\frac{x}{\lambda}\right)^{-\frac{1}{2}\gamma-\frac{1}{2}\lambda}}{2^{\frac{1}{2}\gamma}\Gamma(\frac{1}{2}\gamma)} \quad : x>0$	(8.6.5)
F-ratio	x_6	γ, λ	$\frac{1}{1-\frac{2}{\lambda}}$	$2\frac{\frac{1}{\gamma}+\frac{1}{\lambda-2}}{\left(1-\frac{2}{\lambda}\right)\left(1-\frac{4}{\lambda}\right)}$	$f(x) = \frac{\gamma^{\frac{1}{2}\gamma}\lambda^{\frac{1}{2}\lambda}x^{\frac{1}{2}\gamma-1}}{B(\frac{1}{2}\gamma,\frac{1}{2}\lambda)\{\lambda+\gamma x\}^{\frac{1}{2}(\lambda+\gamma)}} \quad : x>0$	(8.6.6)
Beta-	x_7	m, n	$\frac{m}{m+n}$	$\frac{mn}{[(m+n)^2(m+n+1)]}$	$f(x) = \frac{x^{m-1}(1-x)^{n-1}}{B(m,n)} \quad : 0<x<1$	(8.6.7)
Gamma-	x_8	m	m	m	$f(x) = \frac{x^{m-1} e^{-x}}{\Gamma(m)} \quad : x>0$	(8.6.8)
Dry-ice	x_9	w	$1/w$	$1/w^2$	$f(x) = we^{-wx}$	(8.6.9)

are only newly introduced; the latter name has been derived from the work of Professor Joan Robinson.

$\Gamma(m)$ and $B(m, n)$ are defined as follows:

$$\Gamma(m) = \int_0^\infty x^{m-1}e^{-x}dx, \quad B(m, n) = \int_0^1 x^{m-1}(1-x)^{n-1}dx.$$

Standard results are that

(*i*) $\Gamma(n+1) = n! = 1.2.3.\ldots.n$, whenever n is a positive integer
(*ii*) $\Gamma(1/2) = \sqrt{\pi}$
(*iii*) $B(m, n) = \dfrac{\Gamma(m)\,\Gamma(n)}{\Gamma(m+n)}$

The extension of the definitions of moments to continuous variates may be found in § **8.19** below.

The nine types of variate set out in Table 8.6.1 are not all distinct. For example, the reader will easily verify that the dry-ice variate x_9 with $w = 1$, is the same as the Gamma variate x_8 with parameter $m = 1$. In Exercise 8.19 he will be asked to check the following further simple relationships between pairs of these variates:

(*i*) $2x_8 = x_4$ with $\gamma = 2m$ (between gamma and chi-squared)
(*ii*) $wx_9 = x_8$ with $m = 1$ (between dry-ice and gamma)
(*iii*) $(x_2-a)^2/\sigma^2 = x_4$ with $\gamma = 1$ (between normal and chi-squared)
(*iv*) $x_5 = \gamma x_6$ (between pseudo-chi-squared and F-ratio)
(*v*) $x_3^2 = x_6$ with $\gamma = 1$ (between Student's *t* and *F*-ratio)
(*vi*) $x_6 = \dfrac{\lambda}{\gamma}\dfrac{x}{1-x}$ with $m = \dfrac{\gamma}{2}$, $n = \dfrac{\lambda}{2}$ (between *F*-ratio and beta-)

Again, the density functions for certain of these variates may be regarded as the limits (as a parameter tends to infinity) of those of others. Thus, in exercises 8.20 and 8.21 the reader will be asked to check that in this sense:

(*i*) $x_3 \to x_2$ with $a = 0$ and $\sigma^2 = 1$ as $\lambda \to \infty$ (Student's *t* tends to normal);
(*ii*) $x_5 \to x_4$ as $\lambda \to \infty$ (pseudo-chi-squared tends to chi-squared);
(*iii*) $\gamma^{-1/2}(x_4-\gamma) \to x_1$, with $a = 0$, $\sigma^2 = 2$, as $\gamma \to \infty$ (chi-squared tends to normal).

We may indicate one or two of the very many contexts in which variates with approximately these probability density functions may arise.

Suppose that observations of a continuous variate are measured to the nearest unit. Then the 'error' or discrepancy between the true value and the rounded-off value may be regarded as approximately a rectangular variate, η $(-0{\cdot}5 < \eta < 0{\cdot}5)$, provided the probability density function satisfies certain not very stringent conditions such as that its quartiles are several units apart. Fig. 17 compares the probability density function for the rounding-off error of a normal variate of standard error 0·5, with rounding to the nearest unit, with that of a rectangular variate.

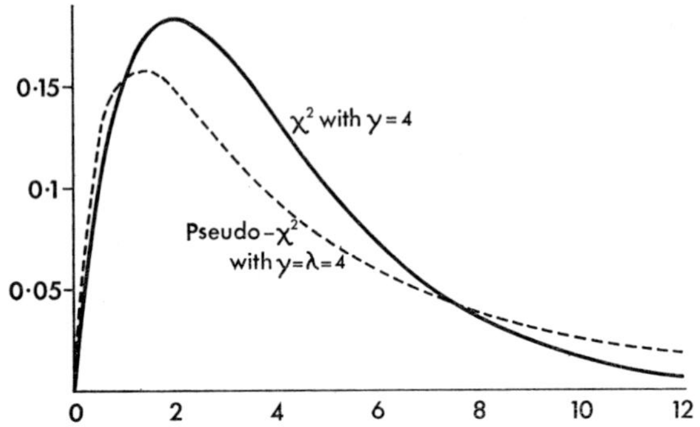

Fig. 18. Comparison of χ^2-density function with pseudo-χ^2-density function

It is always possible, and sometimes advantageous, to transform any continuous variate x with distribution function $F(x)$ to a rectangular variate y $(0 < y < 1)$, by the transformation $y = F(x)$. This is useful in establishing the properties of certain of the distribution-free tests for continuous variates to be described in Chapter 11 of Volume 2.

The normal variate with density function (8.6.2) derives its importance from the fact that for a wide class of statistics t_n based on a sample of size n, the distribution function for $[t_n - E(t_n)]/\sqrt{n}$ tends towards the distribution function of a normal variate of zero mean as n tends to infinity. For example, if T_n denotes the total of n independent observations of a variate x with finite variance σ^2, then $E(T_n) = nE(x)$ and the distribution function of $[T_n - nE(x)]/\sqrt{n}$ tends towards the distribution function of a normal variate of zero mean and variance σ^2. It will be explained in § **27** of this chapter how several of the other variates described in this §**8.6** can be derived by transformation from independent observations of normal variates with parameters (0, 1). In Chapters 11 and 13 we shall see how to construct robust significance tests and confidence intervals by making use of tables of independent observations of normal variates, or by making use of standard scores for ranked data, which are based on the properties of the normal density function.

The chi-squared distribution (8.6.4) derives its importance from the fact that it is possible to test a very wide class of hypotheses H, by constructing a statistical measure χ^2 of the discrepancy of an actual grouped frequency distribution of observations and the 'expected' grouped frequency distribution assuming the hypothesis H, in such a way that the measure χ^2 is approximately a chi-squared variate. On the basis of these measures one may construct chi-squared tests of the type to be described in Chapter 11 of Volume 2.

The F-distribution (8.6.6) derives its importance from the fact that under fairly wide conditions, if the observations are in fact independent of sets of factors X in an analysis of sums of squares (see § **6.24** and Appendix to Chapter 11) the F-ratio found for the set $\boldsymbol{X}$ may be regarded as being approximately an F-variate. (Under conditions of normal variation the F-ratio will be exactly an F-variate.)

The t-variates (8.6.3) are important because for many 'estimator' statistics, the ratio of the (unknown) actual error of estimate to the (known) estimate of the standard error of estimate is approximately a t-variate. Under conditions of normal variation the ratio will be a t-variate exactly.

The β density-function (8.6.7) arises in applications of Bayes' Theorem (see Chapter 9) to estimate the posterior probability density functions for parameter values. The β density function also transforms very simply into the F-density function which we have just discussed.

8.7. Transformation of Variates

Suppose that the continuous variate $\boldsymbol{y}$ is a monotonic differentiable function $y(x)$ of the continuous variate $\boldsymbol{x}$.

Let $F(x)$ and $G(y)$ denote the distribution functions of $\boldsymbol{x}$ and $\boldsymbol{y}$ respectively. Then, since either

$$\boldsymbol{r}(\boldsymbol{x} \leqslant x) = \boldsymbol{r}\{\boldsymbol{y} \leqslant y(x)\}$$

$$\text{or} \quad \boldsymbol{r}(\boldsymbol{x} \leqslant x) = \boldsymbol{r}\{\boldsymbol{y} \geqslant y(x)\}, \tag{8.7.1}$$

we know that either

$$F(x) = G\{y(x)\}$$

$$\text{or} \quad F(x) = 1 - G\{y(x)\}. \tag{8.7.2}$$

Differentiating (8.7.2) we obtain in either case

$$f(x) = g\{y(x)\} \mid dy/dx \mid , \tag{8.7.3}$$

where $f(x)$ and $g(y)$ are the density functions for $\boldsymbol{x}$ and $\boldsymbol{y}$ and where $| dy/dx |$ is the absolute value of dy/dx.

Similarly, where $x(y)$ denotes the inverse function of $y(x)$

$$g(y) = f\{x(y)\} \div | dy/dx |. \tag{8.7.4}$$

If $\boldsymbol{y}$ is *not* a monotonic function of $\boldsymbol{x}$, more care is needed.

Suppose, for example, that

$$\boldsymbol{y} = \boldsymbol{x}^2 \tag{8.7.5}$$

and that $\boldsymbol{x}$ has the density function

$$f(x) = (3/4)(1-x^2) \quad (-1 < x < 1). \tag{8.7.6}$$

Then the distribution function for $\boldsymbol{x}$ is

$$F(x) = (1/2)+(3x/4)-(x^3/4) \tag{8.7.7}$$

when $-1 \leqslant x \leqslant 1$.

Let

$$x(y) = +y^{1/2}; \tag{8.7.8}$$

then the distribution function for $\boldsymbol{y}$ is

$$G(y) = \boldsymbol{p'r}(\boldsymbol{y} \leqslant y) = F\{x(y)\}-F\{-x(y)\} \quad (y \geqslant 0) \tag{8.7.9}$$

$$\therefore \quad G(y) = 2F\{x(y)\}-1 \quad (y \geqslant 0). \tag{8.7.10}$$

By differentiation with respect to x, we find that in this case the density function for $\boldsymbol{y}$ is

$$g(y) = 2(d/dy)F\{x(y)\} = 2f\{x(y)\} \div | dy/dx | \quad (y > 0). \tag{8.7.11}$$

Comparing this with (8.7.3) we note that in (8.7.11) the factor 2 has been introduced because $\boldsymbol{x}$ is a two-valued function $(\pm \boldsymbol{y}^{1/2})$ of the variate $\boldsymbol{y}$. To find the probability density function for $\boldsymbol{y}$, we have to add together the probability densities due to these two values of $\boldsymbol{x}$.

The same principle applies to finding the probability density for $\boldsymbol{y}$ whenever $\boldsymbol{x}$ is a multi-valued function $x(y)$ of $\boldsymbol{y}$. In general

$$g(y) = \sum_{x = x(y)} [f\{x(y)\} \, | (d/dy)x(y) |]. \tag{8.7.12}$$

8.8. Expected Values of Continuous Variates

The expected value of a continuous variate $\boldsymbol{x}$ may be defined in terms of the density function $f(x)$, as

$$E(\boldsymbol{x}) = \int_{-\infty}^{\infty} f(x)x\,dx. \tag{8.8.1}$$

Now, (8.8.1) may be regarded as the limiting form of the equation (5.8.6) which gave the expected value of a discrete variate as

$$E(\boldsymbol{x}) = \sum_{\boldsymbol{x}} f(x)x. \tag{8.8.2}$$

As in the case of the sum (8.8.2) for discrete variates, it is not necessarily the case that the integral (8.8.1) will converge.

In the notation for Stieltjes integrals, both (8.8.1) and (8.8.2) may be expressed as

$$E(\boldsymbol{x}) = \int x dF, \tag{8.8.3}$$

where F is the distribution function for $\boldsymbol{x}$. Those who are conversant with this notation are likely to prefer this form of the definition.

In discussing the expected value of any continuous variate $\boldsymbol{x}$, we shall prefer the notation $E(\boldsymbol{x})$ to the notation $\boldsymbol{p}'\boldsymbol{x}$, since we cannot regard $\boldsymbol{p}'$ and $\boldsymbol{x}$ as vectors in this case.

8.9. Expected Values of Functions of Continuous Variates

Suppose that $\boldsymbol{x}$ is a function, possibly many-valued, $x(y)$, of the variate $\boldsymbol{y}$. Then the probability density for $\boldsymbol{y}$ is given by (8.7.12) as

$$g(y) = \sum_{\boldsymbol{x} = x(y)} [f\{x(y)\} \, . \, |(d/dy)x(y)|]. \tag{8.9.1}$$

Hence

$$E(y) = \int_{-\infty}^{\infty} \sum_{\boldsymbol{x} = x(y)} [f\{x(y)\} \, . \, |(d/dy)x(y)|] \, y dy. \tag{8.9.2}$$

$$\therefore \quad E(y) = \int_{-\infty}^{\infty} f(x) y(x) dx, \tag{8.9.3}$$

where $y(x)$ denotes the inverse function of $x(y)$, provided $\boldsymbol{y}$ is a single-valued function of $\boldsymbol{x}$. If $\boldsymbol{y}$ is not single-valued, we cannot find $E(y)$.

Equation (8.9.3) may be regarded as the limiting form of the equation (5.8.6) which gave the expected value of a function of a discrete variate.

In the notation of Stieltjes integrals (8.9.3) may be written as

$$E(y) = \int y dF, \tag{8.9.4}$$

where F is the distribution function for $\boldsymbol{x}$ and where $\boldsymbol{y}$ is a single-valued function of $\boldsymbol{x}$. This form covers a rather wider class of cases than the earlier forms.

The expected value of a function of two or more continuous variates is discussed in § **A2** of the Appendix to this Chapter.

8.10. Illustration

Consider the variate $\boldsymbol{y} = \boldsymbol{x}^2$ where $\boldsymbol{x}$ has the density function

$$f(x) = (2\pi)^{-1/2}\, e^{-(x^2/2)} \qquad (8.10.1)$$

In this case $\boldsymbol{x}$ is a normal variate (0, 1).
The density function for $\boldsymbol{y}$ is

$$g(y) = \{f(y^{1/2}) + f(-y^{1/2})\} \div |\, dy/dx\,|. \qquad (8.10.2)$$

Now
$$\frac{dy}{dx} = \pm 2y^{1/2} \qquad (y > 0) \qquad (8.10.3)$$

Hence
$$g(y) = (2\pi y)^{-1/2}\, e^{-y/2} \quad (y > 0) \qquad (8.10.4)$$

Also
$$\pi^{1/2} = \Gamma(1/2)$$

$$\therefore \quad g(y) = y^{-1/2} e^{-y/2} \,/\, 2^{1/2}\Gamma(1/2). \qquad (8.10.5)$$

Hence the density function (8.10.4) for $\boldsymbol{y}$ is identical with the χ^2-distribution (8.6.4) with parameter $v = 1$.

We see then that the square of a normal (0, 1) variate is a χ^2-variate with parameter $v = 1$.

The expected value of this variate $\boldsymbol{y}$ is

$$E(\boldsymbol{y}) = \int_0^\infty g(y)\,y\,dy = \int_{-\infty}^{\infty} f(x)\,x^2\,dx = \int_{-\infty}^{\infty} (2\pi)^{-1/2}\{xe^{-x^2/2}\}\,x\,dx$$

$$= (2\pi)^{-1/2} \left[-e^{-x^2/2}x\right]_{-\infty}^{\infty} + \int_{-\infty}^{\infty} f(x)\,dx$$

$$= \int_{-\infty}^{\infty} f(x)\,dx. \qquad (8.10.6)$$

Now, it is known that
$$\int_{-\infty}^{\infty} f(x)\,dx = 1; \qquad (8.10.7)$$

hence $E(\boldsymbol{y}) = 1$. (8.10.8)

Thus the expected value of a χ^2-variate with parameter one is unity. We could have obtained this result directly if we had appealed to the fact that the variance of a normal (0, 1) variate is unity.

8.11. Multivariate Density Functions

It would be unduly tedious to spell out in full detail the extension of the discussion of the last few sections to cover the case of pairs and sets of continuous variates. We need, however, to run briefly over the main points of such a discussion.

If $\boldsymbol{x}(t)$ $(t = 1, 2, \ldots, n)$ denotes a set of continuous variates, then the distribution function for the set is now a function of n variables $x(t)$ $(t = 1, 2, \ldots, n)$, and it may be written as

$$F\{x(1), x(2), \ldots, x(n)\} = \boldsymbol{p}'\left[\bigcap_{t=1}^{n} \boldsymbol{r}\{\boldsymbol{x}(t) \leqslant x(t)\}\right]. \tag{8.11.1}$$

It gives, just as in the case of discrete variates, the probability that no $\boldsymbol{x}(t)$ attains the value $x(t)$ or greater.

We know that when the $\boldsymbol{x}(t)$ are discrete variates, all derivatives of F, when they exist, must be zero. But we now allow the existence of sets of continuous variates, for which the function

$$f\{x(1), x(2), \ldots, x(n)\} = \{\partial^n/\partial x(1)\partial x(2) \ldots \partial x(n)\}F. \qquad (8.11.2$$

exists and is non-negative at all sets of values (apart possibly from extreme values) of the $x(t)$. The function $f\{x(1), x(2), \ldots, x(n)\}$ is called the probability density function for the set of variates $\boldsymbol{x}(1), \boldsymbol{x}(2), \ldots, \boldsymbol{x}(n)$. It must satisfy

$$\int_{-\infty}^{\infty} \cdots \int_{-\infty}^{\infty} f\{x(1) \ldots x(n)\}\, dx(1) \ldots dx(n) = 1, \tag{8.11.3}$$

in order that F may behave correctly when the $x(t)$ all tend to minus infinity or all tend to plus infinity.

The rule for transforming the probability function of a set of variates is similar to the rule (8.7.12) but involves a Jacobian in place of the term $|(d/dy)x(y)|$.

Let $\mathscr{X}$ and $\mathscr{Y}$ stand for two sets of continuous variates $\boldsymbol{x}(t)$, $\boldsymbol{y}(t)$ $(t = 1, 2, \ldots, n)$, and let each $\boldsymbol{x}(t)$ be a function, (possibly many-valued) of all the variates $\boldsymbol{y}(t)$, whereas the $\boldsymbol{y}(t)$ are each single-valued functions of all the $\boldsymbol{x}(t)$.

Let $|\partial \mathrm{X}/\partial \mathrm{Y}|$ denote the absolute value of the determinant of partial derivatives:

$$\boldsymbol{J} = \begin{vmatrix} \partial x(1)/\partial y(1) & \partial x(2)/\partial y(1) & \ldots & \partial x(n)/\partial y(1) \\ \partial x(1)/\partial y(2) & \partial x(2)/\partial y(2) & \ldots & \partial x(n)/\partial y(n) \\ \vdots & \vdots & & \vdots \\ \partial x(1)/\partial y(n) & \partial x(2)/\partial y(n) & \ldots & \partial x(n)/\partial y(n) \end{vmatrix} \tag{8.11.4}$$

then the density function for the variates $\mathscr{Y}$ may be expressed in terms of the density function for the variates $\mathscr{X}$ as

$$g(Y) = \sum_{X(Y) = X} f(X) \mid \partial X/\partial Y \mid \tag{8.11.5}$$

where X denotes $x(1), x(2), \ldots, x(n)$; and Y denotes $y(1), y(2), \ldots, y(n)$.

The expression (8.11.4) is a Jacobian and it follows from the properties of partial derivatives that such expressions obey the general rule:

$$\mid \partial X/\partial Y \mid . \mid \partial Y/\partial X \mid = 1. \tag{8.11.6}$$

Then (8.11.5) can also be written

$$g(Y) = \sum_{X(Y) = X} \{f(X) \div \mid \partial Y/\partial X \mid\} \tag{8.11.7}$$

For a full account of the use of Jacobians in the transformation of functions, the reader may consult Courant, *Differential and Integral Calculus*, Vol. 2, especially Chapter III, § 4-7 and Chapter IV, § 4.

8.12. Illustration. The multivariate normal distribution

$$f(X) = Ce^{-X'AX/2} \tag{8.12.1}$$

Here X stands for a column-vector with elements $x(t)$ $(t = 1, 2, \ldots, n)$ and A stands for a positive definite symmetrical $n \times n$ matrix of constants a_{tu}; $f(X)$ denotes the density function for n variates $\mathbf{x}(t)$ $(t = 1, 2, \ldots, n)$ at the point $\mathbf{x}(t) = x(t)$ $(t = 1, 2, \ldots, n)$, C is a constant chosen to ensure that $f(X)$ satisfies (8.11.3).

The value of the constant C may be found as follows. Since A is a positive definite symmetrical matrix, we can find an $n \times n$ matrix U such that

$$UU' = A. \tag{8.12.2}$$

Now let Y stand for a column vector with elements $y(t)$ $(t = 1, 2, \ldots, n)$ and, similarly, let the sets of variates $\mathbf{x}(t)$, $\mathbf{y}(t)$; $(t = 1, 2, \ldots, n)$ be represented by column vectors $\mathscr{X}, \mathscr{Y}$ of each of which they form the n elements.

Let Y be defined by

$$Y = UX, \tag{8.12.3}$$

and let the variates $\mathscr{Y}$ be similarly defined by

$$\mathscr{Y} = U\mathscr{X}. \tag{8.12.4}$$

The derivatives of the $y(t)$ with respect to the $x(t')$ are given by the elements, $U_{tt'}$, of $\boldsymbol{U}$;

$$\partial y(t)/\partial x(t') = U_{tt'} \tag{8.12.5}$$

so that the determinant $|\partial \boldsymbol{Y}/\partial \boldsymbol{X}|$ is simply the determinant $|\boldsymbol{U}|$ of the matrix $\boldsymbol{U}$.

The density function for the variates may now be obtained by means of (8.11.7) as

$$g(\boldsymbol{Y}) = f(\boldsymbol{X}) \div |\boldsymbol{U}|. \tag{8.12.6}$$

But by (8.12.2)

$$|\boldsymbol{U}| = |\boldsymbol{A}|^{1/2} \tag{8.12.7}$$

and by (8.12.1) (8.12.2) (8.12.3)

$$f(\boldsymbol{X}) = C\mathrm{e}^{-\boldsymbol{Y}'\boldsymbol{Y}/2} \tag{8.12.8}$$

Hence

$$g(\boldsymbol{Y}) = C\,|\boldsymbol{A}|^{-1/2}\mathrm{e}^{-\boldsymbol{Y}'\boldsymbol{Y}/2}. \tag{8.12.9}$$

But $\mathrm{e}^{-\boldsymbol{Y}'\boldsymbol{Y}/2}$ is simply the product of n functions $\mathrm{e}^{-\{y(t)\}^2/2}$ $(t = 1, 2, \ldots, n)$ and it is known (see example 22 at end of Chapter 8) that

$$\int_{-\infty}^{\infty} \mathrm{e}^{-y^2/2} dy = (2\pi)^{1/2}. \tag{8.12.10}$$

Hence

$$\int_{-\infty}^{\infty} \cdots \int_{\infty}^{\infty} g(\boldsymbol{Y})\,dy(1) \ldots dy(n) = C\{|\boldsymbol{A}|\}^{-1/2}(2\pi)^{\pi/2}. \tag{8.12.11}$$

Hence by the general property (8.11.3) of multivariate density functions

$$C\{|\boldsymbol{A}|\}^{-1/2}(2\pi)^{n/2} = 1. \tag{8.12.12}$$

Finally, then, we may evaluate the constant C as

$$C = \{|\boldsymbol{A}|\}^{1/2}(2\pi)^{-n/2}. \tag{8.12.13}$$

We also note the consequence, to be quoted in section 14 below, that by (8.12.9)

$$g(\boldsymbol{Y}) = \prod_{t=1}^{n} [(2\pi)^{-1/2}\,\mathrm{e}^{-\{y(t)\}^2/2}]. \tag{8.12.14}$$

8.13. Independence of Variates

We may take over from our models for events and discrete variates the general property of independence that a pair of variates $\boldsymbol{x}$, $\boldsymbol{y}$ are independent if for all real x, y their distribution function factorises as

$$F(x, y) = G(x)\, H(y), \tag{8.13.1}$$

where $G(x)$, $H(y)$ can then be chosen to be the marginal distribution functions for $\boldsymbol{x}$ and $\boldsymbol{y}$.

Similarly a set of variates $\boldsymbol{x}(t)$ $(t = 1, 2, \ldots, n)$ are independent, if for all $\mathbf{X} = x(t)$ $(t = 1, 2, \ldots, n)$ the distribution function factorises as

$$F(\mathbf{X}) = \prod_{t=1}^{n} G_t\{x(t)\}, \tag{8.13.2}$$

where the $G_t(x)$ can then be chosen so as to be the marginal distribution functions for $\boldsymbol{x}(t)$ $(t = 1, 2, \ldots, n)$ satisfying

$$G_t(x) = \lim_{\substack{x(t') \to \infty \\ \text{all } t' \neq t \\ \alpha(t) = x}} F(\mathbf{X}). \tag{8.13.3}$$

By partial differentiation and considerable care, it can be established that the definition (8.13.2) entails the formally similar property for the density functions. That is to say, the variates $\boldsymbol{x}(t)$ $(t = 1, 2, \ldots, n)$ are independent, if for all $\mathbf{X} = x(t)$ $(t = 1, 2, \ldots, n)$ the density functions factorise as

$$f(\mathbf{X}) = \prod_{t=1}^{n} g_t\{x(t)\}, \tag{8.13.4}$$

in which case $g_t(x)$ can be chosen to be the marginal density function for $\boldsymbol{x}(t)$ given by

$$g_t(x) = (\partial/\partial x)\, G_t(x). \tag{8.13.5}$$

8.14. Special Cases of Sets of Independent Variates

We saw in § **8.12** that, where

$$f(\mathbf{X}) = C e^{-\mathbf{X}'\mathbf{A}\mathbf{X}/2} \tag{8.14.1}$$

($\boldsymbol{A}$ positive definite and symmetric)

and
$$C = |\, \boldsymbol{A}\, |^{1/2} (2\pi)^{-n/2} \tag{8.14.2}$$

and $f(\mathbf{X})$ is the probability density function for variates

$$\mathscr{X} = \boldsymbol{x}(t) \quad (t = 1, 2, \ldots, n), \tag{8.14.3}$$

then we can find a transformation

$$\mathscr{Y} = \boldsymbol{U}\mathscr{X} \tag{8.14.4}$$

such that the variates

$$\mathscr{Y} = y(t) \quad (t = 1, 2, \ldots, n) \tag{8.14.5}$$

have the density function (8.12.14)

$$g(Y) = \prod_{t=1}^{n} \{(2\pi)^{-1/2} e^{-\{y(t)\}^2/2}\}. \tag{8.14.6}$$

Since $g(Y)$ factorises as the product of n individual density functions, we see that $y_1(1), \ldots, y(n)$ are n independent normal variates of zero mean and unit variance. Thus, whenever variates $x(1), x(2), \ldots, x(n)$ have a joint multivariate normal density function, we can find a linear transformation of type (8.14.4) to a set of n independent normal variates of zero mean and unit variance.

We will now consider an example concerning independent chi-squared variates.

EXAMPLE 8.14.1. x and y are independent chi-squared variates of parameters v_1 and v_2. Obtain the bivariate probability density function for (*i*) their sum z, (*ii*) u, the ratio of y to their sum z. Show that z and u are independent and hence show that z is a chi-squared variate of parameter (v_1+v_2) whereas u is a β-variate of parameters $(v_2/2, v_1/2)$.

Solution. From the data of the problem

$$z = x+y, \qquad u = x/(x+y), \tag{8.14.7}$$

where by (8.6.4) the variates x and y have probability density functions

$$\phi(x) = x^{(v_1/2)-1} e^{-x/2}/2^{v_1/2}\Gamma(v_1/2) \tag{8.14.8}$$

$$\psi(y) = y^{(v_2/2)-1} e^{-y/2}/2^{v_2/2}\Gamma(v_2/2) \tag{8.14.9}$$

and, in virtue of their independence, they have the bivariate probability density function

$$f(x, y) = \phi(x)\psi(y) = \frac{x^{(v_1/2)-1}\, y^{(v_2/2)-1}\, e^{-(x+y)/2}}{2^{(v_1+v_2)/2}\,\Gamma(v_1/2)\,\Gamma(v_2/2)}. \tag{8.14.10}$$

The bivariate probability density function for z and u must then be given by

$$g(z, u) = f(x\ y) \begin{vmatrix} \partial x/\partial z & \partial x/\partial u \\ \partial y/\partial z & \partial y/\partial u \end{vmatrix} \tag{8.14.11}$$

where x, y satisfy

$$x+y = z \qquad y/(x+y) = u. \tag{8.14.12}$$

Evidently,

$$y = u\,z; \quad x = (1-u)z, \tag{8.14.13}$$

so that

$$\begin{vmatrix} \partial x/\partial z & \partial x/\partial u \\ \partial y/\partial z & \partial y/\partial u \end{vmatrix} = \begin{vmatrix} 1-u & z \\ u & -z \end{vmatrix} = -z. \tag{8.14.14}$$

Hence by (8.14.10), (8.14.11), (8.14.13), (8.14.14),

$$g(z, u) = \left\{z^{\frac{v_1+v_2}{2}-1} e^{-z/2} \Big/ 2^{\frac{v_1+v_2}{2}} \Gamma\left(\frac{v_1+v_2}{2}\right)\right\}$$

$$\left\{\Gamma\left(\frac{v_1+v_2}{2}\right) u^{\frac{v_2}{2}-1} (1-u)^{\frac{v_1}{2}-1} \Big/ \Gamma\left(\frac{v_1}{2}\right)\Gamma\left(\frac{v_2}{2}\right)\right\}. \tag{8.14.15}$$

But (8.14.15) factorises to the form

$$g(z, u) = h(z)\, j(u) \tag{8.14.16}$$

where

$$h(z) = z^{\frac{1}{2}(v_1+v_2)} e^{-\frac{1}{2}z} / [2^{\frac{1}{2}(v_1+v_2)} \Gamma\{\tfrac{1}{2}(v_1+v_2)\}] \tag{8.14.17}$$

$$j(u) = \left\{\Gamma\left(\frac{v_1+v_2}{2}\right) \Big/ \Gamma\left(\frac{v_1}{2}\right)\Gamma\left(\frac{v_2}{2}\right)\right\} u^{\frac{v_2}{2}-1} (1-u)^{\frac{v_1}{2}-1}. \tag{8.14.18}$$

We see that $\boldsymbol{z}$ and $\boldsymbol{u}$ are independent variates: moreover on reference to Table 8.6.1 we see that $\boldsymbol{z}$ is a χ^2-variate with parameter v_1+v_2 and $\boldsymbol{u}$ is a beta variate with parameters $v_2/2$, $v_1/2$.

8.15. Properties of Expected Values of Functions of Continuous Variates

In our model for discrete variates many important properties of expected variates are written down without proof because they correspond to well-known properties of vectors. For examples, if a_j are constants and $\boldsymbol{x}(j)$ are discrete variates ($j = 1, 2, \ldots, n$), then

$$E\left\{\sum_{j=1}^{n} a_j \boldsymbol{x}(j)\right\} = \boldsymbol{p}' \sum_{j=1}^{n} a_j \boldsymbol{x}(j) = \sum_{j=1}^{n} a_j \boldsymbol{p}'\boldsymbol{x}(j) = \sum_{j=1}^{n} a_j E\{\boldsymbol{x}(j)\}, \tag{8.15.1}$$

and

$$\operatorname{var}\left\{\sum_{j=1}^{n} a_j \boldsymbol{x}(j)\right\} = \sum_{j=1}^{n} \boldsymbol{p}' a_j \widehat{\{\boldsymbol{x}(j) \cdot \mathbf{1}\}} \sum_{j=1}^{n} a_k \{\boldsymbol{x}(k) \cdot \mathbf{1}\}$$

$$= \sum_{j=1}^{n} \sum_{k=1}^{n} a_j a_k \operatorname{cov}\{\boldsymbol{x}(j), \boldsymbol{x}(k)\}$$

$$\therefore \operatorname{var}\left\{\sum_{j=1}^{n} a_j \boldsymbol{x}(j)\right\} = \sum_{j=1}^{n} [a_j^2 \operatorname{var}\{\boldsymbol{x}(j)\} + 2 \sum_{k=j+1}^{n} a_j a_k \operatorname{cov}\{x(j), \boldsymbol{x}(k)\}]. \tag{8.15.2}$$

Provided the right-hand side of (8.15.1) exists, the equality of the L.H.S. and the R.H.S. still holds for continuous variates. But the proof is no longer so straightforward; it is given in the Appendix to this Chapter. In the same appendix a proof is given that formula (8.15.2) for var $\left\{\sum_{j=1}^{n} a_j \mathbf{x}(j)\right\}$ holds also for continuous variates, provided the right-hand side exists.

Another important result concerning the expected values of discrete variates was that where $\mathbf{x}$ and $\mathbf{y}$ were independent variates and where,

$$\mathbf{z} = \hat{\mathbf{x}}\mathbf{y} \tag{8.15.3}$$

then, (5.15.4)

$$E(\mathbf{z}) = E(\mathbf{x}) \,.\, E(\mathbf{y}) \tag{8.15.4}$$

as proved in § **5.15**.

The same result holds for independent continuous variates since, if $\mathbf{x}$, $\mathbf{y}$ are independent, their joint probability density function factorises as

$$f(x, y) = g(x)\, h(y),$$

where $g(x)\, h(y)$ are the marginal probability density functions for $\mathbf{x}$ and $\mathbf{y}$, so that where $\mathbf{z} = \hat{\mathbf{x}}\mathbf{y}$

$$E(\mathbf{z}) = \int_{-\infty}^{\infty} \int_{-\infty}^{\infty} f(x, y) xy\, dx\, dy = \int_{-\infty}^{\infty} \int_{-\infty}^{\infty} g(x) h(y) xy\, dx\, dy$$

$$= \int_{-\infty}^{\infty} g(x) x dx \int_{-\infty}^{\infty} h(y) y dy, \tag{8.15.5}$$

provided these integrals converge.

Hence if $\mathbf{z}$ is the product of two independent continuous variates $\mathbf{x}$ and $\mathbf{y}$

$$E(\mathbf{z}) = E(\mathbf{x})\, E(\mathbf{y}). \tag{8.15.6}$$

8.16. Conditional Distribution Functions

If $\mathbf{x}$ and $\mathbf{y}$ are discrete variates, we have seen in (5.16.5) that the conditional distribution function for $\mathbf{y}$, given that $\mathbf{x} \leqslant x$ is

$$F(y \mid \mathbf{x} \leqslant x) = \mathbf{p}'_{\mid \mathbf{r}(\mathbf{x} \leqslant x)}\, \mathbf{r}(\mathbf{y} \leqslant y). \tag{8.16.1}$$

This may be expanded by (5.13.1) to give

$$F(y \mid \mathbf{x} \leqslant x) = \mathbf{p}'\hat{\mathbf{r}}(\mathbf{x} \leqslant x)\mathbf{r}(\mathbf{y} \leqslant y)/\mathbf{p}'\mathbf{r}(\mathbf{x} \leqslant x), \tag{8.16.2}$$

which reduces to

$$F(y \mid \mathbf{x} \leqslant x) = F(x, y)/G(x), \tag{8.16.3}$$

where $F(x, y)$ is the bivariate distribution function and $G(x)$ is the marginal distribution function for $\boldsymbol{x}$.

In the case of continuous variates $\boldsymbol{x}$ and $\boldsymbol{y}$ we may use (8.16.3) as a *definition* of the conditional distribution function $F(y \mid \boldsymbol{x} \leqslant x)$.

Similarly, as a *definition* of the conditional distribution function for $\boldsymbol{y}$, given that $x-\varepsilon<\boldsymbol{x} \leqslant x$, we may adopt

$$F(y \mid x-\varepsilon<\boldsymbol{x} \leqslant x) = \boldsymbol{p'r}\{(\boldsymbol{y} \leqslant y) \text{ and } (x-\varepsilon<\boldsymbol{x} \leqslant x)\}/\boldsymbol{p'r}\{x-\varepsilon<\boldsymbol{x} \leqslant x\}, \quad (\varepsilon>0)$$

which reduces to

$$F\{y \mid x-\varepsilon<\boldsymbol{x} \leqslant x\} = \{F(x, y)-F(x-\varepsilon, y)\}/\{G(x)-G(x-\varepsilon)\} \quad (\varepsilon>0) \tag{8.16.4}$$

If we are dealing with a pair of continuous variates we may now let $\varepsilon \to 0$ $(\varepsilon>0)$ and obtain the conditional distribution function for $\boldsymbol{y}$, given $\boldsymbol{x} = x$, as

$$F(y \mid \boldsymbol{x}=x) = \left\{\frac{\partial}{\partial x} F(x, y)\right\} \Big/ g(x), \tag{8.16.5}$$

provided $g(x)>0$, where $g(x)$ is the marginal probability density function for $\boldsymbol{x}$.

8.17. Conditional Probability Density Functions

The conditional probability density function for $\boldsymbol{y}$ given $\boldsymbol{x} = x$ may be derived immediately by differentiation (8.16.5) with respect to y.

Thus the conditional probability density function for $\boldsymbol{y}$, given $\boldsymbol{x} = x$, is by definition

$$f\{y \mid \boldsymbol{x}=x\} = (\partial/\partial y) F\{y \mid \boldsymbol{x}=x\}. \tag{8.17.1}$$

It then follows from (8.16.5) and (8.11.2) that

$$f\{y \mid \boldsymbol{x}=x\} = f(x, y)/g(x), \tag{8.17.2}$$

where $f(x, y)$ is the bivariate probability density function and $g(x)$ is the marginal probability density function for $\boldsymbol{x}$.

These concepts may readily be extended from the case of two continuous variates to that of more than two continuous variates.

8.18. Orthogonality and Corrected Variates: Variance and Covariance

Provided we interpret **1** as a variate which always equals 1, the definition D. 26 of §**5.18** may be extended without modification to cover the case of continuous variates; §**5.18** on deviations applies as well to continuous as to discrete variates. The main part of D. 32, the definition of orthogonality, still holds for continuous variates: so does D. 33, the definition of corrected variates, if for $\boldsymbol{x}'\hat{\boldsymbol{p}}(\boldsymbol{y}.\boldsymbol{x})$ we substitute $E\{\hat{\boldsymbol{x}}(\boldsymbol{y}.\boldsymbol{x})\}$. The definitions of variance and covariance (§**5.19**) and of the variance-covariance matrix §**5.22** need no modification, but the expressions (5.20.5 and 5.20.3) must involve integrals in place of sums if they are to apply to continuous variates. They become

$$\text{var}\,\boldsymbol{y} = E\{\boldsymbol{y}-E(y)\}^2 = \int_{-\infty}^{\infty} h(y)y^2dy-\{E(y)\}^2, \tag{8.18.1}$$

where $$E(\boldsymbol{y}) = \int_{-\infty}^{\infty} h(y)ydy$$

and $h(y)$ is the probability density function for $\boldsymbol{y}$.

Similarly

$$\text{cov}\,(\boldsymbol{x}, \boldsymbol{y}) = \int_{-\infty}^{\infty}\int_{-\infty}^{\infty} f(x, y)xy\,dy\,dx-E(\boldsymbol{x})E(\boldsymbol{y}) \tag{8.18.2}$$

where $$E(\boldsymbol{x}) = \int_{-\infty}^{\infty} g(x)x\,dx$$

$$E(\boldsymbol{y}) = \int_{-\infty}^{\infty} h(y)y\,dy \tag{8.18.3}$$

and $f(x, y)$, $g(x)$ and $h(y)$ are the probability density functions for $(\boldsymbol{x}, \boldsymbol{y})$, $\boldsymbol{x}$ and $\boldsymbol{y}$.

8.19. Definitions of Moments Extended to Cover the Case of Continuous Variates

The definitions D. 36, D. 37 of § **5.26** need no modification.

But in the expressions (5.26.1) and (5.26.2) we must use integrals instead of sums when we refer to continuous variates. In this case

$$\mu_j' = E(\boldsymbol{x}^j) = \int_{-\infty}^{\infty} f(x)x^j\,dx \tag{8.19.1}$$

$$\mu_j = E\{(\boldsymbol{x}\cdot\boldsymbol{1})^j\} = \int_{-\infty}^{\infty} f(x)\{x-E(\boldsymbol{x})\}^j\,dx \tag{8.19.2}$$

$$\mu'_{jk} = E\{\hat{\boldsymbol{x}}^j\boldsymbol{y}^k\} = \int_{-\infty}^{\infty}\int_{-\infty}^{\infty} f(x, y)x^jy^k\,dy\,dx \tag{8.19.3}$$

$$\mu_{jk} = E\{\widehat{(\boldsymbol{x}\cdot\boldsymbol{1})^j(\boldsymbol{y}}\cdot\boldsymbol{1})^k\} = \int_{-\infty}^{\infty}\int_{-\infty}^{\infty} f(x, y)\{x-E(\boldsymbol{x})\}^j\{y-E(\boldsymbol{y})\}^k dy\,dx \tag{8.19.4}$$

8.20. Sets of Observations of Continuous Chance Variates

If a set of independent discrete variates $\boldsymbol{z}(j)$ $(j = 1, 2, \ldots, n)$, all had the same probability function $f(x)$, we were able to regard them (D. 25, § **5.17**), as independent observations of a chance variate $\boldsymbol{x}$ with this same probability function $f(x)$. In the case of continuous variates we may similarly regard any set of independent continuous variates $\boldsymbol{z}(j)$ $(j = 1, 2, \ldots, n)$ all with the same probability density function, as being independent observations of a continuous variate $\boldsymbol{x}$ with this density function. We accordingly adopt the definition

D. 54 The set of continuous variates $\boldsymbol{x}(t)$ $(t = 1, 2, \ldots, m)$ constitute m observations of a *chance variate* $\boldsymbol{x}$ with probability density function $g(x)$, if the $\boldsymbol{x}(t)$ $(t = 1, 2, \ldots, m)$ are an independent set and each has the probability density function $g(x)$.

The probability density that the m observations of a chance variate $\boldsymbol{x}$ will take the specified values $x(t)$ $(t = 1, 2, \ldots, m)$ is $\prod_{t=1}^{m} g\{x(t)\}$, where $g(x)$ is the probability density function for $\boldsymbol{x}$.

8.21. Moment-generating Functions and Characteristic Functions

In § **6.25** the moment-generating function for a set of measurements x_t $(t = 1, 2, \ldots, n)$ was defined (D. 49) as

$$M'(\alpha) = \frac{1}{n}\sum_{t=1}^{n} e^{x_t\alpha} = \sum_{r=0}^{\infty} m_r' \frac{\alpha^r}{r!} \tag{8.21.1}$$

where m_r' denoted the r-th moment about the origin. $M'(\alpha)$ was thus the arithmetic mean value of $e^{x_t\alpha}$ taken over the n measurements x_t.

In § **5.27**, we had considered the corresponding formula (5.27.5) giving the moment-generating function for a discrete variate $\boldsymbol{x}$. It was mentioned that the series involved in this formula might fail to converge for some discrete variates. We may adapt the same formula (5.27.5) to provide a definition of the moment-generating-function for a continuous variate $\boldsymbol{x}$, as well as for a discrete variate:

D. 55 The *moment generating function* for a variate $\boldsymbol{x}$ with probability function or with probability density function $f(x)$, is given by

$$\begin{aligned} \mathrm{M}'(\alpha) = E(e^{\alpha x}) &= \sum_x f(x)e^{\alpha x} \text{ for a discrete variate} \\ &= \int_{-\infty}^{\infty} f(x)e^{\alpha x}dx \text{ for a continuous variate.} \end{aligned} \tag{8.21.2}$$

In either case, provided the moments about the origin are finite,

$$M'(\alpha) = \sum_{r=0}^{\infty} \mu_r' \frac{\alpha^r}{r!}. \tag{8.21.3}$$

For some purposes it is convenient to consider, instead, the moment-generating function about the mean. This may be defined as follows:

D. 55*a* If $\boldsymbol{x}$ is any variate with moment generating function $M'(\alpha)$ and with expected value μ_1', the *moment generating function about the mean* is given by

$$M(\alpha) = e^{-\alpha\mu_1'}M'(\alpha) = E(e^{\alpha(x-\mu_1')}).$$

This moment-generating function can, if the moments about the mean are finite be expanded as

$$M(\alpha) = \sum_{r=0}^{\infty} \mu_r \frac{\alpha^r}{r!}. \tag{8.21.4}$$

As in the case of moment-generating functions for discrete variates, the integrals for those of continuous variates may fail to converge.

EXAMPLE 8.21.1. If $\boldsymbol{x}$ is the Cauchy variate with probability density function $f(x) = \dfrac{1}{(1+x^2)\pi}$ $(-\infty < x < \infty)$, show that it has no finite moments except μ_0' and that except in the trivial case, $\alpha = 0$, the integral for the moment generating function fails to converge. Is all this true also for the truncated variate $\boldsymbol{x}'$ with probability function

$$g(x) = \frac{2}{(1+x^2)\pi} \qquad 0 < x < \infty\,?$$

Solution. The integral for μ_1' is $\displaystyle\int_{-\infty}^{\infty} xf(x)xdx = \frac{1}{2\pi} \lim_{x\to\infty} \log \frac{1+k^2x^2}{1+x^2}$ for any constant k, and this clearly is indeterminate: hence, the integral for μ_1' does not converge to a unique limit. Similarly no integral for any higher moment converges.

The integral for $M'(\alpha)$ certainly fails to converge unless $\alpha = 0$, since, for all values of x greater than $\dfrac{4}{\alpha^2}$, $e^{\alpha x}+e^{-\alpha x} > 1+x^2$.

The case of the truncated variate is slightly different.

Again, the moments do not exist, but the expression for the moment-generating function converges for all negative value of α. This follows from the fact that when α is negative and x is positive, $e^{\alpha x} < 1$.

If we consider not the moment generating function $E(e^{\alpha x})$ but instead the function $E(e^{i\alpha x})$ where i denotes the square root of minus one, the integral or sum for this function converges for all variates $\boldsymbol{x}$. This follows from the facts that

$$\left| \int_{-\infty}^{\infty} z(x) f(x)\, dx \right| \leqslant \int_{-\infty}^{\infty} | z(x) | f(x)\, dx, \tag{8.21.5}$$

and that

$$\left| \sum_x z(x) f(x) \right| \leqslant \sum_x | z(x) | f(x), \tag{8.21.6}$$

where $z(x)$ is any real or complex function of x and where $f(x)$ is any real non-negative function of x.

The function

$$\psi(\alpha; \boldsymbol{x}) = E(e^{i\alpha x}) = \int_{-\infty}^{\infty} f(x) e^{i\alpha x} dx \tag{8.21.7}$$

or the equivalent sum in the case of a discrete variate $\boldsymbol{x}$, is known as the *characteristic function* for $\boldsymbol{x}$.

If the moments and cumulants exist

$$\psi(\alpha; \boldsymbol{x}) = \sum_{r=0}^{\infty} \mu_r' \frac{(i\alpha)^r}{r!} = e^{i\alpha\mu_1'} \sum_{r=0}^{\infty} \mu_r \frac{(i\alpha)^r}{r!} \tag{8.21.8}$$

and

$$\log \psi(\alpha; \boldsymbol{x}) = \sum_{r=1}^{\infty} k_r \frac{(i\alpha)^r}{r!} \tag{8.21.9}$$

where μ_r', μ_r are the moments about the origin and about the mean, and where the k_r are the cumulants for $\boldsymbol{x}$.

It is often convenient to use the term cumulant function to refer to the function (8.21.9) instead of to the function (6.26.1).

8.22. The Cumulant Function for the Sum of Two Independent Variates

If $\boldsymbol{x}$ and $\boldsymbol{y}$ are any two independent variates, then any functions of them, $\xi(\boldsymbol{x}), \eta(\boldsymbol{y})$ will also be independent variates. It was shown in § **8.15** that for continuous variates, as for discrete variates, if $\boldsymbol{\xi}, \boldsymbol{\eta}$ are independent then

$$E(\boldsymbol{\xi\eta}) = E(\boldsymbol{\xi}) E(\boldsymbol{\eta}). \tag{8.22.1}$$

It follows immediately that the expected value of the product $e^{\alpha x} e^{\alpha y}$, where $\boldsymbol{x}$ and $\boldsymbol{y}$ are independent, is the product of the expected values

$$E(e^{\alpha x}) \text{ and } E(e^{\alpha y}), \tag{8.22.2}$$

provided these are finite.

Now let

$$z = x + y \tag{8.22.3}$$

then it follows from (8.22.2) that

$$E(e^{\alpha z}) = E(e^{\alpha x})E(e^{\alpha y}), \tag{8.22.4}$$

and hence that

$$\log E(e^{\alpha z}) = \log E(e^{\alpha x}) + \log E(e^{\alpha y}). \tag{8.22.5}$$

But the three expressions in (8.22.5) are simply the cumulant functions for z, x and y. Thus we see that the cumulant function for z, the sum of two *independent* variates x and y, is the sum of the cumulant functions for x and y provided these converge. (8.22.6)

By equating coefficients in the expansions of these cumulant functions where they do converge, it follows further that the cumulants for a variate z, which is the sum of two independent variates x and y, may be found by adding together the corresponding cumulants for x and for y.

8.23. The Reproductive Property of Certain Families of Variates

We saw in the example in § **8.14** that, if x and y were independent χ^2-variates with parameters v_1 and v_2, then their sum z must be a χ^2-variate with parameter $v_1 + v_2$.

Let us consider this fact from another angle. If x is a χ^2-variate with parameter v_1, then its moment-generating function is

$$M'(\alpha; x) = \int_0^\infty f(x)e^{+\alpha x}dx \qquad (\alpha < 0), \tag{8.23.1}$$

where

$$f(x) = x^{(v_1/2)-1}e^{-x/2} \,/\, 2^{v_1/2}\Gamma(v_1/2). \tag{8.23.2}$$

Putting

$$u = (1-2\alpha)x, \tag{8.23.3}$$

it is easily shown that for $0 > \alpha$

$$M'(\alpha, x) = (1-2\alpha)^{-v_1/2}\int_0^\infty f(u)\,du = (1-2\alpha)^{-v_1/2}. \tag{8.23.4}$$

Hence the cumulant function for x is

$$K(\alpha; x) = \log M'(\alpha, x) = -\frac{v_1}{2}\log(1-2\alpha) \tag{8.23.5}$$

and similarly the cumulant function for $\mathbf{y}$ is.

$$K(\alpha;\, \mathbf{y}) = -\frac{v_2}{2}\log(1-2\alpha). \tag{8.23.6}$$

Provided that $\mathbf{x}$ and $\mathbf{y}$ are independent, the cumulant function for their sum $\mathbf{z}$ is obtained by adding together their cumulant functions, so that

$$K(\alpha;\, \mathbf{z}) = -\frac{v_1+v_2}{2}\log(1-2\alpha). \tag{8.23.7}$$

The expression (8.23.7) is the cumulant function for a χ^2-variate with parameter (v_1+v_2). This result thus squares with the conclusion reached in § **8.14** that the sum of two independent χ^2-variates with parameters v_1 and v_2 is itself a χ^2-variate with parameter (v_1+v_2).

Other similar results which may be checked by the same method are the following:

The sum of two independent Poisson variates with parameters v_1, v_2, is itself a Poisson variate with parameter (v_1+v_2).

The sum of two independent binomial variates with parameters (π, n_1) and (π, n_2) is itself a binomial variate with parameter (π, n_1+n_2).

The sum of two independent normal variates with parameters (a_1, σ_1^2) and (a_2, σ_2^2) is itself a normal variate with parameters $(a_1+a_2, \sigma_1^2+\sigma_2^2)$.

In saying that these results could be *checked* by this method, the word *proved* had to be avoided. The fact, for example, that the cumulant function for the sum of two independent Poisson variates with parameters v_1 and v_2 is the same as that for a Poisson variate with parameter (v_1+v_2) does not *by itself* constitute a proof that the sum of these two independent Poisson variates is, in fact, a Poisson variate with parameter (v_1+v_2): for we have not yet excluded the possibility that the sum is some *other* variate which has the same cumulant function but not the same probability function as a Poisson variate. However, it has in fact been established that

THEOREM 8.23.1. *The Uniqueness Theorem.* If two variates $\mathbf{x}$ and $\mathbf{y}$ have the same cumulant function and hence the same characteristic function, then they also have the same distribution function.

For a full discussion of this theorem see Chapter 10 of *Mathematical Methods of Statistics*, by H. Cramer.

With the aid of the Uniqueness Theorem 8.23.1, it *is* possible to establish, through a consideration of their cumulant functions, the reproductive properties given above for chi-squared, Poisson, binomial and normal variates.

8.24. The Cumulant Function for a Linear Function of Independent Variates

Suppose that the variate x has the cumulant function $K(\alpha, x)$ and that

$$y = x\beta, \tag{8.24.1}$$

where β is a constant. Then the cumulant function for y must satisfy

$$K(\alpha; y) = K(\alpha\beta; x)$$

since

$$K(\alpha; y) = \log E(e^{i\alpha y}) = \log E(e^{i\alpha\beta x}). \tag{8.24.2}$$

By repeated application of the result (8.22.6) it then follows that if $x(t)$ $(t = 1, 2, \ldots, n)$ are a set of n independent variates, and if

$$z = \sum_{t=1}^{n} \beta_t x(t) \tag{8.24.3}$$

then the cumulant function for z satisfies

$$K(\alpha; z) = \sum_{t=1}^{n} K\{\alpha\beta_t; x(t)\}. \tag{8.24.4}$$

Now let $k_r(z)$; $k_r\{x(t)\}$ $(t = 1, 2, \ldots, n)$, denote the r-th cumulants for z and for the $x(t)$; then by equating coefficients of α^r in (8.24.4) we discover that

$$k_r(z) = \sum_{t=1}^{n} \beta_t^r k_r\{x(t)\}. \tag{8.24.5}$$

One case of particular interest is the one where $x(t)$ $(t = 1, 2, \ldots, n)$ are independent observations of a chance variate x and where

$$z = n^{-1} \sum_{t=1}^{n} x(t) \tag{8.24.6}$$

is the arithmetic mean.

The cumulant function for z is then

$$K(\alpha; z) = nK(\alpha/n; x) \tag{8.24.7}$$

and the r-th cumulant for z is

$$k_r(z) = n^{1-r} k_r(x). \tag{8.24.8}$$

In particular the expected value of z equals the expected value of x whereas the variance of z is to be obtained by dividing the variance of x by n.

8.25. The Expected Value of the Variance of a Set of Observations

We have seen, in §**7.6**, that the expected value of the variance of a sample of n members taken at random with replacement from a population of variance σ^2 is $\frac{n-1}{n}\sigma^2$. This is a special case of the general result that the expected value of the variance of n independent observations of a chance variate $\boldsymbol{x}$ is $\frac{n-1}{n}$ var $\boldsymbol{x}$.

We may choose our origin of measurement so that $E(\boldsymbol{x}) = 0$. Then let $\boldsymbol{z}_1, \boldsymbol{z}_2, \ldots, \boldsymbol{z}_n$ denote the n independent observations of $\boldsymbol{x}$, and let $\bar{\boldsymbol{z}}$ and $\boldsymbol{v}$ denote their arithmetic mean and variance. Then

$$E(\boldsymbol{z}_i) = 0 \quad (i = 1, 2, \ldots, n) \tag{8.25.1}$$

$$E(\boldsymbol{z}_i^2) = \text{var}\,(\boldsymbol{z}_i) = \text{var}\,\boldsymbol{x} \quad (i = 1, 2, \ldots, n) \tag{8.25.2}$$

$$E(\bar{\boldsymbol{z}}) = 0, \quad \text{and var}\,\bar{\boldsymbol{z}} = \frac{1}{n}\,\text{var}\,\boldsymbol{x} \text{ by (8.24.7)} \tag{8.25.3}$$

$$n\boldsymbol{v} = \sum_{i=1}^{n} (\boldsymbol{z}_i - \bar{\boldsymbol{z}})^2 = \sum_{i=1}^{n} \boldsymbol{z}_i^2 - n\bar{\boldsymbol{z}}^2$$

$$\therefore \quad nE(\boldsymbol{v}) = \sum_{i=1}^{n} \{E(\boldsymbol{z}_i^2)\} - n\,\text{var}\,\bar{\boldsymbol{z}}. \tag{8.25.4}$$

Hence, by (8.25.2), (8.25.3) and (8.25.4)

$$E(\boldsymbol{v}) = \frac{n-1}{n}\,\text{var}\,\boldsymbol{x}. \tag{8.25.5}$$

8.26. Linear Functions of Independent Normal Variates

We have seen (8.24.3), (8.24.5) that if $\boldsymbol{x}(1), \boldsymbol{x}(2), \ldots, \boldsymbol{x}(n)$ are independent and

$$\boldsymbol{z} = \sum_{t=1}^{n} \beta_t \boldsymbol{x}(t) \tag{8.26.1}$$

then the cumulants for $\boldsymbol{z}$ are given in terms of the cumulants for $\boldsymbol{x}(1), \boldsymbol{x}(2), \ldots, \boldsymbol{x}(n)$ by

$$k_r(\boldsymbol{z}) = \sum_{t=1}^{n} \beta_t^r k_r\{\boldsymbol{x}(t)\}. \tag{8.26.2}$$

Hence, if $\boldsymbol{x}(1), \boldsymbol{x}(2), \ldots, \boldsymbol{x}(n)$ are normal variates so that

$$k_r\{\boldsymbol{x}(t)\} = 0 \text{ for } r > 2 \text{ and } t = 1, 2, \ldots, n, \tag{8.26.3}$$

then

$$k_r(z) = 0 \text{ for } r > 2; \tag{8.26.4}$$

which implies that z is a normal variate also. The expected value and variance of z are given as $k_1(z)$ and $k_2(z)$ by (8.26.2) in terms of the expected values and variances of the $x(t)$.

8.27. The Family of Variates Related to the Normal Variates

Many of the variates described in §**8.6** can be regarded as functions of independent normal variates of zero mean and unit variance. To avoid waste of words, we shall refer to such normal variates as independent normal (0,1) variates.

In the example of § **8.10**, for instance, we showed that a chi-squared variate with parameter unity may be regarded as the square of a normal (0,1) variate. In § **8.14** we showed that the sum of two independent chi-squared variates with parameters v_1 and v_2 is itself a chi-squared variate with parameter (v_1+v_2). It follows that a chi-squared variate with parameter 2 may be regarded as the sum of two independent chi-squared variates each with parameter unity. By repeated application, a chi-squared variate with parameter m may be regarded as the sum of m independent chi-squared variates each with parameter unity. Hence, also, a chi-squared variate with parameter m may be regarded as the sum of squares of m independent normal (0, 1) variates.

It was also shown in § **8.14** that the ratio of a chi-squared variate with parameter v_2 to the sum of itself and an independent chi-squared variate with parameter v_1 is a β-variate with parameters $(v_2/2, v_1/2)$. Accordingly, where γ_1, γ_2 are integers, we may regard y, a β-variate with parameters $(\gamma_1/2, \gamma_2/2)$ as given by

$$y = \frac{x_1^2+x_2^2+\ldots+x_{\gamma_1}^2}{x_1^2+x_2^2+\ldots+x_{\gamma_1+\gamma_2}^2}, \tag{8.27.1}$$

where $x_1, x_2, \ldots, x_{\gamma_1+\gamma_2}$ are independent normal (0,1) variates.

If y is a β-variate with parameters $(m/2, n/2)$, and if

$$z = ny \,/\, m(1-y), \tag{8.27.2}$$

then z is an F-variate with parameters (m, n). The proof runs as follows.

Since

$$z = ny \,/\, m(1-y),$$

$$y = mz \,/\, (n+mz). \tag{8.27.3}$$

The probability density function for y is, by (8.6.6)

$$f(y) = \Gamma\left(\frac{m+n}{2}\right) y^{(m-2)/2}(1-y)^{(n-2)/2} \Big/ \Gamma\left(\frac{m}{2}\right)\Gamma\left(\frac{n}{2}\right). \tag{8.27.4}$$

Hence the probability density function for z is

$$g(z) = f(y)\left|\frac{dy}{dz}\right| = mnf(y) \,/\, (n+mz)^2 \qquad (8.27.5)$$

$$\therefore\ g(z) = mn\Gamma\left(\frac{m+n}{2}\right)y^{(m-2)/2}(1-y)^{(n-2)/2} \,/\, \Gamma\left(\frac{m}{2}\right)\Gamma\left(\frac{n}{2}\right)(n+mz)^2 \qquad (8.27.6)$$

$$\therefore\ g(z) = mn\Gamma\left(\frac{m+n}{2}\right)(mz)^{(m-2)/2}\,n^{(n-2)/2} \,/\, \Gamma\left(\frac{m}{2}\right)\Gamma\left(\frac{n}{2}\right)(n+mz)^{\frac{m+n}{2}} \qquad (8.27.7)$$

$$\therefore\ g(z) = \Gamma\left(\frac{m+n}{2}\right)m^{m/2}\,n^{n/2}\,z^{(m-2)/2} \,/\, \Gamma\left(\frac{m}{2}\right)\Gamma\left(\frac{n}{2}\right)(n+mz)^{(m+n)/2} \qquad (8.27.8)$$

so that by (8.6.5), z is an F-variate with parameters (m, n).

Now if

$$y = \frac{x_1^2+x_2^2+\ \ldots\ +x_m^2}{x_1^2+x_2^2+\ \ldots\ldots\ x_{m+n}^2}. \qquad (8.27.9)$$

and

$$z = ny \,/\, m(1-y), \qquad (8.27.10)$$

then

$$z = \frac{(x_1^2+x_2^2+\ \ldots\ x_m^2)/m}{(x_{m+1}^2+x_{m+2}^2+\ \ldots\ +x_{m+n}^2)/n}. \qquad (8.27.11)$$

Hence, if z is an *F-variate* with parameters (m, n), we may regard z as the ratio of the mean of squares of m independent normal (0,1) variates to the mean of squares of another n independent normal (0,1) variates.

Suppose now that y is a t-variate with parameter n, and let

$$z = y^2. \qquad (8.27.12)$$

We shall show that z is an F-variate with parameters $(1, n)$.

Proof

$$f(y) = \Gamma(n+1)/2]/\Gamma(n/2)\sqrt{(n\pi)}[1+(y^2/n)]^{(n+1)/2} \qquad (8.27.13)$$

Now $\sqrt{\pi} = \Gamma(1/2)$ (8.27.14)

$$g(z) = 2f(y) \,/\left|\frac{dz}{dy}\right| = f(y) \,/\, |\,y\,| \qquad (8.27.15)$$

$$\therefore \quad g(z) = \Gamma[(n+1)/2] \,/\, \Gamma(n/2)\Gamma(1/2)\sqrt{(nz)}[1+(z/n)]^{(n+1)/2} \tag{8.27.16}$$

$$= \Gamma[(n+1)/2]n^{n/2}z^{+1/2} \,/\, \Gamma(n/2)\Gamma(1/2)(n+z)^{(n+1)/2} \tag{8.27.17}$$

Hence by (8.6.5) $\boldsymbol{z}$ is an F-variate with parameters $(1, n)$.

$$\therefore \quad \boldsymbol{y}^2 = \boldsymbol{z} = \boldsymbol{x}_1^2 \,/\, (\boldsymbol{x}_2^2+\boldsymbol{x}_3^2+ \ldots +\boldsymbol{x}_{n+1}^2) \tag{8.27.18}$$

where $\boldsymbol{x}_1, \boldsymbol{x}_2, \ldots, \boldsymbol{x}_{n+1}$ are independent normal (0,1) variates.

It can indeed be shown that, when $\boldsymbol{y}$ is a t-variate with parameter n, then

$$\boldsymbol{y} = \boldsymbol{x}_1 \,/\, \sqrt{(\boldsymbol{x}_2^2+\boldsymbol{x}_3^2+ \ldots +\boldsymbol{x}_{n+1}^2)} \tag{8.27.19}$$

where $\boldsymbol{x}_1, \boldsymbol{x}_2, \ldots, \boldsymbol{x}_{n+1}$ are independent normal (0,1) variates.

Appendix to Chapter 8

A8.1. Continuous Variates Regarded as the Limiting Cases of Sequences of Discrete Variates

One good reason for considering continuous variates is that their distribution functions represent the limiting case of a sequence of distribution functions of discrete variates. In fact it will be shown that any specified distribution function $G(x)$ of a continuous variate can be obtained as the limit of a sequence of conditional distribution functions in a discrete variate probability model.

Let the variate $\boldsymbol{\alpha}$ have integer elements α_t satisfying

$$2^{\alpha_t} \leqslant t+2 < 2^{\alpha_t+1} \quad (t = 0, 1, 2, 3, \ldots), \tag{A8.1.1}$$

and let the probability vector $\boldsymbol{p}'$ have elements p_t such that

$$p_t = 4^{-\alpha_t} \tag{A8.1.2}$$

Then α has elements 1, 1, 2, 2, 2, 2, 3, 3, 3, 3, 3, 3, 3, 3, 4, . . . and $\boldsymbol{p}'$ has elements 1/4, 1/4, 1/16, 1/16, 1/16, 1/16, 1/64, . . . and it is easily verified that

$$\boldsymbol{p}'\mathbf{1} = \mathbf{1}. \tag{A8.1.3}$$

Now, let $\boldsymbol{x}$ be the vector with elements x_t satisfying

$$x_t = 2^{-\alpha_t}(t+2)-1, \tag{A8.1.4}$$

so that in binary notation, the elements of $\boldsymbol{x}$ will be

$$0{\cdot}0,\ 0{\cdot}1,\ 0{\cdot}00,\ 0{\cdot}01,\ 0{\cdot}10,\ 0{\cdot}11,\ 0{\cdot}000,\ 0{\cdot}001,\ 0{\cdot}010, \ldots$$

Now consider the conditional distributional function

$$F(x \mid \boldsymbol{\alpha} = n) = \boldsymbol{p}'_{\mid \boldsymbol{r}(\alpha = n)}\boldsymbol{r}(\boldsymbol{x} \leqslant x). \tag{A8.1.5}$$

All the elements of the row vector $\boldsymbol{p}'_{\mid \boldsymbol{r}(\alpha = n)}$ are zeros except for the elements with suffices t such that

$$2^n - 2 \leqslant t < 2^{n+1} - 2, \tag{A8.1.6}$$

and each of these non-zero elements will have the value 2^{-n} because

$$\boldsymbol{p}'_{\mid \boldsymbol{r}(\alpha = n)} = \frac{\boldsymbol{p}'\hat{\boldsymbol{r}}(\alpha = n)}{\boldsymbol{p}'\boldsymbol{r}(\alpha = n)} = 2^n\boldsymbol{p}'\hat{\boldsymbol{r}}(\alpha = n) = 2^{-n}\boldsymbol{r}'(\alpha = n) \text{ by (A8.1.2).}$$

The 2^n corresponding elements x_t of $\boldsymbol{x}$ satisfy

$$x_t = (t+2)2^{-n} - 1 \tag{A8.1.7}$$

in view of (A8.1.4), since for these elements, $\alpha_t = n$.

It follows that the expression (A8.1.5) is the sum of a set of products all of which are zero except for a subset of products each equal to 2^{-n}. These will be products of the t-th element of $\boldsymbol{p}'_{\mid \boldsymbol{r}(\alpha = n)}$ with the t-th element of $\boldsymbol{r}(\boldsymbol{x} \leqslant x)$ for the values of t which by (A8.1.4) satisfy

$$0 \leqslant (t+2)2^{-n} - 1 \leqslant x. \tag{A8.1.8}$$

When $x < 1$ the number of integers t satisfying (A8.1.8) must be $[2^n x + 1]$ where $[y]$ denotes the integral part of y. Hence, when $0 < x < 1$, the conditional distribution function of (A8.1.5) is to be obtained by multiplying $[2^n x + 1]$ by 2^{-n} to obtain

$$F(x \mid \alpha = n) = 2^{-n}[2^n x + 1] = x + \theta \quad (0 \leqslant \theta \leqslant 2^{-n},\ 0 \leqslant x \leqslant 1). \tag{A8.1.9}$$

As n is increased through the integer values, we obtain a sequence of distributions functions $F(x \mid \alpha = n)$ which, for given x $(0 \leqslant x \leqslant 1)$, approach the limit

$$F(x \mid \alpha \to \infty) = x \quad (0 \leqslant x \leqslant 1). \tag{A8.1.10}$$

Now let $G(x)$ be any specified distribution function of a continuous variate, and let $H(y)$ be the inverse function such that

$$H(y) = \min x \quad (G(x) = y). \tag{A8.1.11}$$

Let $\boldsymbol{u}$ be the variate whose elements u_t are given by

$$u_t = H(x_t), \tag{A8.1.12}$$

so that

$$x_t = G(u_t). \tag{A8.1.13}$$

Since $G(u)$ must be a monotonic increasing function of u,

$$\boldsymbol{r}(\boldsymbol{u} \leqslant u) = \boldsymbol{r}[\boldsymbol{x} \leqslant G(u)], \tag{A8.1.14}$$

and the conditional distribution function for $\boldsymbol{u}$ for fixed α is

$$F^*(u \mid \alpha = n) = F[G(u) \mid \alpha = n)], \tag{A8.1.15}$$

where F is the function (A8.1.9). Hence

$$F^*(u \mid \alpha = n) = G(u) + \theta' \qquad \left(0 < \theta' < 2^{-n}\frac{dG}{du}\right). \tag{A8.1.16}$$

Provided that $\dfrac{dG}{du}$ is bounded above, it follows that as $n \to \infty$ the conditional distribution function for the variate $\boldsymbol{u}$ defined by (A8.1.12) satisfies

$$F^*(u \mid \alpha = n) \to G(u). \tag{A8.1.17}$$

A8.2. The Expected Value of a Function of Two or More Continuous Variates

Let the variate $\boldsymbol{u}$ be given by

$$\boldsymbol{u} = \phi[\boldsymbol{x}(1), \boldsymbol{x}(2), \ldots, \boldsymbol{x}(n)] \tag{A8.2.1}$$

a function of the n variates $\boldsymbol{x}(j)$, $(j = 1, 2, \ldots, n)$, whose multivariate probability density function is $f[x(1), x(2), \ldots, x(n)]$.

Then, by analogy with (5.17.2) we may define the expected value of the variate $\boldsymbol{u}$ as

$$E(\boldsymbol{u}) = \int\int\ldots\int f[x(1), \ldots, x(m)]\phi[x(1), \ldots, x(m)]dx(1)\ldots dx(m). \tag{A8.2.2}$$

We must first check that this definition is consistent with the definition already given in (8.9.3). If ϕ is a function involving $x(1)$ only, say $\phi = \psi[x(1)]$,

$$\begin{aligned} E[\psi\{\boldsymbol{x}(1)\}] = E(\boldsymbol{u}) &= \int\int\ldots\int f[x(1)\ldots x(m)]\psi[x(1)]dx(1)\ldots dx(m) \\ &= \int \psi[x(1)]\left[\int\ldots\int f[x(1), \ldots, x(m)]\,dx(2), \ldots, dx(m)\right]dx(1) \\ &= \int \psi[x(1)]g[x(1)]dx(1), \end{aligned} \tag{A8.2.3}$$

where $g[x(1)]$ is the marginal probability density function for $\boldsymbol{x}(1)$. Equation (A8.2.3) is in agreement with equation (8.9.3).

Now consider the case where

$$\boldsymbol{u} = \phi[\boldsymbol{x}(1), \boldsymbol{x}(2), \ldots, \boldsymbol{x}(m)] = \sum_{t=1}^{m} a_t \boldsymbol{x}(t); \qquad \text{(A8.2.4)}$$

then by (A8.2.2)

$$E(u) = \int\int \ldots \int f[x(1), \ldots, x(m)] \sum_{t=1}^{m} a_t x(t) dx(1) dx(2) \ldots dx(m) \qquad \text{(A8.2.5)}$$

$$\therefore \quad E(\boldsymbol{u}) = \sum_{t=1}^{m} a_t I_t, \qquad \text{(A8.2.6)}$$

$$\text{where} \quad I_t = \int \ldots \int f[x(1), \ldots, x(m)] x(t) dx(1) \ldots dx(m). \qquad \text{(A8.2.7)}$$

But by (A8.2.3)

$$I_t = E[\boldsymbol{x}(t)]. \qquad \text{(A8.2.8)}$$

Hence

$$E\left[\sum_{t=1}^{m} a_t \boldsymbol{x}(t)\right] = E(\boldsymbol{u}) = \sum_{t=1}^{m} a_t E[\boldsymbol{x}(t)]. \qquad \text{(A8.2.9)}$$

Equation (A8.2.9) generalises equation (8.15.1) to the case of continuous variates. It will be noted that the proof does not assume that the variates $\boldsymbol{x}(t)$ are independent.

Now consider the variance of the linear function (A8.2.4).

$$\operatorname{var} \boldsymbol{u} = E[\boldsymbol{u} - E(\boldsymbol{u})]^2 = E\left[\sum_{t=1}^{m} a_t \{\boldsymbol{x}(t) - E[\boldsymbol{x}(t)]\}\right]^2 \qquad \text{(A8.2.10)}$$

$$\therefore \quad \operatorname{var} \boldsymbol{u} = \sum_{s=1}^{m} \sum_{t=1}^{m} a_s a_t E[\{\boldsymbol{x}(s) - E[\boldsymbol{x}(s)]\}\{\boldsymbol{x}(t) - E[\boldsymbol{x}(t)]\}]$$

$$= \sum_{s=1}^{m} \sum_{t=1}^{m} a_s a_t \operatorname{cov}[\boldsymbol{x}(s), \boldsymbol{x}(t)]. \qquad \text{(A8.2.11)}$$

Hence, we may write

$$\operatorname{var}\left[\sum_{t=1}^{m} a_t \boldsymbol{x}(t)\right] = \sum_{t=1}^{m} \left[a_t^2 \operatorname{var} \boldsymbol{x}(t) + 2 \sum_{s=t+1}^{m} a_s a_t \operatorname{cov}\{\boldsymbol{x}(s), \boldsymbol{x}(t)\}\right] \qquad \text{(A8.2.12)}$$

This generalises the result (8.15.2) to the case of continuous variates.

A8.3. The Spherical Multivariate Normal Distribution

Let $\boldsymbol{X}'$ denote the row of m independent normal variates $\boldsymbol{x}_i$ $(i = 1, 2, \ldots, m)$ each with parameters (0,1); then the probability density function for $\boldsymbol{X}'$ may be written in the form

$$g(X') = (2\pi)^{-m/2}\, \mathrm{e}^{-(1/2)X'X}, \qquad \text{(A8.3.1)}$$

where X' denotes a row-vector of m elements.

Now, let $\boldsymbol{Y}'$ denote the row of m variates $\boldsymbol{y}_j$ $(j = 1, 2, \ldots, m)$ obtained by an orthogonal transformation of the variates $\boldsymbol{x}_i$ given by

$$\boldsymbol{Y}' = \boldsymbol{X}'\boldsymbol{A} \qquad \text{(A8.3.2)}$$

with

$$\boldsymbol{A}'\boldsymbol{A} = \hat{\mathbf{1}}. \qquad \text{(A8.3.3)}$$

Then, by (A8.3.3) $|\, \boldsymbol{A} \,|^2 = 1$, and by (A8.3.2), the Jacobian of the transformation is

$$\left|\frac{\partial Y}{\partial X'}\right| = |\, A \,| = \pm 1. \qquad \text{(A8.3.4)}$$

It also follows from (A8.3.2, (A8.3.3) that

$$\boldsymbol{Y}'\boldsymbol{Y} = \boldsymbol{X}'\boldsymbol{X} \qquad \text{(A8.3.5)}$$

so that the density function for the variates $\boldsymbol{y}_j$ composing the row-vector $\boldsymbol{Y}'$ is

$$h(Y') = (2\pi)^{-m/2}\, \mathrm{e}^{-(1/2)Y'Y}, \qquad \text{(A8.3.6)}$$

where $\boldsymbol{Y}'$ is the row-vector of values of the $\boldsymbol{y}_j$; this implies that the $\boldsymbol{y}_j$ are m independent normal variates, each with parameters (0,1).

We can use this result to establish the following theorem.

THEOREM A8.3.1. If $\boldsymbol{X}'$ is the row-vector of m independent normal variate $\boldsymbol{x}_i$ $(i = 1, 2, \ldots, m)$ each with parameters (0,1), and if $\boldsymbol{y}$ is the single variate given by $\boldsymbol{y} = \boldsymbol{X}'\boldsymbol{a}$, where $\boldsymbol{a}$ is a vector satisfying $\boldsymbol{a}'\boldsymbol{a} = 1$, then the variate $\boldsymbol{z} = \boldsymbol{X}'\boldsymbol{X} - \boldsymbol{y}'\boldsymbol{y}$ is a chi-squared variate with parameter $(m-1)$.

Proof. Using the result of section A4.9, we can find an orthogonal matrix $\boldsymbol{A}$ of which $\boldsymbol{a}$ is the first column. Then let

$$\boldsymbol{Y}' = \boldsymbol{X}'\boldsymbol{A}. \qquad \text{(A8.3.7)}$$

Since $\boldsymbol{y} = \boldsymbol{X}'\boldsymbol{a}$ and $\boldsymbol{a}$ is the first column of $\boldsymbol{A}$, $\boldsymbol{y}$ must be the first element of $\boldsymbol{Y}'$; let $\boldsymbol{y}_2, \boldsymbol{y}_3, \ldots, \boldsymbol{y}_m$ then denote the remaining elements of $\boldsymbol{Y}'$. In this case, by the result following (A8.3.6), $\boldsymbol{y}_1, \boldsymbol{y}_2, \ldots, \boldsymbol{y}_m$ are m independent normal variates with parameters (0,1) and

$$z = X'X - y'y = Y'Y - y'y = \sum_{j=2}^{m} y_j' y_j \qquad \text{(A8.3.8)}$$

so that z is the sum of the squares of $(m-1)$ independent normal variates each with parameters (0,1): hence (see § **8.27**) z is a chi-squared variate with parameter $(m-1)$. This proves the theorem.

In particular, if $\bar{x}$ denotes the arithmetic mean of the m variates x_i and if

$$z = \sum_{i=1}^{m} (x_i - \bar{x})^2 = X'X - y'y, \qquad \text{(A8.3.9)}$$

where $y = \bar{x}m^{1/2}$, z is a chi-squared variate with parameter $(m-1)$.

These results are particular cases of a theorem due to Cochran (1934). (For further discussion of this theorem, the reader may consult Kendall and Stuart, *Advanced Theory of Statistics*, Volume I, Chapter 15, p. 360.)

Now suppose that x_i $(i = 1, 2, \ldots, m)$ are independent normal variates, each with parameters a, σ^2; by applying the Theorem A8.3.1 to the variates $t_i = (x_i - a)/\sigma$ $(i = 1, 2, \ldots, m)$ which are independent normal variates with parameters (0,1), we find that where

$$z = \sum_{i=1}^{m} (x_i - \bar{x})^2/\sigma^2 \qquad \text{(A8.3.10)}$$

z is a chi-squared variate with parameter $(m-1)$.

If we further suppose that x_i' $(i = 1, 2, \ldots, m')$ are an independent set of m' independent normal variates each with parameters (a', σ^2) with the *same* variance σ^2, (but not necessarily with the same expected value as the x_i have) then, where

$$z' = \sum_{i=1}^{m} (x_i' - \bar{x}')^2/\sigma^2 \qquad \text{(A8.3.11)}$$

and $\bar{x}'$ denotes the arithmetic mean of the m' variates x_i', z' must be a chi-squared variate with parameter $(m'-1)$ and independent of the variate z. Dividing z by z', and cancelling out σ^2, we see that where

$$F = \frac{1}{m} \sum_{i=1}^{m} (x_i - \bar{x})^2 \Bigg/ \frac{1}{m'} \sum_{i=1}^{m'} (x_i' - \bar{x}')^2 \qquad \text{(A8.3.12)}$$

and where

$$G = mF = \sum_{i=1}^{m} (x_i - \bar{x})^2 \Bigg/ \frac{1}{m'} \sum_{i=1}^{m'} (x_i' - \bar{x}_1')^2, \qquad \text{(A8.3.13)}$$

F is an F-variate with parameters (m, m') and G is a pseudo-chi-squared variate with parameters (m, m'). These results will be used frequently in Chapter 11 (Volume 2) on significance tests.

The results just obtained may be regarded as a special case of the general principle that where $\boldsymbol{G}$ is m' times the ratio of two orthogonal sums of squares in an analysis of sums-of-squares (or analysis of variance) table, where the numerator and denominator are quadratic forms of ranks m and m', $\boldsymbol{G}$ is a pseudo-chi-squared variate with parameters (m, m') and $\boldsymbol{G}/m$ is an $\boldsymbol{F}$-variate with parameters (m, m'). The reader is again referred to the discussion of Cochrane's theorem in Kendall and Stuart, Vol. 1, p. 360, for more detail.

Exercises 8

8.1. Obtain the moment-generating functions for the variates with the following probability density functions and expand them as far as the term in α^2,

(*i*) Normal $f(x) = \dfrac{1}{\sqrt{(2\pi)}\sigma} \mathrm{e}^{-(x-a)^2/2\sigma^2}$,

(*ii*) Rectangular $f(x) = 1/(b-a) \quad (a < x < b)$,

(*iii*) Gamma $f(x) = \mathrm{e}^{-x}x^t/(t!) \quad (0 < x,\ t$ a positive integer).

8.2. Obtain the first two cumulants for each variate in question 1.

8.3. Use the fact that if $\boldsymbol{x}$, $\boldsymbol{y}$ are independent variates, so are $\mathrm{e}^{\alpha\boldsymbol{x}}$ and $\mathrm{e}^{\alpha\boldsymbol{y}}$, to show that in this case, if $\boldsymbol{z} = \boldsymbol{x}+\boldsymbol{y}$, the cumulant generating-function for $\boldsymbol{z}$ is the sum of those for $\boldsymbol{x}$ and for $\boldsymbol{y}$.

8.4. If $K(\alpha, \boldsymbol{x})$ denotes the cumulant-generating function for $\boldsymbol{x}$, prove that $K(\alpha, \boldsymbol{x}\beta) = K(\alpha\beta, \boldsymbol{x})$ where β is any real number. Derive a formula giving k_r^*, the r-th cumulant for $\boldsymbol{x}\beta$, in terms of β and k_r the $\boldsymbol{r}$-th cumulant for $\boldsymbol{x}$.

8.5. If $\boldsymbol{x}(1), \boldsymbol{x}(2), \ldots, \boldsymbol{x}(n)$ are independent, and each have the same cumulant-generating function $K(\alpha, \boldsymbol{x})$, obtain from the result of question 3 the cumulant-generating function for z, where $\boldsymbol{z} = \sum_{r=1}^{n} \boldsymbol{x}(r)$.

8.6. Investigate the cumulants for $\boldsymbol{y} = \bar{\boldsymbol{x}} = \boldsymbol{z}/n$ and for $\boldsymbol{t} = (\bar{\boldsymbol{x}} - \mathbf{1}k_1)\sqrt{(n/k_2)}$ where k_1 and k_2 are the first two cumulants for $\boldsymbol{x}(1)$.

8.7. (Tchebycheff's Theorem). Prove that if θ has expected value a and finite variance σ^2, then

$$\text{Prob}\,[|\,\theta - a\,| > \lambda\sigma] \leqslant (1/\lambda^2).$$

Hint. Where $f(x)$ is the probability density function for θ, show that

$$\int_{|x-a|>\lambda\sigma} \sigma^2\lambda^2 f(x)\,dx < \sigma^2.$$

8.8. Show that if x_i^2 is distributed like chi-squared with one degree of freedom, for $i = 1, 2, \ldots, n$, then $\sum_{i=1}^{n} x_i^2$ is distributed like chi-squared with n degrees of freedom.
[Cambridge Economics Tripos Part 2, 1961]
'Chi-squared with n degrees of freedom' may be taken to mean a chi-squared variate with parameter n.

8.9. Explain how moment-generating functions can be used to determine the moments of multivariate distributions.

x and **y** are distributed jointly according to the bivariate norma distribution. Write down the most general form of their joint density function, and derive their covariance.
[Cambridge Economics Tripos Part 2, 1961]

8.10. Prove that the mean and variance of a Poisson variate are equal and obtain a formula for all the cumulants of a variate with density function

$$f(x) = e^{-L} L^x/x! \quad (x = 0, 1, 2, \ldots)$$

How would you prove that the sum of two independent Poisson variates is itself a Poisson variate?
[Cambridge Economics Tripos Part 2, 1962]

8.11. If income, **y**, has a log-normal distribution such that

$$f(y) = \frac{1}{\sigma y\sqrt{(2\pi)}} \exp\left[-\frac{1}{2\sigma^2}(\log y-\mu)^2\right] \quad \text{for} \quad 0<y<\infty$$

find the ratios between mean, median and modal incomes.

The expenditure, x_i, of the i-th individual on a good is related to his income y_i, as

$$x_i = \alpha+\beta \log y_i, \quad \text{for all } i.$$

Discuss some of the properties of this Engel curve.

If income has the log-normal distribution given above, what is the relationship between total expenditure on the good, X, and mean income Y?
[Cambridge Economics Tripos Part 2, 1963]

8.12. (*a*) A player throws a pair of dice and continues throwing until a total of 7 or of 8 is thrown. What is the expected duration of play (number of times dice thrown)? What is the probability that the total of the final throw is 7?

(b) In a carton of two dozen component parts, three are defective. A sample of five parts is drawn at random (with replacement). What is the chance of getting (*i*) no defective, and (*ii*) just one defective? Compare these exact chances with the approximate chances obtained by using an appropriate Poisson distribution. [Cambridge Economics Tripos Part 2, 1964]

8.13. (a) If X is a random variable with positive mean, show that Prob $[X>0] > 1/(1+c^2)$ where c is the coefficient of variation of X.

(b) Show that the correlation coefficient between X/Y and Y/Z is negative, where X, Y and Z are independent random variables with finite variance, taking positive values. What is the reason for this in general terms?
[Cambridge Economics Tripos Part 2, 1964]

8.14. (a) A random variable x has probability density function: $f(x) = ce^{-\lambda x}$ $(0 < x < \infty)$. Find the constant c in terms of the parameter λ. Evaluate the mean and variance. Calculate the probability that two observations of x are both greater than a given value a. Hence obtain the probability density function for the *smaller* of the two observations.

(b) The joint probability density function of two variables x and y is: $f(x, y) = e^{-(x+y)}$; $(0 < x < \infty;\ 0 < y < \infty)$. Show that the joint moment-generating function is

$$M(t_1, t_2) = 1/(1-t_1)(1-t_2).$$

Hence, or otherwise, show that $E(xy) = 1$. Determine the probability density function of the variate $z = x+y$.
[Cambridge Economics Tripos Part 2, 1964]

8.15. $(x_1, x_2, \ldots, x_n)$ are n independent drawings from a population which has the normal distribution with mean μ and variance σ^2. Show that

$$u = \sum_{i=1}^{n} \frac{1}{\sigma^2}(x_i-\mu)^2$$

is distributed like χ_n^2 with the density function

$$f(u) = u^{(n/2)-1}e^{-u/2} \,/\, 2^{n/2}\Gamma(n/2).$$

[Cambridge Economics Tripos Part 2, 1965]

8.16. A die is tossed 90 times. Find the probability of getting just 15 aces, (*i*) using the binomial formula and Stirling's approximation, and (*ii*) using the normal curve approximation.
Stirling's approximation: $n! \sim \sqrt{(2\pi n)} n^n e^{-n}$.
[Cambridge Economics Tripos, Part 2, 1965]

8.17. (a) If x and y are independent random variables with the same continuous distribution function F, and an expression for $P[x \leqslant t, y \leqslant t]$, use this to find the distribution function $G(z)$ of the variable $z = \max(x, y)$.

(b) If

$$f(x) = \mu^x e^{-\mu}/x! \quad (x = 0, 1, 2, \ldots, \infty),$$

and

$$f(y, x) = \frac{x}{y} p^y (1-p)^{x-y} \quad (y = 0, 1, 2, \ldots, x),$$

show that the marginal frequency function of y is given by

$$g(y) = (\mu p)^y e^{-\mu p}/y!.$$

[Cambridge Economics Tripos Part 2, 1965]

8.18. A sociologist has classified children on the basis of their fathers' occupations, into 'middle-class' and 'other', 20% of all children falling into the first category. An intelligence test has been standardised on a large random sample of all children to give normally distributed scores with mean 100 and variance 225; when it is given to a large random sample of 'middle-class' children, their scores are found to have mean 110 and variance 169.

If all children with scores of 112 or more are admitted to grammar schools, what proportion of children in each class will be admitted? What will be the average score (a) of the 'middle-class' children admitted, (b) of the other children admitted? (You may assume that the distribution of scores within each class is approximately normal.)

If in a particular town, 30 'middle-class' and 70 'other' children take the test, what will be the mean score of those admitted?
[Cambridge Economics Tripos Part 2, 1965]

8.19. Use the density functions given in Table 8.6.1 and the formula (8.7.12) for the transformation of variates to show that:

(*i*) If $\boldsymbol{x}$ is a normal variate with parameters a, σ^2, and $\boldsymbol{y} = (\boldsymbol{x}-a)/\sigma$ then $\boldsymbol{y}$ is a chi-squared variate with parameter 2.
(*ii*) If $\boldsymbol{x}$ is a gamma-variate with parameter m, and $\boldsymbol{y} = 2\boldsymbol{x}$, then $\boldsymbol{y}$ is a chi-squared variate with parameter $2m$,

(*iii*) If $\boldsymbol{x}$ is a dry-ice variate with parameter w, and $\boldsymbol{y} = wx$, then $\boldsymbol{y}$ is a gamma-variate with parameter 1.

(*iv*) If $\boldsymbol{x}$ is an F-ratio variate with parameters γ, λ then, if $\boldsymbol{y} = \gamma\boldsymbol{x}$, $\boldsymbol{y}$ is a pseudo-chi-squared variate with parameters γ, λ.

(*v*) If $\boldsymbol{x}$ is a Student's t-variate with parameter λ and $\boldsymbol{y} = \boldsymbol{x}^2$, then $\boldsymbol{y}$ is an F-variate with parameters $(1, \lambda)$.

(*vi*) If $\boldsymbol{x}$ is an F-variate with parameters γ, λ and if $y = \frac{\lambda}{\gamma}\frac{x}{1-x}$, then $\boldsymbol{y}$ is a beta-variate with parameters $\frac{1}{2}\gamma, \frac{1}{2}\lambda$.

8.20. $u(x;\lambda)$ and $v(x;\lambda)$ are the density functions of a t-variate with parameters λ and of a pseudo-chi-squared variate with parameters γ, λ; verify that, where $u(x) = \lim_{\lambda\to\infty} u(x;\lambda)$ and $v(x) = \lim_{\lambda\to\infty} v(x;\lambda)$, $u(x)$ and $v(x)$ are the density functions respectively of

(*i*) a normal variate with parameters 0,1, and

(*ii*) a chi-squared variate with parameter γ.

8.21. If $\boldsymbol{x}$ is a chi-squared variate with parameter γ, find the density function $g(y;\gamma)$ for $\boldsymbol{y}$, where $\boldsymbol{y} = \gamma^{-1/2}(\boldsymbol{x}-\gamma)$.

Verify that where $g(y) = \lim_{\gamma\to\infty} g(y;\gamma)$, $g(y)$ is the density function for a normal variate of zero expected value and variance 2.

8.22. $\boldsymbol{x}$ and $\boldsymbol{y}$ are independent variates with probability density functions

$$g(x) = Ge^{-x^2/2};\; h(y) = Ge^{-y^2/2} \quad (-\infty < x < \infty) \quad (-\infty < y < \infty)$$

The two variates $\boldsymbol{r}, \boldsymbol{\theta}$ satisfy

$$\boldsymbol{x} = \boldsymbol{r}\cos\boldsymbol{\theta};\; \boldsymbol{y} = \boldsymbol{r}\sin\boldsymbol{\theta} \quad (-\pi < \boldsymbol{\theta} \leqslant \pi;\; \boldsymbol{r} \geqslant 0).$$

Obtain the bivariate probability density function for $(\boldsymbol{r}, \boldsymbol{\theta})$ and derive the marginal probability density functions for $\boldsymbol{r}$ and for $\boldsymbol{\theta}$. Hence obtain the value $G = 1/\sqrt{2\pi}$, without assuming the value of $\int_{-\infty}^{\infty} e^{-x^2/2}dx$. Hence establish equation (8.12.10) of Chapter 8.

Chapter 9

BAYES' PROCEDURES FOR INTERPRETING STATISTICAL EVIDENCE

9.1. General Discussion of Bayes' Procedures

Whenever a situation is *completely* represented by a probability model of the types discussed in Chapters 4, 5 and 8 of this book, the natural way to use fresh statistical evidence for estimation is to substitute conditional probabilities (given the evidence) for the original probabilities provided by the model. This is the method of estimation associated with the name of Bayes, and the formulae for applying it are provided by the various forms of *Bayes' Theorem*. The various forms of this method may be described as *Bayes' procedures* and this chapter will discuss them at considerable length.

The reason that Bayes' procedures are not more widely used than they are, is that it is unusual for a situation to be *completely* represented by a probability model. This is because of the prevalence of *uncertainty*. As a rule any probability model is likely to contain some chance variates whose probability functions are unknown. Even if the forms of these functions are specified in the model, they are likely to contain one or more unknown parameters; indeed the purpose of the statistical analysis of evidence will often be precisely that of estimating the values of such parameters.

The difficulty about *completely* representing the situation by a probability model does not arise merely from the fact that the values of such parameters in the model are unknown. It arises whenever there is *uncertainty* about the values of such parameters. By 'uncertainty', we here mean an inability to provide even a probability function for the values of these parameters. When the model does contain such a probability function, then the main difficulty vanishes, and the way is clear for the application of Bayes' procedures. In the early pages of this chapter we shall consider how to proceed in this easier case. In its later pages we shall consider various suggestions for ascribing probability functions for the parameters, even in cases where there is uncertainty. We leave until Volume 2 the discussion of devices for avoiding altogether the need to provide any probability function for parameters about whose values there is great uncertainty.

9.2. Arguments in Favour of the Use of Bayes' Procedures when Interpreting Economic Statistics

Bayes' procedures are particularly convenient for use in the estimation of the kinds of parameters, such as price- or income-elasticities of demand, which interest economists. For very often, even *before* the start of a statistical enquiry, some fairly definite views will be held as to whereabouts the true value of the parameter is likely to lie, and whereabouts, on the other hand, it is most unlikely to lie. It is very proper to wish to take some account of such general knowledge, and, if one uses a Bayes' procedure, one can take account of the general knowledge by incorporating such beliefs in a probability function for the parameter value, a probability function which then forms a part of the probability model itself. Other methods of estimation, which will be discussed in Volume 2, deliberately make no use at all of these preliminary ideas about the value of the parameter. This may be a sensible precaution, if there is no general consensus of opinion with regard to these preliminary ideas; for at least it will remove the temptation to allow one's own pet theories unduly to bias one's estimates. But when all concerned are fairly well agreed beforehand, as to where, very roughly, the true parameter value is likely to lie, it is wasteful *not* to make use of this kind of evidence, along with the statistical evidence to be collected later. In economic applications of statistical methods there is accordingly a strong prima facie case for the use of Bayes' procedures.

9.3. Bayes' Theorem applied to Events

Bayes' procedures all consist in the application of Bayes' Theorem in one of its forms. The simplest form of the theorem is that which revises the probability judgement between two possible states of affairs, $\boldsymbol{j}(1)$ and $\boldsymbol{j}(2)$, in the light of fresh evidence that an event $\boldsymbol{k}$ has occurred. The theorem gives a formula for the ratio of the conditional probabilities of $\boldsymbol{j}(1)$ and $\boldsymbol{j}(2)$, given the occurrence of $\boldsymbol{k}$. The formula is

$$\frac{\boldsymbol{p}'_{|\boldsymbol{k}}\boldsymbol{j}(1)}{\boldsymbol{p}'_{|\boldsymbol{k}}\boldsymbol{j}(2)} = \frac{\boldsymbol{p}'_{|\boldsymbol{j}(1)}\boldsymbol{k} \cdot \boldsymbol{p}'\boldsymbol{j}(1)}{\boldsymbol{p}'_{|\boldsymbol{j}(2)}\boldsymbol{k} \cdot \boldsymbol{p}'\boldsymbol{j}(2)} \tag{9.3.1}$$

where, in the model, the three event vectors $\boldsymbol{j}(1)$, $\boldsymbol{j}(2)$ and $\boldsymbol{k}$ represent the states of affairs $\boldsymbol{j}(1)$ and $\boldsymbol{j}(2)$ and the event $\boldsymbol{k}$.

It will be seen that the theorem in the form (9.3.1) states that the ratio of the *conditional* probabilities of the two states of affair, given the knowledge of $\boldsymbol{k}$, is equal to the product of the ratio of their original probabilities and the ratio of the conditional probabilities of the occurrence of $\boldsymbol{k}$, given the two states of affairs.

This is sometimes expressed by stating that 'the conditional *odds*, given $\boldsymbol{k}$, in favour of $\boldsymbol{j}(1)$ against $\boldsymbol{j}(2)$, is equal to the original odds in favour of $\boldsymbol{j}(1)$ against $\boldsymbol{j}(2)$, multiplied by the *likelihood-ratio* of $\boldsymbol{j}(1)$ to $\boldsymbol{j}(2)$ in the light of $\boldsymbol{k}$'.

The theorem expressed in (9.3.1) enables one to revise one's estimate of the odds in favour of one state of affairs against another, in the light of fresh evidence $\boldsymbol{k}$.

The proof of the theorem in this form is very direct.

It follows immediately from the definitions given in section **4.14** that

$$\boldsymbol{p}'_{|\,\boldsymbol{k}}\boldsymbol{j}(1)\,\boldsymbol{p}'\boldsymbol{k} = \boldsymbol{p}'\hat{\boldsymbol{k}}\boldsymbol{j}(1) \quad \text{and} \quad \boldsymbol{p}'_{|\,\boldsymbol{k}}\boldsymbol{j}(2)\,.\,\boldsymbol{p}'\boldsymbol{k} = \boldsymbol{p}'\hat{\boldsymbol{k}}\boldsymbol{j}(2) \qquad (9.3.2)$$

and that $$\boldsymbol{p}'_{|\,\boldsymbol{j}(1)}\boldsymbol{k}\boldsymbol{p}'\boldsymbol{j}(1) = \boldsymbol{p}'\hat{\boldsymbol{k}}\boldsymbol{j}(1) \quad \text{and} \quad \boldsymbol{p}'_{|\,\boldsymbol{j}(2)}\boldsymbol{k}\boldsymbol{p}'\boldsymbol{j}(2) = \boldsymbol{p}'\hat{\boldsymbol{k}}\boldsymbol{j}(2). \qquad (9.3.3)$$

(9.3.1) follows from (9.3.2) and (9.3.3), by elementary algebra.

9.4. An Example of the Use of Bayes' Theorem for Events

In illustrating the use of Bayes' Theorem, it is best to start with very simple artificial examples taken from games of chance; otherwise, one's attention is likely to be diverted by irrelevant doubts as to the correspondence between the probability model and the real-world situation which it is supposed to represent.

Consider a bag which is believed to contain *either*

(*a*) 2 white balls and 3 black balls *or*
(*b*) 4 white balls and 1 black ball.

At first we are to regard possibility (*a*) as being twice as probable as possibility (*b*). We are now to draw one ball at random from the bag, observe the ball's colour and revise our estimate of the relative probabilities of (*a*) and (*b*).

To do this, we may set up a probability model in which the two possibilities (*a*) and (*b*) are represented by event-vectors $\boldsymbol{a}$ and $\boldsymbol{b}$ such that

$$\frac{\boldsymbol{p}'\boldsymbol{a}}{\boldsymbol{p}'\boldsymbol{b}} = 2. \qquad (9.4.1)$$

We may represent the event that the ball drawn is white by the event-vector $\boldsymbol{k}$, in which case the event that it is not white will be represented by the event vector $(\mathbf{1}-\boldsymbol{k})$. It is obvious from the conditions of the problem that

$$\boldsymbol{p'}_{|a}\boldsymbol{k} = 0{\cdot}4 \qquad \boldsymbol{p'}_{|a}(\boldsymbol{1}-\boldsymbol{k}) = 0{\cdot}6,$$

$$\boldsymbol{p'}_{|b}\boldsymbol{k} = 0{\cdot}8 \quad \text{and} \quad \boldsymbol{p'}_{|b}(\boldsymbol{1}-\boldsymbol{k}) = 0{\cdot}2. \tag{9.4.2}$$

Now suppose that the colour of the ball drawn is found to be white. Then $\boldsymbol{k}$ is known, and (9.3.1) tells us that

$$\frac{\boldsymbol{p'}_{|k}\boldsymbol{a}}{\boldsymbol{p'}_{|k}\boldsymbol{b}} = \frac{\boldsymbol{p'}_{|a}\boldsymbol{k}\boldsymbol{p'a}}{\boldsymbol{p'}_{|b}\boldsymbol{k}\boldsymbol{p'b}}. \tag{9.4.3}$$

Substituting the numerical values given in (9.4.2) and (9.4.1), we find

$$\frac{\boldsymbol{p'}_{|k}\boldsymbol{a}}{\boldsymbol{p'}_{|k}\boldsymbol{b}} = \boldsymbol{1}. \tag{9.4.4}$$

Thus the observation that the drawn ball is white has made the odds even between the possibilities (*a*) and (*b*), whereas before this, the odds were 2 to 1 in favour of possibility (*a*).

Now suppose, instead, that the ball drawn was found to be black; then we know the event $(\boldsymbol{1}-\boldsymbol{k})$. Bayes' Theorem in the form (9.3.1) now tells us

$$\frac{\boldsymbol{p'}_{|(1-k)}\boldsymbol{a}}{\boldsymbol{p'}_{|(1-k)}\boldsymbol{b}} = \frac{\boldsymbol{p'}_{|a}(\boldsymbol{1}-\boldsymbol{k})\,\boldsymbol{p'a}}{\boldsymbol{p'}_{|b}(\boldsymbol{1}-\boldsymbol{k})\,\boldsymbol{p'b}} = 6 \quad \text{by (9.4.1) and (9.4.2).} \tag{9.4.5}$$

Thus the draw of a black ball strengthens the odds in favour of possibility (*a*) from 2:1 to 6:1 on account of the likelihood ratio in favour of (*a*) being 3:1, given that the drawn ball was black.

This very elementary example has been set out so fully, because of the fundamental importance of this form of reasoning, which, we shall see, can be extended to deal with very elaborate situations as well.

9.5. Bayes' Theorem for the Conditional Probability Function of a Discrete Variate

By regarding the taking of two alternative values y_1 and y_2 by a variate $\boldsymbol{y}$ as being the two events $\boldsymbol{r}(\boldsymbol{y} = y_1)$ and $\boldsymbol{r}(\boldsymbol{y} = y_2)$, we may use Bayes' theorem, in the form (9.3.1), to obtain the conditional probability function for $\boldsymbol{y}$, given $\boldsymbol{k}$ (some event).

For it follows from (9.3.1) that

$$\frac{\boldsymbol{p'}_{|k}\boldsymbol{r}(\boldsymbol{y} = y_1)}{\boldsymbol{p'}_{|k}\boldsymbol{r}(\boldsymbol{y} = y_2)} = \frac{\boldsymbol{p'}_{|r(y=y_1)}\boldsymbol{k}\boldsymbol{p'r}(\boldsymbol{y} = y_1)}{\boldsymbol{p'}_{|r(y=y_1)}\boldsymbol{k}\boldsymbol{p'r}(\boldsymbol{y} = y_2)}. \tag{9.5.1}$$

Let $h(y)$ and $f_{|k}(y)$ denote the probability function for $\boldsymbol{y}$ and the conditional probability function for $\boldsymbol{y}$, given $\boldsymbol{k}$; then (9.5.1) becomes

$$\frac{f_{|k}(y_1)}{f_{|k}(y_2)} = \frac{p'_{|r(y=y_1)}k\,h(y_1)}{p'_{|r(y=y_2)}k\,h(y_2)}. \qquad (9.5.2)$$

More generally, if the values which $\boldsymbol{y}$ can take are y_t ($t = 1, 2, \ldots, n$), then

$$f_{|k}(y_t) = \lambda p'_{|r(y=y_t)}k\,h(y_t) \quad (t = 1, 2, \ldots, n),$$

where λ is a constant. (9.5.3)

Since
$$\sum_{t=1}^{n} f_{|k}(y_t) = 1, \qquad (9.5.4)$$

the constant λ may be evaluated, so as to establish

$$f_{|k}(y) = p'_{|r(y=y)}k\,h(y) \Big/ \sum_{u=1}^{n} [p'_{|r(y=y_u)}k\,h(y_u)]. \qquad (9.5.5)$$

An important special case of (9.5.5) arises when $\boldsymbol{k}$ is the event, $\boldsymbol{r}(\boldsymbol{x} = x)$, that another discrete variate $\boldsymbol{x}$ takes a particular value x. Substituting for $\boldsymbol{k}$ in (9.5.5) we obtain

$$f_{|r(x=x)}(y) = \phi_{|r(y=y)}(x)\,h(y) \Big/ \sum_{u=1}^{n} [\phi_{|r(y=y_u)}(x)\,h(y_u)]. \qquad (9.5.6)$$

where $\phi_{|r(y=y)}(x)$ denotes the conditional probability function for x, given that $\boldsymbol{y} = y$.

In the notation of § **5.13**, equation (9.5.6) may be more conveniently expressed as

$$h(y \mid \boldsymbol{x}=x) = \phi(x \mid \boldsymbol{y}=y)h(y) \Big/ \sum_{u=1}^{n} [\phi(x \mid \boldsymbol{y}=y_u)h(y_u)]. \qquad (9.5.7)$$

This is a form of Bayes' Theorem which it is often useful to quote.

9.6. An Example of the Use of Bayes' Theorem to Find a Posterior Probability Function

A bag contains 10 balls of which an unknown number $\boldsymbol{y}$ are white. The probability function for y is given as

$$h(y) = (y+1)/66 \qquad (y = 0, 1, 2, \ldots, 10). \qquad (9.6.1)$$

Three balls are drawn at random (without replacement) and the number $\boldsymbol{x}$ of these which are white is observed. If none of the three is white, find the posterior probability function for $\boldsymbol{y}$.

Solution. Bayes' Theorem establishes that

$$h(y \mid \boldsymbol{x}=0) = g(0 \mid \boldsymbol{y}=y)h(y) \Big/ \sum_{v=0}^{10} [g(0 \mid \boldsymbol{y}=v)h(v)]$$

where $g(x \mid \boldsymbol{y}=y)$ is the conditional probability function for $\boldsymbol{x}$ given that $\boldsymbol{y}=y$.

From the conditions of the problem

$$g(0 \mid \boldsymbol{y}=y) = (10-y)(9-y)(8-y)/720.$$

Hence we may calculate the posterior probability function $h(y \mid \boldsymbol{x}=0)$ as follows:

y	0	1	2	3	4	5	6	7	8	9	10
$66h(y)$	1	2	3	4	5	6	7	8	9	10	11
$120g(0 \mid \boldsymbol{y}=y)$	120	84	56	35	20	10	4	1	0	0	0
$7920h(y)g(0 \mid \boldsymbol{y}=y)$	120	168	168	140	100	60	28	8	0	0	0
$h(y \mid \boldsymbol{x}=0)$	·1515	·2121	·2121	·1768	·1263	·0758	·0354	·0101	0	0	0

In Fig. 19 the posterior probability function is compared with the prior probability function.

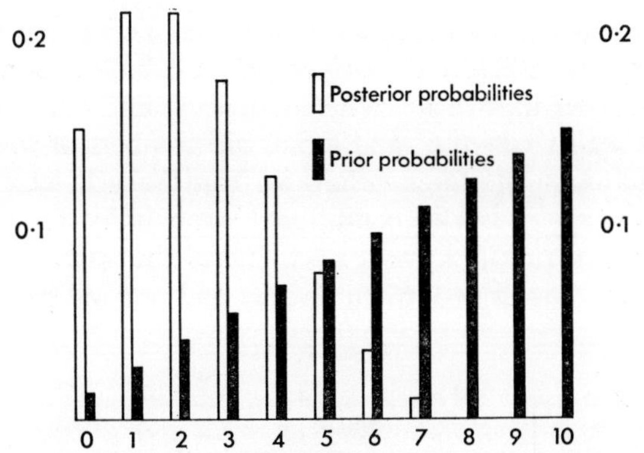

Fig. 19. Posterior and prior probability functions compared

9.7. An Example of the Use of Bayes' Theorem to Find the Posterior Probability Function for a Pair of Parameters

The logic of the argument is not essentially affected, if we apply Bayes' theorem to find the posterior probability function for a pair of variates, instead of for a single variate. Analogously to equation (9.5.7) we may then write

$$f(y, z \mid \boldsymbol{x}=x) = \phi(x \mid \boldsymbol{y}=y, \boldsymbol{z}=z)h(y, z) \Big/ \sum_{u}\sum_{v} [\phi(x \mid \boldsymbol{y}=y_u, \boldsymbol{z}=z_v) h(y_u, z_v)] \qquad (9.7.1)$$

where $h(y, z)$ denotes the prior probability function for the variate-pair $(\boldsymbol{y}, \boldsymbol{z})$.

EXAMPLE 9.7.1. A bag contains 6 red balls, 6 white balls and 6 black balls. Of these, six have been removed at random in the dark and placed in an otherwise empty sack. Three balls are drawn at random from the sack, and when light is restored, it is found that two of them are white and one is red. Find the posterior probability function for the numbers of red and white balls among the six placed in the sack.

Solution. Let $\boldsymbol{y}$ and $\boldsymbol{z}$ be variates representing the numbers of red and white balls placed in the sack. Then from the conditions of the problem it follows that the prior probability function for the variates $\boldsymbol{y}$ and $\boldsymbol{z}$ is

$$h(y, z) = \binom{6}{y}\binom{6}{z}\binom{6}{6-y-z} \Big/ \binom{18}{6}$$

whence $$h(y, z) = [y!\, z!\, (6-y-z)!]^{-1} \lambda$$

where λ is a constant independent of y and z. This is the probability function for $\boldsymbol{y}$ and $\boldsymbol{z}$ before the colours of the 3 balls drawn have been observed. In order to save space by condensing into one variate $\boldsymbol{x}$, the two variates $\boldsymbol{n}_w$, $\boldsymbol{n}_r$, where $\boldsymbol{n}_w$ and $\boldsymbol{n}_r$ are the numbers of white and red balls observed among the three withdrawn from the sack, let $x = 4\boldsymbol{n}_w + \boldsymbol{n}_r$. If $\phi(x \mid \boldsymbol{y}=y, \boldsymbol{z}=z)$ is the conditional probability function for $\boldsymbol{x}$, given y and z, then $\phi(9 \mid \boldsymbol{y}=y, \boldsymbol{z}=z)$ takes the values given in the following table, since $\boldsymbol{x} = 9$ when 2 white and one red ball are drawn.

TABLE 9.7.1

$\boldsymbol{y}$	0	1	2	3	4	5	6
	$20\phi(9 \mid \boldsymbol{y} = y, \boldsymbol{z} = z)$						
$\boldsymbol{z}$							
0	0	0	0	0	0	0	0
1	0	0	1	3	6	10	
2	0	0	2	6	12		
3	0	0	3	9			
4	0	0	4				
5	0	0					
6	0						

We may use this table and equations (9.7.1) and (9.7.3) to obtain the following table of $f(y, z \mid \boldsymbol{x}=9)$.

TABLE 9.7.2

y	0	1	2	3	4	5	6
			$27f(y, z \mid x = 9)$				
z							
0	0	0	0	0	0	0	0
1	0	0	1	3	3	1	
2	0	0	3	6	3		
3	0	0	3	3			
4	0	0	1				
5	0	0					
6	0						

From Table 9.7.2 we can read off the posterior probability function $f(y, z \mid \boldsymbol{x} = 9) = f(y, z \mid n_w = 2, n_r = 1)$, for the pair of variates $\boldsymbol{y}$, $\boldsymbol{z}$, in the light of evidence that 2 white and 1 red ball have been drawn.

9.8. The Use of Bayes' Theorem in Estimating the Value of a Parameter

Suppose that $\boldsymbol{y}$ is an unknown parameter governing the conditional probability function $f(z \mid \boldsymbol{y} = y)$ for a variate $\boldsymbol{z}$. Suppose that the parameter $\boldsymbol{y}$ has itself a probability function $h(y)$.

Suppose that we obtain n observations z_t $(t = 1, 2, \ldots, n)$ of $\boldsymbol{z}$. We assume these to be independent, *given* y, the value of $\boldsymbol{y}$. Let us pause to consider the exact meaning of this assumption: it means that, given y, the probability function for any set of values z_t $(t = 1, 2, \ldots, n)$ is the product of their individual probabilities, given y. This condition may be written as

$$\boldsymbol{p}'_{\mid \boldsymbol{r}(\boldsymbol{y} = y)} \bigcap_{t=1}^{n} \boldsymbol{r}[\boldsymbol{z}_t = z_t] = \prod_{t=1}^{n} [\boldsymbol{p}'_{\mid \boldsymbol{r}(\boldsymbol{y} = y)} \boldsymbol{r}(\boldsymbol{z}_t = z_t)]. \tag{9.8.1}$$

Let us denote the event that we obtain n particular values $z_1, z_2, \ldots, z_n$ by the event vector $\boldsymbol{k}$. Then

$$\boldsymbol{k} = \bigcap_{t=1}^{n} \boldsymbol{r}[\boldsymbol{z}_t = z_t], \tag{9.8.2}$$

and by the assumption (9.8.1),

$$\boldsymbol{p}'_{\mid \boldsymbol{r}(\boldsymbol{y} = y)} \boldsymbol{k} = \prod_{t=1}^{n} [\boldsymbol{p}'_{\mid \boldsymbol{r}(\boldsymbol{y} = y)} \boldsymbol{r}(\boldsymbol{z}_t = z_t)] = \prod_{t=1}^{n} [f(z_t \mid \boldsymbol{y} = y)]. \tag{9.8.3}$$

Hence, we may apply Bayes' Theorem, in the form (9.5.5), to show that

$$f_{|k}(y) = h(y) \prod_{t=1}^{n} [f(z_t \mid \mathbf{y} = y)] \Big/ \sum_{Y} \left[h(Y) \prod_{t=1}^{n} f(z_t \mid \mathbf{y} = Y) \right]. \tag{9.8.4}$$

EXAMPLE 9.8.1. A sack contains 5 red balls and an unknown number $\mathbf{y}$ of white balls. The probability function for $\mathbf{y}$ is assumed to be

$$h(y) = 1/(y+1)(y+2) \qquad (y = 0, 1, 2, \ldots). \tag{9.8.5}$$

5 balls are drawn at random without replacement, and the number, x_1, of red balls among them, is observed. The 5 balls are then replaced and the experiment is repeated, to give a second observation x_2. Altogether, 20 independent observations, $x_1, x_2, \ldots, x_{20}$, are obtained in this way, and the following frequency distribution is found for them.

TABLE 9.8.1

$x =$	0	1	2	3	4	5
Number of cases	1	3	5	6	4	1

Obtain the posterior probability function for $\mathbf{y}$.

Solution. Let $\mathbf{k}$ denote the event of obtaining the particular sequence of observations $x_1, x_2, \ldots, x_{20}$, that was obtained. Then the conditional probability of $\mathbf{k}$, given y, is

$$\prod_{t=1}^{20} [f(x_t \mid \mathbf{y} = y] = \mathbf{p}'_{|r(\mathbf{y}=y)}\mathbf{k}, \tag{9.8.6}$$

where

$$f(x \mid \mathbf{y} = y) = \binom{5}{5-x}\binom{5}{y-5+x} \Big/ \binom{y+5}{5}. \tag{9.8.7}$$

Hence

$$\mathbf{p}'_{|(r\mathbf{y}=y)}\mathbf{k} = \prod_{t=0}^{5} [f(t \mid \mathbf{y} = y)]^{n_t} \tag{9.8.8}$$

where $\quad n_0 = 1,\ n_1 = 3,\ n_2 = 5,\ n_3 = 6,\ n_4 = 4,\ n_5 = 1. \qquad (9.8.9)$

By the application of Bayes' Theorem, (9.8.7) and (9.8.8), the posterior probability function for $\mathbf{y}$ is

$$f(y \mid \mathbf{k}) \propto (y+1)^{-1}(y+2)^{-1} \prod_{t=0}^{5} [(y!)^2\{(y+5)!(y-5+t)!\}^{-1}] \tag{9.8.10}$$

where the n_t are given in (9.8.9).

By substituting numerical values, the first few $f(y \mid \mathbf{k})$ may be found as:

TABLE 9.8.1

y	<5	5	6	7	8	9	10
$f(y \mid \boldsymbol{k})$	0	0·641	0·305	0·0485	0·0049	0·0004	0·00004

9.9. Bayes' Theorem for Continuous Variates

Bayes' Theorem for continuous variates may be written down in exactly the same symbols as Bayes' Theorem for discrete variates, except that integrals must replace sums: moreover, symbols which previously represented probability functions will now represent probability density functions.

Two examples should make the correspondence perfectly clear.

Corresponding to equation (9.5.7) for discrete variates, we may write for continuous variates $\boldsymbol{x}$ and $\boldsymbol{y}$,

$$h(y \mid \boldsymbol{x}=x) = \phi(x \mid \boldsymbol{y}=y)h(y) \bigg/ \int_{-\infty}^{\infty} \phi(x \mid \boldsymbol{y}=u)h(u)\,du \quad (9.9.1)$$

However, in (9.9.1), $h(y)$ represents the marginal probability density function for $\boldsymbol{y}$ and both $h(y \mid \boldsymbol{x}=x)$ and $\phi(x \mid \boldsymbol{y}=y)$ are conditional marginal probability density functions.

Again, corresponding to equation (9.8.4) for discrete variates we may write for continuous variates $\boldsymbol{y}$ and $\boldsymbol{z}$,

$$f_{\mid \boldsymbol{k}}(y) = h(y) \prod_{t=1}^{n} [f(z_t \mid \boldsymbol{y}=y)] \bigg/ \int_{-\infty}^{\infty} h(u) \prod_{t=1}^{n} [f(z_t \mid \boldsymbol{y}=u)]\,du \quad (9.9.2)$$

The proof of these formulae is similar to that of the corresponding formulae for discrete variates and is based on equation (8.17.2), which gives the conditional probability density function for one continuous variate, given the value of another.

Since most of the parameters which economists are likely to wish to estimate can most conveniently be represented by continuous variates, it is Bayes' Theorem applying to continuous variates which is likely to be the form most often useful to them.

9.10. An Example of the Application of Bayes' Theorem for Continuous Variates

EXAMPLE 9.10.1. The variate $\boldsymbol{x}$ has the conditional probability density function

$$\phi(x \mid \boldsymbol{y}=y) = y\mathrm{e}^{-xy} \quad (x>0, y>0), \qquad (9.10.1)$$

where the value y is unknown. 5 independent observations, x_1 to x_5 are obtained for $\boldsymbol{x}$. If the original probability density function for $\boldsymbol{y}$ was

$$h(y) = \mathrm{e}^{-y}, \tag{9.10.2}$$

find the posterior probability density function in the light of the 5 observations.

Solution. By equation (9.9.2) we have for the posterior probability density

$$f_{\mid \boldsymbol{k}}(y) = h(y) \prod_{t=1}^{5} [\phi(x_t \mid \boldsymbol{y} = y)] \bigg/ \int_0^\infty h(u) \prod_{t=1}^{5} [\phi(x_t \mid \boldsymbol{y} = u)]\, du. \tag{9.10.3}$$

On substituting for $\phi(x \mid \boldsymbol{y} = y)$ and for $h(y)$, we obtain

$$f_{\mid \boldsymbol{k}}(y) = y^5 \mathrm{e}^{-Xy} \bigg/ \int_0^\infty u^5 \mathrm{e}^{-Xu} du, \text{ where } X = 1 + \sum_{t=1}^{5} x_t \tag{9.10.4}$$

and on integrating the denominator we finally obtain

$$f_{\mid \boldsymbol{k}}(y) = (X^6/120)\, y^5 \mathrm{e}^{-Xy}. \tag{9.10.5}$$

EXERCISE. Find the mean and variance of the posterior density function (9.10.5) in the case where the arithmetic mean of the 5 observations of $\boldsymbol{x}$ is 2·00.

Solution. Mean = 6/11 Variance = 6/121.

9.11. The Simultaneous Estimate of Several Parameters

There is no difficulty, in principle, in extending Bayes' Theorem to provide posterior probability density functions for sets of parameters. Thus if the variates $\boldsymbol{y}_j$ ($j = 1, 2, \ldots, m$) have the joint probability density function $h(y_1, \ldots, y_m)$, and if evidence $\boldsymbol{k}$ is collected which has the conditional probability, or probability density

$$\phi[\boldsymbol{k} \mid \bigcap_{j=1}^{m} \boldsymbol{r}(\boldsymbol{y}_j = y_j)]$$

depending on the $\boldsymbol{y}_j$, then the posterior probability density for the variates $\boldsymbol{y}_j$, in the light of the evidence $\boldsymbol{k}$, is

$$f[y_1, \ldots, y_m \mid \boldsymbol{k}] = h(y_1, \ldots, y_m)\phi\left[\boldsymbol{k} \,\middle|\, \bigcap_{j=1}^{m} \boldsymbol{r}(\boldsymbol{y}_j = y_j)\right] \bigg/ I \tag{9.11.1}$$

where

$$I = \int_{-\infty}^{\infty} \ldots \int h(u_1, \ldots, u_m)\phi \left[\boldsymbol{k} \middle| \bigcap_{j=1}^{m} \boldsymbol{r}(\boldsymbol{y}_j = u_j) \right] du_1 \ldots du_m. \tag{9.11.2}$$

EXAMPLE 9.11.1. $\boldsymbol{x}_1$ and $\boldsymbol{x}_2$ have the probability density function

$$\phi \left[x_1, x_2 \middle| \bigcap_{j=1}^{3} \boldsymbol{r}(\boldsymbol{y}_j = y_j) \right] = \mathrm{e}^{-[y_1x_1^2 + 2y_2x_1x_2 + y_3x_2^2]}\sqrt{(y_1y_3 - y_2^2)}/\pi. \tag{9.11.3}$$

If the parameters $\boldsymbol{y}_1$, $\boldsymbol{y}_2$ and $\boldsymbol{y}_3$ have the prior probability density function

$$h(y_1, y_2, y_3) = \mathrm{e}^{-(y_1^2 + y_2^2 + y_3^2)}\pi^{-3/2} \tag{9.11.4}$$

obtain the posterior probability density function for $(\boldsymbol{y}_1, \boldsymbol{y}_2, \boldsymbol{y}_3)$ after n observation-pairs (x_{1t}, x_{2t}) $(t = 1, 2, \ldots, n)$ have been taken of $\boldsymbol{x}_1$ and $\boldsymbol{x}_2$.

Solution. Let $\boldsymbol{k}$ denote the evidence collected and let

$$S_{11} = \sum_{t=1}^{n} x_{1t}^2; \quad S_{12} = \sum_{t=1}^{n} x_{1t}x_{2t}; \quad S_{22} = \sum_{t=1}^{n} x_{2t}^2. \tag{9.11.5}$$

Then, by Bayes' Theorem,

$$(y_1, y_2, y_3 \mid \boldsymbol{k}) = (y_1y_3 - y_2^2)^{n/2}\,\mathrm{e}^{-[y_1^2 + S_{11}y_1 + y_2^2 + S_{12}y_2 + y_3^3 + S_{22}y_3]}/I, \tag{9.11.6}$$

where, with $\sqrt{(v_1v_3 - v_2^2)} = v$,

$$I = \int_0^{\infty}\int_0^{\infty}\int_{-v}^{v} (v_1v_3 - v_2^2)^{n/2}\,\mathrm{e}^{-[v_1^2 + S_{11}v_1 + v_2^2 + S_{12}v_2 + v_3^2 + S_{22}v_3]}\,dv_2dv_1dv_3 \tag{9.11.7}$$

The result (9.11.6) illustrates the fact that the posterior density functions obtained by means of Bayes' Theorem are usually fairly elaborate. This need not be any considerable disadvantage, since what one needs in any specific case are numerical results, and with the aid of a computer these can be tabulated easily from a functional specification such as (9.11.6)-(9.11.7).

In the days when it was always necessary to rely on desk calculating machines, it was often better to use methods, other than Bayes' procedures, which yielded answers of simpler functional form and enabled summary statistics and estimates of their variance to be more quickly provided. Such rule-of-thumb methods are to be described in chapters in Volume 2.

9.12. Improper Prior Probability Density Functions

In applying Bayes' Theorem to find the posterior probability function for a variate **y** which can take any one of an infinite set of values, for example the negative and positive integers, including zero, it is often convenient to suppose that originally all these values were equally probable. Strictly speaking it is impossible to include such a probability function in our model since the sum of the probabilities has to equal unity, and it is therefore impossible for them all to be the same strictly positive number. However, in applying Bayes' theorem, one can drop the requirement that the sum of the prior probabilities should equal unity, and one can set them all equal to some strictly positive number and yet get sensible results. This is because the effect of increasing all the prior probabilities by the same proportion cancels out in the numerator and denominator of the expression for the posterior probabilities. By proceeding in this way, one gets the same result as by spreading the prior probability equally between the cases $-N$ to $+N$ inclusive, and finding the posterior probabilities and then allowing N to tend to infinity.

Much the same considerations apply when the parameter or parameters concerned are continuous. We may then apply Bayes' Theorem to find a posterior probability density function, after attributing a prior probability density function which is uniform over an infinite region, and yet get sensible results. Here again, the result is as though we had applied a uniform prior probability density function over a finite region, calculated the posterior probability function, and had then found to what limit this function tended as we allowed the extent of the region to tend to infinity.

EXAMPLE 9.12.1. The variate **x** has the conditional probability density function

$$f(x \mid \mathbf{y}=y) = e^{-(x-y)^2/2} / \sqrt{(2\pi)},$$

where **y** has a uniform prior probability density function extending along the whole range of real values.

Find the posterior probability density function for **y**, when 3 independent observations of **x** have been found to have an arithmetic mean a.

Solution. Take as prior probability density function for **y**, $h(y) = 1$, (9.12.1) and let $\boldsymbol{k}$ denote the evidence of the set of three observations of $\boldsymbol{x}$. The conditional probability density for $\boldsymbol{k}$, given the value y of $\boldsymbol{y}$, is

$$\phi(\boldsymbol{k} \mid \boldsymbol{y} = y) = \frac{1}{(2\pi)^{3/2}} \exp\{-\tfrac{1}{2} \sum_{i=1}^{3} (x_i - a)^2\} = C \exp\{-\tfrac{3}{2}(a-y)^2\} \tag{9.12.2}$$

where C is a constant independent of the value of y.

By Bayes' Theorem, the posterior probability density function for $\boldsymbol{y}$ is

$$f(y \mid \boldsymbol{k}) = \exp\{-\tfrac{3}{2}(y-a)^2\} \bigg/ \int_{-\infty}^{\infty} \exp\{-\tfrac{3}{2}(u-a)^2\} du = \sqrt{\left(\frac{3}{2\pi}\right)} \exp\{-\tfrac{3}{2}(y-a)^2\} \tag{9.12.3}$$

9.13. The Effects of Varying the Prior Probability Density Function

Reluctance to use Bayes' procedures of estimation often arises because of the lack of agreed principles on which to base the choice of prior probability density functions. A considerable subjective and arbitrary element is felt to affect this choice. Advocates of the procedures sometimes argue that this objection is of very little weight, since the difference made by the variation between sensible alternative prior probability density functions is, as a rule, no greater than would be made by taking just one observation.

This latter contention would be very difficult to establish, since there is no obvious way of deciding what are sensible choices of prior probability density functions and what are not. In any given example, it is always possible to point to two possible prior probability density functions which are so different that it would take a million (or any other specified large number) observations to outweigh the effects of choosing one rather than the other. Yet a few examples will show how variations in the choice of prior probability density function, although on the face of things they appear to be quite substantial, yet cause as little variation in the posterior probability density function as would the taking of just one more observation.

Consider the example 9.12.1. If instead of choosing $h(y) = 1$ we had chosen $h(y) = \dfrac{1}{\sqrt{(2\pi)}} e^{-y^2/2}$, then the posterior probability density function for $\boldsymbol{y}$, given the three observations, would have been

$$f(y \mid \boldsymbol{k}) = (2\pi)^{-2} e^{-2[y-(3a/4)]^2}, \tag{9.13.1}$$

which has a smaller mean and a smaller variance than the density function found in (9.12.3). But if we had retained the earlier prior probability density function $h(y) = 1$ and had taken one more observation of x and it had been a zero, then we should have again obtained the posterior probability density function (9.13.1).

The suggestion that one observation may make as much difference to the posterior probability density function as would the difference between any two sensible prior probability density functions, is, however, less reassuring than appears at first. For one observation, if it is sufficiently atypical and misleading, can cause an extremely serious modification to the posterior probability density function. Consider again the example of the previous paragraph. Suppose the true value of y to be 100 and suppose that $a = 100{\cdot}40$. Then, if we were using the prior probability density function $h(y) = 1$, we should obtain for y the posterior probability density function (9.12.3) with $a = 100{\cdot}40$. This is a normal density function with mean 100·40 and variance 0·3333. If we then obtained a fourth and highly misleading observation of zero, the posterior probability density function would be modified to (9.13.1) with $a = 100{\cdot}40$. This would be a normal density function with mean 75·30 and variance 0·2500. On the basis of this probability density function, we should be led to assert that there was a probability of less than one in 10^{1000} that y exceeded 99·0000. We should have been led to the same wild assertion if we had used only the three good observations with their average $a = 100{\cdot}40$ but had used the prior probability density function, $h(y) = \dfrac{1}{\sqrt{2\pi}}\, e^{-y^2/2}$, which led to the posterior probability density function (9.13.1).

Thus the very Example 9.12.1, which we used in order to show that a change in prior probability density function will typically make no more difference to our conclusions than a single observation, shows also that such a difference may be catastrophic.

There is, despite what was said in favour in § **9.2**, a formidable case against the use of Bayes' procedures. For the prior probability density function typically reflects a state of personal opinion, or in the absence of any definite opinion, the application of some conventional rule for choosing a prior probability density function. Different enquirers are likely to have different opinions and to respect different conventions. We have just seen that an unfortunate choice of prior probability density function can lead to absurd conclusions; and it is clear that, if they are allowed to use Bayes' procedures, different scientists may be led to interpret the same observations in widely different fashions.

This formidable case rests on the supposition that the scientists will in fact choose prior probability density functions sufficiently diverse to have appreciably different effects on the posterior probability density functions. The advocates of Bayes' procedures claim that such diversity in choice of prior probability density functions can be avoided by the observation of various precautions in making the choice. This claim has not as yet been firmly established and would be hotly disputed by many qualified statisticians. The remainder of this chapter will be concerned mainly with a discussion of the various precautions and conventions which have been suggested in order to prevent undue diversification in the choice of prior probability density functions.

9.14. Bayes' Postulate

Before considering the choice of prior probability density functions for continuous parameters, it is instructive to return to the simpler problem of choosing prior probabilities for a set of discrete alternatives. We may start with the choice of prior probabilities for two alternative hypotheses $\boldsymbol{h}$ and $\boldsymbol{k}$, where we assume that one or other of them must be true. In the typical case, we shall already have some opinion as to which is true, for example that $\boldsymbol{h}$ is rather more probable than $\boldsymbol{k}$, but this opinion will not be so precise that we can go on to assert that, although $\boldsymbol{h}$ has a probability as great at 0·6034, it has not got so great a probability as 0·6036. In such a case we should be hard put to it to know what probability to choose for $\boldsymbol{h}$ between, say, 0·55 and 0·85. This is one of the cases where it would not, in fact, much affect our conclusions where we chose the prior probability within this range, provided we were able to collect a fair amount of evidence to discriminate between the two hypotheses.

The typical case includes *some* prior evidence or opinion, but it is instructive to consider the case where there is absolutely no indication which of the two hypotheses should be regarded as the more probable. Two different examples of this case may be distinguished. There may be a great deal of evidence concerning the plausibility of each hypothesis, but it may be exactly symmetrical between the two. For example we may be considering the two hypotheses that in the throw of a certain die, an even number is more likely to result, and that an odd number is more likely to result; and we may know that in 12 300 throws of the die, just 6150 even numbers and 6150 odd numbers have been thrown. On the other hand, we may know absolutely nothing about the die, beyond the fact that it is possible to throw it, and that an odd or an even number can be thrown; we may not even know what shape it is.

In the former case where there is a great deal of evidence, which is exactly symmetrical between the two hypotheses, it seems very reasonable to give the hypotheses *equal* probability. This is the procedure which was suggested by Bayes. The attribution of equal probabilities to such hypotheses, in the absence of evidence favouring one above the other is known as Bayes' postulate. Strictly speaking it is not a postulate but a procedure or convention, since it concerns the choice of prior probabilities, and there is no sense in which a prior probability can be said to have any particular *true* value in such a case, and we are certainly not concerned with any postulate about what is the *true* value of the probability.

Bayes' postulate or convention, may also be applied when there is a complete absence of evidence about the two hypotheses. We could, for example, attribute prior probabilities of a half each, to the theories that an even throw was more likely and that an odd throw was more likely, in the case of a die about which we knew absolutely nothing, except that a throw might result in an odd or an even number. This is the application of Bayes' postulate or convention which has aroused the more controversy. In support of it, it may be said that the adoption of any other convention, such as giving odds of two to one in favour of the hypothesis first specified, offends common sense even more because of its asymmetry. If Bayes' Theorem is to be used at all in this case, then Bayes' postulate or convention is the only procedure conceivable to the writer, which is not thoroughly objectionable on grounds of asymmetry or irrelevance.

The above discussion of the case of two alternative hypotheses extends to the case of several hypotheses. If there is no evidence or opinion about their probability ordering, or if there is only symmetrical evidence, then Bayes' procedure is to give all the hypotheses the same prior probability. As in the case of two hypothesis, this is the only procedure conceivable to the writer which is not objectionable on grounds of asymmetry or irrelevance.

Properly interpreted, the rule for this procedure cannot lead to inconsistency. Arguments designed to show that it can lead to inconsistency usually introduce asymmetry into the evidence in some such fashion as the following.

'The United Kingdom consists of two regions, E and not-E. The Q was born in the United Kingdom. Choose a prior probability for the hypothesis that the Q was born in E.' In the absence of further evidence, Bayes' procedure is to give this hypothesis a probability of one half. But suppose another sociologist has been confronted with much the same problem in another form: 'The United Kingdom contains four parts, E, S, U and W. The Q was born in the United Kingdom. Choose a prior probability for the hypothesis that the Q was born in E.' Having no further evidence, this less ill-informed sociologist correctly used Bayes' procedure and gives the hypothesis a probability of one quarter. In one case the procedure leads to the answer 'one half' and in the other case to the answer 'one quarter'. This is perfectly consistent, since the second sociologist was given the evidence, denied to the former, that not-E consisted of three parts, S, U and W. This tipped the scales and resulted in the attribution of a considerably higher prior probability to the hypothesis about not-E than to the hypothesis about E. The evidence available to the second sociologist about the two hypotheses was not *quite* symmetrical.

Bayes' procedure may also be used to choose a set of prior probabilities for an unlimited number of hypotheses between which opinion and evidence is completely symmetrical. It may however be observed that it is extremely difficult to imagine any such set of hypotheses. But if we *are* confronted with such a set of hypotheses, the procedure is to give them all equal probability. This will of course involve the same kind of improper prior distribution as we considered in § **9.12**: since the number of hypotheses is unlimited, if we give each the same non-zero numerical probability, their total probability will not be finite. Nevertheless we may adopt this procedure and then use Bayes' theorem after collecting further evidence, without this leading to ridiculous posterior probability functions, because, as in the continuous case discussed in § **9.12**, the effect of any constant multiplying the prior probability function cancels out, because of its presence in both the numerator and denominator of the expression for the posterior probability. The effect of the procedure is equivalent to giving equal probability to just N of the hypotheses, and working out the posterior probabilities, and then allowing N to tend to infinity.

EXAMPLE 9.14.1. A scientist sincerely believes that where N is the total number of worlds in the observable universe which are inhabited by beings more intelligent than those on Earth; then the probability of his receiving and recognising signals from at least one of them within T listening hours is given by

$$f(T, N) = 1-\beta^{TN} \tag{9.14.1}$$

where β is a *known* positive constant slightly less than unity. So far, he has listened for 30,000 hours without positive result. What should he regard the posterior probability function for N to be?

Solution. Suppose he is unaware of any evidence as to what N may be.

He decides to adopt Bayes' procedure. Accordingly, he attributes an 'improper' prior probability function for N, namely

$$h(n) = 1 \quad (n = 0, 1, 2, \ldots). \tag{9.14.2}$$

Applying Bayes' Theorem and using (9.14.1) he accordingly obtains the posterior probability function in the light of the evidence E,

$$f(n \mid E) \propto \beta^{30\,000n} \quad (n = 0, 1, 2, \ldots). \tag{9.14.3}$$

This is, of course, a case where it is highly questionable whether the evidence can be regarded as symmetrical between all the different possible values of N. The procedure is very dubious in this case. As has already been observed, this is probably a common feature of all examples involving an unlimited number of discrete hypotheses. In this particular case, one might regard the smaller numbers as individually more probable than the larger, a priori, but one would be hard put to it to decide by how much more probable. If instead of $h(n) = 1$, our scientist had taken $h(n) = 1/(n+1)(n+2)$, then he would have obtained the posterior probability function

$$f(n \mid E) \propto \beta^{30\,000n}/(n+1)(n+2) \tag{9.14.4}$$

instead of the function (9.14.3).

9.15. Bayes' Postulate for Continuous Variates

If one attempts to extend Bayes' procedure, so as to provide a prior probability density function for a parameter or set of parameters, about whose value one professes to be completely at a loss, one must expect some difficulty. One will be led to the adoption of the improper prior probability density function which is uniform over all possible values of the parameter or parameters. Before considering objections to this procedure we will run through an example.

EXAMPLE 9.15.1. $\boldsymbol{\theta}$ is the unknown probability that an experiment will have a specified result $\boldsymbol{k}$. In n experiments the result $\boldsymbol{k}$ occurs in r cases. Find the posterior probability density function for $\boldsymbol{\theta}$.

Solution. Since we are told absolutely nothing about $\boldsymbol{\theta}$ we will adopt the Bayes' procedure of giving $\boldsymbol{\theta}$ the prior probability density function

$$h(t) = 1 \qquad (0 < t < 1). \tag{9.15.1}$$

The probability, given the value t of $\boldsymbol{\theta}$, of getting r results of type $\boldsymbol{k}$ is

$$\phi(r \mid \boldsymbol{\theta} = t) = \binom{n}{r} t^r (1-t)^{n-r}. \tag{9.15.2}$$

Hence by Bayes' theorem the posterior probability density function for θ is

$$f(t \mid r) = t^r(1-t)^{n-r} \bigg/ \int_0^1 u^r(1-u)^{n-r}\,du \tag{9.15.3}$$

namely

$$f(t \mid r) = t^r(1-t)^{n-r} \,/\, B(r+1, n-r+1). \tag{9.15.4}$$

The mean and variance of this probability density function are

$$\mu_1{}' = \frac{r+1}{n+2} \qquad \mu_2 = \frac{(n-r+1)(r+1)}{(n+2)^2(n+3)}. \tag{9.15.5}$$

A critic of Bayes' postulate might well draw attention to the following awkward paradox. Had we been told that the unknown probability of the experiment having the result k was γ^2, where $0 < \gamma < 1$, we might have applied Bayes' postulate by giving γ the prior probability density function

$$h(g) = 1, \tag{9.15.6}$$

and we could have used (9.15.2) in the form

$$\phi(r \mid \gamma = g) = \binom{n}{r} g^{2r}(1-g^2)^{n-r} \tag{9.15.7}$$

to obtain the posterior probability density function for γ as

$$v(g \mid r) = g^{2r}(1-g^2)^{n-r} \bigg/ \int_0^1 u^{2r}(1-u^2)^{n-r}\,du \tag{9.15.8}$$

namely

$$v(g \mid r) = 2g^{2r}(1-g^2)^{n-r} \,/\, B(r+\tfrac{1}{2}, n-r+1). \tag{9.15.9}$$

But we can obtain from (9.15.9) the posterior probability density function for θ, where $\theta = \gamma^2$, the probability that the experiment will have the result k. This probability density function must satisfy

$$f(t \mid r) = v(g \mid r)\frac{dg}{dt}, \quad \text{where } g^2 = t, \tag{9.15.10}$$

by the usual rule (see § **8.7**) for transformation of variates. Hence

$$f(t \mid r) = t^{r-1/2}(1-t)^{n-r} \,/\, B(r+\tfrac{1}{2}, n-r+1). \tag{9.15.11}$$

But this result differs from the result (9.15.4) which we obtained earlier by applying Bayes' procedure to find the prior probability density function for θ.

This is very unsatisfactory; the trouble arises from the fact that by assuming a uniform prior probability density function for γ we imply a prior probability density function for $\boldsymbol{\theta}$ which is of the form

$$j(t) = \frac{dg}{dt} \tag{9.15.12}$$

and which will therefore differ from the form (9.15.1) obtained by applying Bayes' procedure directly to $\boldsymbol{\theta}$.

The difficulty is general. A different prior probability density function, and hence also a different posterior probability density function, will be obtained for a parameter, according as we apply Bayes' procedure directly to that parameter or to some non-linear monotonic function of it. In cases where it is not obvious whether to apply the procedure to one parameter or to another, which is a monotonic function of it, different workers may derive different conclusions from the same observations and from the application of Bayes' procedures. In cases where two or more parameters are to be estimated simultaneously, the difficulty is even more likely to be serious.

The upshot of the discussion of this section is that Bayes' postulate does provide a suitable procedure for attributing a prior probability function to a *discrete* variate or set of variates, in the case where the evidence between them is completely symmetrical or entirely absent; but it does not provide a suitable procedure for attributing a probability density function to one or more *continuous* variates, since by applying the same procedure to monotonic functions of the variates, we should be led to different answers. We shall see later in the chapter how to get over this difficulty.

9.16. The Use of Vague Evidence about a Set of Hypotheses

The cases in which it is appropriate to use Bayes' postulate are exceptional ones. As a rule, before we collect the observations in a statistical enquiry, we already have some notion as to roughly where the correct value of an unknown parameter lies; or if we are trying to choose between a set of hypotheses, there is likely to be already some degree of agreement as to which of the hypotheses are the more plausible. In such cases, there is no advantage to be gained by ignoring these beliefs, and there is nothing to be lost by allowing them to influence our choice of the prior probabilities. In this section we shall discuss how this may be done in the case where we are choosing between a set of hypotheses, or estimating a discrete parameter.

If we are considering a set of disjoint hypotheses $H_1, H_2, \ldots,$ we may first attempt to *rank* them in probability. As a rule the probability ordering which we arrive at, whilst it may be consistent, will contain a great many probability relations of type $H_i \parallel H_j$ exhibiting complete uncertainty with regard to which of two hypotheses is more likely. Hence there will be a very wide variety of different *illustrations* available to us of the probability ordering. (See § **2.9**, for an explanation of such illustrations.)

In deciding which illustration of the ordering to use when selecting our prior probabilities, we may adopt a modified form of Bayes' postulate. We may substitute $H_i = H_j$ for $H_i \parallel H_j$, in all cases where this can be done without inconsistency. As a result of such substitutions further substitutions such as $H_m > H_n$ for $H_m \parallel H_n$ will have to be made to avoid inconsistency; and eventually we shall have eliminated by substitutions all relations of type $H_i \parallel H_j$. We shall thus have obtained an illustration of the probability ordering, in which the hypotheses fall into a number of sets $S_1, S_2, \ldots$, such that within any one set all hypotheses are equal in probability, whereas every hypotheses in any one set is more probable than every hypothesis in any later set. The probability ordering will as usual extend to unions of the discrete disjoint hypotheses, as well as to the disjoint hypotheses themselves.

Let us first suppose that we have a strict probability ordering of the disjoint hypothesis and that we have numbered them so that it is

$$H_1 > H_2 > H_3 > H_4 \ldots \tag{9.16.1}$$

If we have further probability comparisons, such as

$$(H_2 \cup H_4) > H_1, \tag{9.16.2}$$

these will impose quite useful limits on the numerical probabilities which we can consistently attribute to the various hypotheses. But, in general, there will remain quite a lot of latitude in our choice. This latitude reflects our state of uncertainty, and it is often instructive to choose various different sets of numerical probabilities within this scope and to see what posterior probabilities each of them leads to. But for purposes of discussion, let us suppose that we have only the information provided by (9.16.1) and no further information of type (9.16.2). This leaves very considerable latitude in the choice of prior probabilities, and it is really a matter of choosing a convention, when we come to select one particular method of attributing the prior probabilities.

The danger involved in the use of Bayes' procedures is that by the dogmatic virtual exclusion of a hypothesis due to some personal quirk or prejudice, the power of evidence to secure near-acceptance for a true hypothesis will be frustrated for an unduly long time. For this reason, it is prudent not to choose for H_1, H_2, etc. a series of numerical prior probabilities which converge very rapidly towards zero; we should beware, in fact, of attributing near-certainty, a priori, to the first few hypotheses alone. On the other hand, it would be foolish to ignore the probability judgements almost altogether, by attributing some very slowly converging series of prior probabilities, such as

$$10^{-9}, 10^{-9}\eta, 10^{-9}\eta^2, \ldots, \quad \text{where } \eta = 1-10^{-9},$$

since this might make it necessary to collect a great deal of further evidence merely to rule out hypotheses which we were already almost sure would be non-starters. Exactly where to draw the line between these two extremes is a matter of convention; and, since the choice is needed if the Bayes Theorem is to be applied, we suggest the following compromise convention.

If a set of hypotheses are arranged in order of probability so that (9.16.1) is satisfied, and if we have no further information such, for example as that given in (9.16.2), then we should allocate the prior probabilities

$$\phi[H_r] = \frac{1}{r(r+1)} \qquad (r = 1, 2, 3, \ldots). \qquad (9.16.3)$$

If the number of disjoint hypotheses is finite, N say, then the prior probabilities should be raised to

$$\phi[H_r] = \frac{N+1}{Nr(r+1)} \qquad (r = 1, 2, 3, \ldots, N). \qquad (9.16.4)$$

A slight further modification of the rule is required when it is only possible to rank the hypotheses into classes,

$$H_1 \text{ to } H_{n_1}, H_{n_1+1} \text{ to } H_{n_2}, \ldots, H_{n_s+1} \text{ to } H_{n_{s+1}}, \text{ etc.}$$

where the hypotheses within any one class are ranked equal in probability, and those in any class are ranked higher in probability than those in any later class. If there are altogether k classes, containing n hypotheses in all, then the prior probabilities should be taken as

$$\phi(H_{n_s}) = (n_k+1) / [n_k(n_s+1)(n_{s-1}+1)] \qquad (9.16.5)$$

for each hypothesis in class s.

EXAMPLE 9.16.1. Seven hypotheses are ranked in probability as

$$H_1 = H_2 > H_3 = H_4 = H_5 = H_6 > H_7.$$

Apply the suggested procedure so as to attribute prior probabilities to them.

Solution. We may write $n_0 = 0$, $n_1 = 2$, $n_2 = 6$, $n_3 = 7$, $k = 3$, and apply (9.16.5) to obtain

$$\phi(H_1) = \phi(H_2) = \tfrac{8}{21}$$

$$\phi(H_3) = \phi(H_4) = \phi(H_5) = \phi(H_6) = \tfrac{8}{147}$$

$$\phi(H_7) = \tfrac{1}{49}.$$

9.17. Prior Probabilities for the Values of Discrete Parameters

Suppose that a parameter θ may take any value 0, 1, 2, 3, . . .; Bayes' postulate would suggest that we attach the same probability to each value, if we have no evidence supporting some values more than others. But in this case, it is difficult to regard the evidence as wholly symmetrical between the values. The value 0, for example, has a prominence and simplicity which is not fully shared by the value 824 077 159 645. This is a proper ground for attributing higher prior probability to $\theta = 0$ than to $\theta = 824\,077\,159\,645$. It is indeed suggested that where θ may take the values 0, 1, 2, 3, . . . , we should regard 0 as the simplest value and give it the highest prior probability, regard 1 as the next most simple and give it the next highest prior probability, regard 2 as the next, and so on, ranking the numbers in probability according to their natural order. If this suggestion is followed and the procedure leading to (9.16.3) is used, the following prior probability function for θ results:

$$\phi(x)) = 1/[(x+1)(x+2)] \quad (x = 0, 1, 2, \ldots). \qquad (9.17.1)$$

If, on the other hand, θ can take the value of any *positive or negative* integer, then we might regard 0 as the most simple value, 1 and -1 as equal second and third, 2 and -2 as equal fourth and fifth, and so on, and rank them in probability accordingly. If we then used the rule of (9.16.5) we should obtain the following prior probability function for θ.

$$\phi(x) = 1\,/\,4x(x+1) \quad (x = \pm 1, \pm 2, \pm 3, \ldots)$$
$$\phi(0) = \tfrac{1}{2}. \qquad (9.17.2)$$

It is emphasised that the above rules are no more than suggestions for use, when all other methods of deciding on the probabilities of the different values of the parameter break down. If, in any particular case, it is found that the outcome of using these rules seems clearly wrong, then this indicates that something else *can* be said about the subjective probabilities of the various hypotheses or parameter values considered. In such a case, prior probabilities which seem less wrong should certainly be substituted for those obtained by use of the conventional rules.

9.18. Prior Probabilities Based on Common Sense

So far, we have supposed that only one person is concerned with assigning probabilities to different hypotheses or parameter values. More usually a number of persons will be concerned. It is unlikely that the probability orderings of the different individuals will exactly agree.

If they are to collaborate in using a prior probability function to combine with statistical evidence in forming estimates, then this prior probability function should somehow reflect their common sense of the prior probabilities, and yet give some weight to peculiar individual points of view. A procedure which should have this effect is the following. Each individual should assign his own prior probability function to the parameter values (or hypotheses) and a composite prior probability function should be formed, by taking the unweighted arithmetic mean of the individual functions.

9.19. Prior Probability Functions Paying Due Weight to Various Schools of Thought

An individual who has a fairly clear subjective probability ordering for a set of parameter-values or hypotheses, may yet wish to give some weight to schools of thought with which he personally disagrees. In this case he may work out prior probability functions for the parameters according to the judgements he supposes each school of thought to subscribe to. He may then combine them by taking an unweighted arithmetic mean of the prior probabilities appropriate to each school of thought.

If, as is probable, he cannot quite bring himself to allow the other schools of thought to have equal weight with his own, he may first use the rules of § **9.16** to attribute weights to the schools of thought in accordance with his own probability ranking of the schools of thought. He may then form a prior probability function for the hypotheses, by working out the appropriately weighted arithmetic means of the prior probability functions ascribed on the basis of the various schools of thought. This procedure allows a healthy preponderance of his own views over those of other schools, but allows for the remote probability that subsequent evidence will favour other views against his own. In this unfortunate case, he will be led to a graceful revision of his own published probability judgements more quickly than if he had been faithful to his own disposition to ignore the dissenting schools of thought altogether.

EXAMPLE 9.19.1. There are n alternative hypotheses H_1 to H_n, and *m* rival schools of thought S_1 to S_m, numbered in order of my judgement of their merit. I suppose the probability ranking of hypothesis H_x by school S_y to be R_{xy}. What probabilities would the rules suggested above attribute to the *n* hypotheses?

Solution. The prior probability function attributed to school S_y would be

$$\phi_y[H_x] = (n+1)\,/\,nR_{xy}(R_{xy}+1) \tag{9.19.1}$$

and the weight attached to this function would be

$$w_y = (m+1)\,/\,my(y+1), \tag{9.19.2}$$

so that the prior probabilities which the rule would ascribe to the hypotheses would be

$$\phi[H_x] = \sum_{y=1}^{m} w_Y \phi_y[H_x] = \frac{(m+1)(n+1)}{mn} \sum_{y=1}^{m} \frac{1}{y(y+1)R_{xy}(R_{xy}+1)}. \tag{9.19.3}$$

9.20. The Use of Vague Evidence about Continuous Parameters

In the §§ **9.16** *et seq.* we considered the use of vague evidence in attributing prior probability functions to a set of alternative hypotheses or to the different values which might be taken by a discrete parameter. In this section, we shall consider the more common case, where we wish to take account of vague evidence, in attributing a prior probability density function to a continuous parameter.

Suppose, for example, that we are concerned with the parameters α and β in the model

$$\boldsymbol{Y}_t = \alpha + \beta \boldsymbol{X}_t + \boldsymbol{\eta}_t, \tag{9.20.1}$$

where the $\boldsymbol{\eta}_t$ are independent observations of a normal (0,1) variate, and are independent of $\boldsymbol{X}_t$.

We have the vague information that β is almost certainly positive and less than 2, and that it probably lies between 0·5 and 1·5; we are also told that when $\boldsymbol{X}_t = 5$, $\boldsymbol{Y}_t$ will probably lie somewhere between 80 and 120.

We might first quantify this vague information in the form

$$\begin{aligned} &p'r[0{\cdot}5 < \beta < 1{\cdot}5] \simeq 2/3 \\ &p'r[0 < \beta < 2] > 0{\cdot}95 \\ &p'r[80 < \alpha + 5\beta < 120] \simeq 2/3. \end{aligned} \tag{9.20.2}$$

(9.20.2) still leaves open a very wide choice of bivariate probability density functions for α and β. We may take advantage of this wide scope of choice, to find a suitable prior probability density function, which will yet be of such a form as to enable the calculation and interpretation of the posterior probability function to be carried out with the minimum of trouble.

A sensible method for finding such a prior probability density function for α and β, is to choose one of the same, or very similar, functional form as the likelihood function. Having chosen the form of the prior probability density function in this way, one may then settle the details so as to fit in with the vague evidence as set out in (9.20.2).

In our example, if we had observations (X_t, Y_t), $(t = 1, 2, \ldots, n)$, the likelihood function would be

$$L(\alpha, \beta) = (2\pi)^{-n/2} \exp\left[-\sum_{t=1}^{n} (Y_t - \alpha - \beta X_t)^2/2\right]$$
$$= \exp\left[-A\alpha^2 - 2H\alpha\beta - B\beta^2 - 2G\alpha - 2F\beta - C\right] \quad (9.20.3)$$

where A, B, C, F, G, and H depend on the observations.

Our method prescribes that we should try a prior probability density function of the form

$$\phi(\alpha, \beta) = \exp\left[-a\alpha^2 - 2h\alpha\beta - b\beta^2 - 2g\alpha - 2f\beta - c\right]. \quad (9.20.4)$$

Now (9.20.4) is simply the most general form of a bivariate normal density function, provided the quadratic is negative definite. Accordingly we may look for a bivariate normal prior probability density function for α and β which will satisfy (9.20.2).

Let us put

$$\gamma = \alpha + 5\beta; \quad (9.20.5)$$

then we may give β and γ prior probability density functions such that β is a normal variate with mean 1 and variance 0·25 and γ is an independent normal variate with mean 100 and variance 400. This will meet the conditions (9.20.2). Accordingly, we take the prior probability density function

$$\psi(\beta, \gamma) = \frac{1}{20\pi} \exp\left[-2(\beta-1)^2 - \frac{(\gamma-100)^2}{800}\right] \quad (9.20.6)$$

and, since the Jacobian $\dfrac{\partial(\beta, \gamma)}{\partial(\beta, \alpha)} = 1$, we may transform to

$$\phi(\alpha, \beta) = \frac{1}{20\pi} \exp\left[-\tfrac{65}{32}(\beta-1)^2 - \tfrac{1}{80}(\beta-1)(\alpha-95) - \tfrac{1}{800}(\alpha-95)^2\right]. \quad (9.20.7)$$

The prior probability density function (9.20.7) has the advantage that it gives expression to the vague evidence incorporated in (9.20.2), and is of such a functional form that any posterior probability density function for α and β which may result from observations (X_t, Y_t) $(t = 1, 2, 3, \ldots)$ will be of the same functional form, namely the bivariate normal. (See Exercise 6 at the end of this chapter.)

The above example is of course considerably simpler than any which is likely to be met in practice, but just the same general principle can always be applied with advantage, where there is vague evidence about the values of continuous parameters. The evidence should be absorbed into a prior probability density function which is of the same functional form as the likelihood function, for this will result in the posterior probability density function also being of that functional form.

Exercises 5, 6 and 12 at the end of this chapter will throw some further light on the use of this method. The reader who wishes for a more exhaustive treatment should consult Raiffa and Schlaifer.†

In the next few sections we shall consider the more difficult case, where we are in complete ignorance as to the values of the continuous parameters.

9.21. Inefficiency of a Prior Probability Density Function

We have seen that Bayes' postulate does not give clear guidance how to choose a prior probability density function for one or more parameter values. In seeking some other rule we shall be guided by the wish to avoid certain forms of inefficiency in the prior probability density function.

The extreme form of inefficiency is to rule out completely a range of possible values of the parameter. For example, if the probability density function for a variate $\boldsymbol{x}$ is

$$f(x \mid \theta = t) = te^{-tx} \quad (x > 0) \quad (t > 0), \tag{9.21.1}$$

it would be extremely inefficient to take for the prior probability density function for θ

$$\phi(t) = 1/t^2 \quad (t > 1) \qquad \phi(t) = 0 \quad (t < 1), \tag{9.21.2}$$

unless it was *known* that θ could not take values less than unity.

There is a lesser form of inefficiency which we shall also seek to avoid. This form reveals itself, when we have practically ruled out a range of values of the parameter, and then find, after collecting some

† *Applied Statistical Decision Theory*, Howard Raiffa and Robert Schlaifer. Published 1961 by Research Division of Harvard Business School.

observations, that the (posterior) probability that the parameter value lies, after all, in this range, is very high. This kind of inefficiency turns out to be an elusive concept and one which it is hard to define precisely.

Suppose that we have chosen a prior probability density function $\phi(t)$ for a parameter value θ, over a finite range of values, $-T$ to T. We collect a large number, N, of observations, and drive the posterior probability density function for θ, which we may denote by $\beta(t)$, taken over the range $-T$ to T. Inefficiency in $\phi(t)$ would be revealed, by there being a range R of values of θ, such that both

$$\int_R \phi(t)\,dt \quad \text{and} \quad \left[1-\int_R \beta(t)\,dt\right] \quad \text{were very small.}$$

In order to define a measure of this kind of inefficiency, we choose arbitrarily a quantity slightly less than unity, say 0·99. We measure the *revealed inefficiency* $i[\phi, \beta, N]$ as follows, by comparing $\beta(t)$ and $\phi(t)$

$i[\phi, \beta, N]$ = the least p, such that for some interval R,

$$\int_R \beta(t)\,dt = 0{\cdot}99 \quad \text{and} \quad \sqrt{N}\int_R \phi(t)\,dt \leqslant p.$$

Thus, if with $N = 10\,000$, we could find an interval R such that the posterior probability that θ lay in R was 0·99, whereas the prior probability was only 0·002, and if there was no interval with posterior probability 0·99 and prior probability less than 0·002, we should find $i(\phi, \beta, N) = 0{\cdot}2$.

Thus defined, $i(\phi, \beta, N)$ is a variate whose value will depend on the observations. But, with N, ϕ and θ given, $i[\phi, \beta, N]$ will have an expected value which we may denote by $J(\theta, \phi)$. This measure $J(\theta, \phi)$ will depend to some extent on N, but it can be shown that it will converge rapidly to a limit as N becomes large. We may therefore refer to this limiting value by the notation $J(\theta, \phi)$.

It is *small* values of $i[\phi, \beta, N)$ which reveal the kind of inefficiency we wish to avoid, and it is therefore small values of $J(\theta, \phi)$ which we should try to rule out when choosing ϕ. For given ϕ, $J(\theta, \phi)$ will have some greatest lower bound (i.e. roughly speaking, some minimum value), over the interval $-T < \theta < T$ and we may denote this by $M[\phi(t)]$. This may be taken as reflecting the inefficiency of the prior probability function $\phi(t)$ over the interval $-T$ to T. Small values of $M[\phi(t)]$ correspond to great inefficiency.

A criterion for choosing the prior probability density function $\phi(t)$ is to try to make $M[\phi(t)]$ as large as possible. It can be shown (see Appendix 9) that this can be done by making $\phi(t)$ proportional to the square root of a function usually denoted by I, which is commonly known as the *Amount of Information*. (See Kendall and Stuart, *Advanced Theory of Statistics*, Vol. 2, p. 10.) I may be defined as follows:

$$I(t) = E\left[\left(\frac{\partial \log f}{\partial t}\right)^2\right] = \int_{-\infty}^{\infty} f(x \mid \theta = t)\left(\frac{\partial \log f}{\partial t}\right)^2 dx, \qquad (9.21.3)$$

where $f(x \mid \theta = t)$ is the probability density function, the value, t, of whose parameter, θ, we are concerned to estimate. If $\boldsymbol{x}$ is a discrete and not a continuous variate, then the integral in (9.20.3) should be replaced by the corresponding sum.

The solution

$$\phi(t) \propto \sqrt{[I(t)]} \qquad (9.21.4)$$

is the same as was proposed by Professor Harold Jeffreys in the second and later editions of *The Theory of Probability*. It can be expressed in other ways, one of which is the following.

Let k_N denote a set of N observations of $\boldsymbol{x}$. Then the maximum likelihood estimator $t^+(k_N)$ of θ, based on the observations k_N, is that value of θ which maximises the probability, given θ, of the observations k_N. (For a fuller explanation see § **12.2** (in Volume 2).) The maximum likelihood estimator $t^+(k_N)$ is itself a variate, since it depends on the observations; and it has its own probability density function $g(t^+, t)$ depending on t, the true value of θ. Taking N as very large, and ignoring terms of order $(1/N)$, it may be shown that rule (9.21.4) may also be expressed as

$$\phi(t) \propto g(t, t). \qquad (9.21.5)$$

Thus for each value t, of θ, the prior probability density for θ should be taken as proportional to the probability density (given that value of θ) of the maximum likelihood estimator, at the correct value, in large samples.

A great advantage of the rule embodied in (9.21.4) or (9.21.5) is that, unlike Bayes' postulate, its operation is unaffected by a transformation from θ to some monotonic function of θ. Furthermore the rule generalises without difficulty to the case where several parameters are involved in place of only one. Before proceeding to this generalisation, it may be useful to consider some examples of the procedure for attributing a prior probability density function for a single parameter.

9.22. Examples Involving the Choice of a Prior Probability Density Function for a Single Parameter

EXAMPLE 9.22.1. The variate $\boldsymbol{x}$ has the probability density function

$$f(x \mid \theta = t) = te^{-tx} \qquad x > 0 \quad t > 0. \qquad (9.22.1)$$

Choose a prior probability density function for θ by the method of section **21** of this chapter.

Solution. The amount of information I is given by

$$I = \int_0^\infty f(x \mid \theta = t)\left(\frac{\partial \log f}{\partial t}\right)^2 dx = \int_0^\infty tx^2 e^{-tx} dx = t^{-2}\int_0^\infty y^2 e^{-y} dy$$

$$\therefore \quad I \propto t^{-2}. \tag{9.22.2}$$

Hence, by rule (9.21.4), we put

$$\phi(t) \propto \frac{1}{t}. \tag{9.22.3}$$

Such a prior probability density function is of course 'improper' in the sense that it is impossible to choose any constant C such that $\int_0^\infty Ct^{-1} dt = 1$. As we have illustrated in sections **9.14** and **9.15**, the fact that a probability function or probability density function is 'improper' need not deter us from using it as a prior probability function or density function, since the multiplying constant cancels out in the expression for the posterior probability function or density function.

EXAMPLE 9.22.2. The discrete variate $\boldsymbol{x}$ has the probability function

$$f(0 \mid \theta = t) = 1-t;\ f(1 \mid \theta = t) = t;\ f(x \mid \theta = t) = 0 \text{ otherwise}$$
$$(0 < t < 1). \tag{9.22.4}$$

Suggest a prior density function for θ.

Solution. The expression for I will be a sum rather than an integral, since $\boldsymbol{x}$ is a discrete variate.

$$I = (1-t)\left[\frac{\partial \log f}{\partial t}\right]^2_{x=0} + t\left[\frac{\partial \log f}{\partial t}\right]^2_{x=1}. \tag{9.22.5}$$

$$I = \frac{1}{1-t} + \frac{1}{t} = \frac{1}{t(1-t)}. \tag{9.22.6}$$

Hence by rule (9.21.4) we put

$$\phi(t) = [t(1-t)]^{-1/2} \Big/ \int_0^1 [t'(1-t')]^{-1/2} dt', \tag{9.22.7}$$

and on integrating the denominator, we obtain

$$\phi(t) = \frac{1}{\pi\sqrt{[t(1-t)]}}. \qquad (9.22.7)$$

EXAMPLE 9.22.3. The variate-pair $(\mathbf{x}, \mathbf{y})$ has the probability density function

$$f(x, y \mid \theta = t) = e^{-(x^2+2xy\sin t+y^2)}(\cos t) / \sqrt{\pi}. \qquad (9.22.8)$$

Suggest a prior probability density function for θ.

Solution.

$$I = \int_{-\infty}^{\infty}\int_{-\infty}^{\infty} f\left(\frac{\partial \log f}{\partial t}\right)^2 dx\, dy. \qquad (9.22.9)$$

$$\therefore\; I = -\int_{-\infty}^{\infty}\int_{-\infty}^{\infty} \frac{1}{\sqrt{\pi}} e^{-(x^2+2xy\sin t+y^2)}(\tan t+2xy\cos t)\, dx\, dy$$

$$= \int_{-\infty}^{\infty}\int_{-\infty}^{\infty} \frac{1}{\sqrt{\pi}} e^{-(u^2+v^2)}[\tan t+2v(u-v\tan t)]^2\, du\, dv,$$

where $\quad u = x+y\sin t \quad$ and $\quad v = y\cos t. \qquad (9.22.10)$

$$\therefore\; I = 1+2\tan^2 t. \qquad (9.22.11)$$

Hence, applying rule (9.21.4), we put

$$\phi(t) = \sqrt{(1+2\tan^2 t)} \bigg/ \int_{\pi/2}^{-\pi/2} \sqrt{(1+2\tan^2 u)}\, du \qquad (9.22.11)$$

$$(-\pi/2 < t < \pi/2).$$

9.23. The Choice of a Prior Probability Density Function for Two or More Parameters

In § **9.21** we saw that the maximum inefficiency could be kept to a minimum, by taking as the prior density function for a single parameter θ, the function

$$\phi(t) \propto \sqrt{I(t)}. \qquad (9.23.1)$$

The same formula holds for the case where two or more parameters $\theta_1, \theta_2, \ldots, \theta_m$ are involved. The general case can be fully illustrated by the case when $m = 2$. In order to keep the maximum inefficiency to a minimum, we should use the prior probability density function for θ_1, θ_2:

$$\phi(t_1, t_2) \propto \sqrt{I(t_1, t_2)}, \qquad (9.23.2)$$

where the amount of information $I(t_1, t_2)$ is defined as the determinant

$$I(t_1, t_2) = \begin{vmatrix} E\left[\dfrac{\partial \log f}{\partial t_1}\dfrac{\partial \log f}{\partial t_1}\right], & E\left[\dfrac{\partial \log f}{\partial t_1}\dfrac{\partial \log f}{\partial t_2}\right] \\ E\left[\dfrac{\partial \log f}{\partial t_2}\dfrac{\partial \log f}{\partial t_1}\right], & E\left[\dfrac{\partial \log f}{\partial t_2}\dfrac{\partial \log f}{\partial t_2}\right] \end{vmatrix} \tag{9.23.3}$$

where $f(x \mid \theta_1 = t_1, \theta_2 = t_2)$ is the probability density function, for whose parameters we seek a prior probability density function.

The separate elements of the determinant (9.23.3) may be evaluated by means of the formula

$$E\left[\frac{\partial \log f}{\partial t_r}\frac{\partial \log f}{\partial t_s}\right] = \int_{-\infty}^{\infty} f[x \mid \theta_1 = t_1, \theta_2 = t_2]\frac{\partial \log f}{\partial t_r}\frac{\partial \log f}{\partial t_s}\,dx. \tag{9.23.4}$$

EXAMPLE 9.23.1. The variate $\boldsymbol{x}$ has the probability density function

$$f(x \mid \theta_1 = t_1, \theta_2 = t_2) = t_2 g(y), \tag{9.23.5}$$

where $g(y)$ is a known probability density function and where

$$y = t_1 + t_2 x. \tag{9.23.6}$$

Suggest a prior probability density function for the parameters θ_1 and θ_2.

Solution. Let $\boldsymbol{u}_1$ and $\boldsymbol{u}_2$ denote $\dfrac{\partial \log f}{\partial t_1}$ and $\dfrac{\partial \log f}{\partial t_2}$ and let $\boldsymbol{u}$ denote $\dfrac{d \log g}{dy}$.

Then, by (9.23.5)

$$\log f = \log t_2 + \log g(y), \tag{9.23.7}$$

so that by (9.23.6)

$$\boldsymbol{u}_1 = \boldsymbol{u} \quad \text{and} \quad \boldsymbol{u}_2 = \frac{1}{t_2} + \boldsymbol{u} t_2. \tag{9.23.8}$$

Now,

$$E(\boldsymbol{u}) = \int_{-\infty}^{\infty} g(y)\frac{d \log g}{dy}\,dy, \tag{9.23.9}$$

since $g(y)$ is the density function for $\boldsymbol{y}$.

Hence

$$E(\boldsymbol{u}) = \int_{-\infty}^{\infty} \frac{dg}{dy}\,dy = \lim_{Y\to\infty} [g(Y)-g(-Y)] = 0, \tag{9.23.10}$$

since $g(y)$ is a density function.

$$\therefore\quad E(\boldsymbol{u}_1) = 0,\ E(\boldsymbol{u}_2) = \frac{1}{t_2},\ E(\boldsymbol{u}_1^2) = E(\boldsymbol{u}^2)$$

$$E(\boldsymbol{u}_1\boldsymbol{u}_2) = E(\boldsymbol{u}_2\boldsymbol{u}_1) = t_2E(\boldsymbol{u}^2);\ E(\boldsymbol{u}_2^2) = \frac{1}{t_2^2}+t_2^2E(\boldsymbol{u}^2). \tag{9.23.11}$$

Now the amount of information is by (9.23.3)

$$I(t_1, t_2) = E(\boldsymbol{u}_1^2)E(\boldsymbol{u}_2^2)-[E(\boldsymbol{u}_1\boldsymbol{u}_2)]^2 \tag{9.23.12}$$

$$\therefore\quad I(t_1, t_2) = \frac{E(\boldsymbol{u}^2)}{t_2^2}. \tag{9.23.13}$$

Since $E(\boldsymbol{u}^2)$ depends only on the density function for $\boldsymbol{y}$, namely $g(y)$, it is independent of t_1 and t_2. Hence the rule (9.23.2) leads us to the prior density function for $\boldsymbol{\theta}_1$ and $\boldsymbol{\theta}_2$

$$\phi(t_1, t_2) \propto \frac{1}{t_2}, \tag{9.23.14}$$

which is improper in the sense of § **9.12**, but no less applicable for that reason.

In this example, $\boldsymbol{\theta}_1$ was a *location parameter* and $\boldsymbol{\theta}_2$ was a *scale parameter*, since the effect of a change in the value of $\boldsymbol{\theta}_1$ is to alter the origin of measurement, and the effect of a change in the value of $\boldsymbol{\theta}_2$ is to alter the scale of measurement, for the variate $\boldsymbol{x}$. Accordingly, (9.23.14) suggests that whenever we are faced with the choice of a prior probability density function for a location parameter $\boldsymbol{\theta}_1$ and a scale parameter $\boldsymbol{\theta}_2$, we should take the prior probability density as inversely proportional to the scale parameter and independent of the location parameter.

9.24. Choice of Prior Probability Density Function for Several Parameters

The only modification required to equations (9.23.2) and (9.23.3) when we are concerned with more than two parameters is that the determinant for the amount of information will now be

$$I(t_1, t_2, \ldots, t_m) = \left| E\left[\frac{\partial \log f}{\partial t_r}\frac{\partial \log f}{\partial t_s}\right]\right| \tag{9.24.1}$$

consisting of m rows and m columns of elements, of which that in the r-th row and s-th column is $E\left[\frac{\partial \log f}{\partial t_r}\frac{\partial \log f}{\partial t_s}\right]$.

(9.24.1) may also be written in the form

$$I(t_1, t_2, \ldots, t_m) = \left| -E\left[-\frac{\partial^2 \log f}{\partial t_r \partial t_s}\right]\right|. \tag{9.24.2}$$

To verify the equivalence of (9.24.1) and (9.24.2), it is sufficient to show that

$$E\left[\frac{\partial \log f}{\partial t_r}\frac{\partial \log f}{\partial t_s}+\frac{\partial^2 \log f}{\partial t_r \partial t_s}\right] = 0. \tag{9.24.3}$$

To prove (9.24.3) we note that since

$$\int_{-\infty}^{\infty} f(x)\,dx = 1 \tag{9.24.4}$$

$$\int_{-\infty}^{\infty} \frac{\partial^2 f}{\partial t_r \partial t_s}\,dx = \frac{\partial^2}{\partial t_r \partial t_s}\int_{-\infty}^{\infty} f(x)\,dx = 0. \tag{9.24.5}$$

Now

$$\frac{\partial^2 f}{\partial t_r \partial t_s} = \frac{\partial}{\partial t_r}\left[f\frac{\partial \log f}{\partial t_s}\right] = f\left[\frac{\partial \log f}{\partial t_r}\frac{\partial \log f}{\partial t_s}+\frac{\partial^2 \log f}{\partial t_r \partial t_s}\right], \tag{9.24.6}$$

so that by (9.24.5)

$$\int_{-\infty}^{\infty} f\left[\frac{\partial \log f}{\partial t_r}\frac{\partial \log f}{\partial t_s}+\frac{\partial^2 \log f}{\partial t_r \partial t_s}\right]dx = 0, \tag{9.24.7}$$

which establishes (9.24.3) and hence that (9.24.1) and (9.24.2) are equivalent.

EXAMPLE 9.24.1. The variate y is assumed to be normally distributed with variance θ_4^2 about the regression plane

$$y = \theta_1+\theta_2 x_2+\theta_3 x_3. \tag{9.24.8}$$

Observations Y_i, X_{2i}, X_{3i} ($i = 1, 2, \ldots, n$) are obtained. Suggest a prior probability density function for θ_1, θ_2, θ_3 and θ_4, and hence obtain their posterior probability density function in the light of the observations.

Solution. Let

$$z = y-t_1-t_2 x_2-t_3 x_3; \tag{9.24.9}$$

then, when $\theta_1 = t_1$, $\theta_2 = t_2$, $\theta_3 = t_3$ and $\theta_4 = t_4$, z is normally distributed, independent of x_2 and x_3 about mean zero with variance t_4^2. Hence the probability density function $g(z)$ for z satisfies

$$\log g(z) = -\tfrac{1}{2}\log(2\pi) - \log t_4 - \frac{z^2}{2t_4^2}. \qquad (9.24.10)$$

Let $f(Y_1, \ldots, Y_n)$ denote the probability density function for the n observations of y, then, where

$$Z_i = Y_i - t_1 - t_2 X_{2i} - t_3 X_{3i} \quad (i = 1, 2, \ldots, n), \qquad (9.24.11)$$

$$\log f = -\sum_{i=1}^{n} \log g(Z_i) = -\frac{n}{2}\log(2\pi) - n\log t_4 - \frac{1}{2t_4^2}\sum_{i=1}^{n} Z_i^2 \qquad (9.24.12)$$

Let $\quad X_{1i} = 1 \quad (i = 1, 2, \ldots, n)$,

and let $\quad c_{rs} = \sum_{i=1}^{n} X_{ri}X_{si} \quad (r, s = 1, 2, 3);$ then $\qquad (9.24.13)$

$$-\frac{\partial^2 \log f}{\partial t_r \partial t_s} = \frac{c_{rs}}{t_4^2} \quad (r, s = 1, 2, 3)$$

$$-\frac{\partial \log f}{\partial t_4} = \frac{n}{t_4} - \frac{1}{t_4^3}\sum_{i=1}^{n} Z_i^2$$

$$-\frac{\partial^2 \log f}{\partial t_r \partial t_4} = \frac{2}{t_4^3}\sum_{i=1}^{n} X_{ri}Z_i \quad (r = 1, 2, 3)$$

$$-\frac{\partial^2 \log f}{\partial t_4^2} = -\frac{n}{t_4^2} + \frac{3}{t_4^4}\sum_{i=1}^{n} Z_i^2. \qquad (9.24.14)$$

Since z has the probability density function (9.24.10),

$$E(Z_i^2) = t_4^2;\; E(X_{ri}Z_i) = 0 \quad (i = 1, 2, \ldots, n), \; (r = 1, 2, 3). \qquad (9.24.15)$$

Hence by (9.24.1) and (9.24.14)

$$I(t_1, t_2, t_3, t_4) = \begin{vmatrix} \frac{c_{11}}{t_4^2} & \frac{c_{12}}{t_4^2} & \frac{c_{13}}{t_4^2} & 0 \\ \frac{c_{21}}{t_4^2} & \frac{c_{22}}{t_4^2} & \frac{c_{23}}{t_4^2} & 0 \\ \frac{c_{31}}{t_4^2} & \frac{c_{32}}{t_4^2} & \frac{c_{33}}{t_4^2} & 0 \\ 0 & 0 & 0 & \frac{2n}{t_4^2} \end{vmatrix} \qquad (9.24.16)$$

$$\therefore \quad I(t_1, t_2, t_3, t_4) = Cnt_4^{-8}, \tag{9.24.17}$$

where C is a determinant independent of t_1, t_2, t_3, and t_4.

Hence we take for the prior probability density function for $(\theta_1, \theta_2, \theta_3, \theta_4)$

$$\phi(t_1, t_2, t_3, t_4) \propto t_4^{-4}. \tag{9.24.18}$$

By applying Bayes' Theorem, we find the posterior probability density function as

$$p(t_1, t_2, t_3, t_4) \propto t_4^{-2(n+2)} \exp\left[-\sum_{i=1}^{n} (Y_i - t_1 - t_2 X_{2i} - t_3 X_{3i})^2 \Big/ 2t_4^2 \right] \tag{9.24.19}$$

9.25. Marginal Posterior Probability Density Functions

The typical estimation problem will involve the estimation of several parameters. Whilst, in principle, the task is completed when the posterior probability function has been provided, as for example in solution (9.24.19), such a manner of presentation is too generalised to be altogether informative. One also wishes to know the (marginal) posterior probability density function for each parameter separately.

If we have found the overall posterior probability density function $\beta(t_1, t_2, t_3, t_4)$ for four parameters $\theta_1, \theta_2, \theta_3$ and θ_4, the marginal probability density function for one of them, θ_1 say, will be found as

$$\beta_1(t) \propto \int_{-\infty}^{\infty} \int_{-\infty}^{\infty} \int_{-\infty}^{\infty} \beta(t_1, t_2, t_3, t_4)\, dt_2\, dt_3\, dt_4. \tag{9.25.1}$$

The task of evaluating such integrals exactly is often very heavy. Usually, however, numerical tables of the marginal probability density functions can be obtained quite quickly by means of an electronic computer, to as high a degree of accuracy as can be useful. In some cases the integration can be carried out by hand with the aid of some approximations.

A number of examples of more elaborate applications of the various procedures discussed in this chapter will now be considered.

9.26. Further Examples of the Application of Bayes' Theorem

In this section a fairly wide range of examples of the applications of the procedures already described will be provided.

EXAMPLE 9.26.1. The variate $\boldsymbol{x}$ has the probability density function

$$f(x \mid \theta_1 = t_1, \theta_2 = t_2) = t_2^{t_1} x^{t_1 - 1} \exp(-t_2 x) / \Gamma(t_1) \quad (0 < x < 1/t_2). \tag{9.26.1}$$

(*i*) Suggest a prior probability density function for $\boldsymbol{\theta}_1$ if it is known that $\boldsymbol{\theta}_2 = 1$.
(*ii*) Suggest a prior probability density function for $(\boldsymbol{\theta}_1, \boldsymbol{\theta}_2)$, when both parameters are unknown.
(*iii*) Derive the marginal prior probability density function for $\boldsymbol{\theta}_1$ from (*ii*) and compare it with the solution (*i*).

Solution. (*i*) In this case

$$f(x \mid \theta_1 = t_1) = x^{t_1 - 1} e^{-x} / \Gamma(t_1) \quad (0 < x < 1) \tag{9.26.2}$$

$$\log f = (t_1 - 1) \log x - x - \log(\Gamma(t_1)). \tag{9.26.3}$$

By (9.24.2) the amount of information is

$$I(t_1) = -E\left[\frac{\partial^2 \log f}{\partial t_1^2}\right] = \frac{\partial^2 \log \Gamma(t_1)}{\partial t_1^2}.$$

Accordingly, we take as the prior probability density function for $\boldsymbol{\theta}_1$

$$\phi(t_1) \propto \sqrt{\left[\frac{\partial^2 \log \Gamma(t_1)}{\partial t_1^2}\right]}. \tag{9.26.4}$$

Whilst it is not possible to express $\dfrac{\partial^2 \log \Gamma(t_1)}{\partial t_1^2}$ simply, in terms of more elementary functions, it is possible to expand it in series in various simple manners.

Thus

$$\frac{\partial^2 \log \Gamma(t_1)}{\partial t_1^2} = \sum_{r=0}^{\infty} \left[\frac{1}{(t_1 + r)^2}\right], \tag{9.26.5}$$

whence also

$$\frac{\partial^2 \log \Gamma(t_1)}{\partial t_1^2} = \frac{2}{2t - 1} - \sum_{r=0}^{\infty} \left[\frac{1}{(t_1 + r)^2 \{4(t_1 + r)^2 - 1\}}\right]. \tag{9.26.6}$$

We can see from equation (9.26.6) that the prior probability density function (9.26.4) is 'improper' in the sense explained in § **9.12**. Equation (9.26.6) would enable $\phi(t_1)$ to be very easily tabulated on a computer.

(*ii*) From equation (9.26.1) we obtain

$$\log f = t_1 \log t_2 + (t_1 - 1) \log x - \log[\Gamma(t_1)]. \tag{9.26.7}$$

Hence by equation (9.24.2), the amount of information is

$$I(t_1, t_2) = \begin{vmatrix} \dfrac{\partial^2 \log \Gamma(t_1)}{\partial t_1^2} & -\dfrac{1}{t_2} \\ -\dfrac{1}{t_2} & \dfrac{t_1}{t_2^2} \end{vmatrix}. \tag{9.26.8}$$

$$\therefore \quad I(t_1, t_2) = \frac{1}{t_2^2}\left[t_1 \frac{\partial^2 \log \Gamma(t_1)}{\partial t_1^2} - 1\right]. \tag{9.26.9}$$

Hence we take for the prior probability density function for $\boldsymbol{\theta}_1$ and $\boldsymbol{\theta}_2$,

$$\phi(t_1, t_2) \propto \frac{1}{t_2} \sqrt{\left[t_1 \frac{\partial^2 \log \Gamma(t_1)}{\partial t_1^2} - 1\right]}. \tag{9.26.10}$$

This factorises into the product of a function of t_1 and a function of t_2; which implies that, according to this prior probability density function, $\boldsymbol{\theta}_1$ and $\boldsymbol{\theta}_2$ are independent variates. It is therefore natural to take the marginal probability density functions for $\boldsymbol{\theta}_1$ and $\boldsymbol{\theta}_2$ separately as

$$\phi_1(t_1) \propto \sqrt{\left[t_1 \frac{\partial^2 \log \Gamma(t_1)}{\partial t_1^2} - 1\right]} \qquad \phi_2(t_2) \propto \frac{1}{t_2}. \tag{9.26.11}$$

However since the prior probability density function $\phi(t_1, t_2)$ is an 'improper' one, it is not legitimate to calculate the marginal density functions in this manner. The choice of marginal density functions will depend on the way we allow the area (in the $t_1 - t_2$ plane) over which we integrate $\phi(t_1, t_2)$ to tend to infinity. If we keep the area rectangular, with sides parallel to the t_1- and t_2-axes, then, when we allow the sides of the rectangle to tend to infinity, we shall be led to choose the marginal prior probability density functions $\phi_1(t_1)$ and $\phi_2(t_2)$ of (9.26.11). But, had we chosen some other shape of area over which to integrate, we could have been led to adopt other marginal prior probability density functions. This need cause no inconvenience, however, because, if there are two parameters involved, then it is only the bivariate prior probability density function that is needed. From this, and from the evidence, the *posterior* marginal probability density functions for each parameter can be unambiguously determined provided at least two observations are available. For with two or more observations available, the posterior probability density function will be 'proper', although the prior probability density function is 'improper'.

It can be shown, by reductio ad absurdum, that the marginal prior probability density function $\phi_1(t_1)$ of (9.26.11) is not proportional to the prior probability density function $\phi(t_1)$ of (9.26.4). For the proportionality of $\phi_1(t_1)$ and $\phi(t_1)$ would imply the existence of a constant λ satisfying

$$[\phi(t_1)]^2(t_1-\lambda) = 1 \quad \text{for all positive } t_1; \tag{9.26.12}$$

whence

$$[\phi(t_1)]^2-[\phi(t_1+1)^2 = 1/[(t_1-\lambda)(t_1+1-\lambda)], \tag{9.26.13}$$

whereas by (9.26.4)

$$[\phi(t_1)]^2-[\phi(t_1+1)]^2 \propto 1/t_1^2 \tag{9.26.14}$$

and (9.26.13) and (9.26.14) cannot both hold exactly.

9.27. A More General Multi-Variate Model

The following multi-parameter estimation problem can arise in many contexts. n sets of observations of variates $\boldsymbol{y}_1, \boldsymbol{y}_2, \ldots, \boldsymbol{y}_k$, under conditions $\boldsymbol{x}_1, \boldsymbol{x}_2, \ldots, \boldsymbol{x}_m$, are obtained, and it is required to estimate the parameters c_{rs} and t_{ij} in the model

$$\boldsymbol{Y} = \boldsymbol{XC}+\mathbf{Z} \tag{9.27.1}$$

$$\mathbf{Z}\boldsymbol{T} = \boldsymbol{\eta}. \tag{9.27.2}$$

$\boldsymbol{Y}$ denotes the $n\times k$ matrix of observations of the variates $\boldsymbol{y}_s$, $\boldsymbol{X}$ denotes the $n\times m$ matrix of the m conditions $\boldsymbol{x}_r$ and $\boldsymbol{C}$ denotes the $m\times k$ matrix of unknown regression coefficients. $\mathbf{Z}$ denotes an $n\times k$ matrix of deviations from the true regression and these are accounted for by (9.27.2) in terms of $\boldsymbol{\eta}$, an $n\times k$ matrix each of whose elements is an independent observation of a normal variate of zero mean and unit variance. $\boldsymbol{T}$ is an upper triangular matrix whose terms t_{ij} are zero when $i>j$, and unknown when $i\leqslant j$. The elements of $\mathbf{Z}$ may then be regarded as n sets of observations of k (possibly) correlated normal variates.

We shall suppose that the number n of sets of observations is as great as the number, $k[2m+k+1]/2$, of parameters c_{rs} and t_{ij} which we are required to estimate. We may then apply the usual procedures for selecting a prior probability density function for the parameters, and then use Bayes' Theorem to find the posterior probability density function for them. In order that this should be reasonably informative, it is necessary to obtain marginal posterior probability density functions for single parameters, and possibly for pairs of parameters, by integrating out the other parameters. If k and m exceed 2 or 3, this may be a formidable task, even with the aid of an electronic computer, unless some use is made of approximations.

EXAMPLE 9.27.1. We shall consider the case where $k=3$ and $m=2$. We shall also suppose that the condition $\boldsymbol{x}_1$ is a dummy with value always set equal to unity. The general case with unrestricted values of k and m differs only in complexity from this case.

First, we find the prior probability density function for the 12 unknown parameters. Let us denote them by

$$\theta_1 = c_{11},\ \theta_2 = c_{12},\ \theta_3 = c_{13},\ \theta_4 = c_{21},\ \theta_5 = c_{22},\ \theta_6 = c_{23},$$

$$\theta_7 = t_{11},\ \theta_8 = t_{12},\ \theta_9 = t_{22},\ \theta_{10} = t_{13},\ \theta_{11} = t_{23},\ \theta_{12} = t_{33}. \qquad (9.27.3)$$

As usual we set the prior probability density function for the 12 parameters proportional to the square root of the amount of information, I, where in this case I is the determinant $|\,A\,|$ of the 12×12 matrix A whose general term is given by

$$a_{uv} = -E\left[\frac{\partial^2 \log L}{\partial\theta_u \partial\theta_v}\right], \qquad (9.27.4)$$

where L is the likelihood.

Fortunately, it can be proved without much difficulty that all the terms a_{uv} for which $u \leqslant 6$ and $v > 6$ are zero, as also are all the terms a_{uv} for which $u > 6$ and $v \leqslant 6$. Assuming this result, it follows that the determinant $|\,A\,|$ factorises as

$$|\,A\,| = |\,A_1\,| \times |\,A_2\,| \qquad (9.27.5)$$

where A_1 and A_2 are the 6×6 matrices,

$$\{a_{uv}\} \text{ with } u \leqslant 6 \text{ and } v \leqslant 6$$

and $$\{a_{uv}\} \text{ with } 12 \geqslant u > 6 \text{ and } 12 \geqslant v > 6. \qquad (9.27.6)$$

The general term of the matrix A_1 is of the form

$$a_{uv} = -E\left[\frac{\partial^2 \log L}{\partial c_{pq} \partial c_{rs}}\right] \qquad (9.27.7)$$

where the logarithm of the likelihood is

$$\log L = -\frac{3n}{2}\log(2\pi) - \sum_{i=1}^{n}\sum_{j=1}^{3}(\eta_{ij}^2/2) + n \log\left|\frac{\partial(\eta_1, \eta_2, \eta_3)}{\partial(y_1, y_2, y_3)}\right| \qquad (9.27.8)$$

where the final term is the logarithm of the Jacobian of the transformation from $\boldsymbol{y}$ to η, and where by (9.27.2), this Jacobian is equal to $t_{11}t_{22}t_{33}$.

It follows that

$$-\frac{\partial^2 \log L}{\partial c_{pq}\partial c_{rs}} = \tfrac{1}{2}\sum_{i=1}^{n}\sum_{j=1}^{3}\frac{\partial^2(\eta_{ij}^2)}{\partial c_{pq}\partial c_{rs}}. \tag{9.27.9}$$

Now

$$\frac{\partial \eta_{ij}}{\partial c_{rs}} = \frac{\partial \eta_{ij}}{\partial z_{is}}\frac{\partial z_{is}}{\partial c_{rs}} = -t_{sj}x_{ir} \tag{9.27.10}$$

$$\therefore \quad \frac{\partial^2(\eta_{ij}^2)}{\partial c_{pq}\partial c_{rs}} = 2t_{qj}t_{sj}x_{ip}x_{ir} \tag{9.27.11}$$

$$\therefore \quad -\frac{\partial^2 \log L}{\partial c_{pq}\partial c_{rs}} = \sum_{i=1}^{n} x_{ip}x_{ir}\sum_{j=1}^{3} t_{qj}t_{sj} = \beta_{pr}\gamma_{qs} \quad \text{(say)} \tag{9.27.12}$$

where β_{pr} is independent of the values of the parameters, and where both β_{pr} and γ_{qs} are independent of what values the variates may take. (The x_{ir} are not here regarded as variates but as predetermined conditions.)

Let $\boldsymbol{T}$ denote the 3×3 (upper triangular) matrix of the t_{pr} and let Γ denote the 3×3 matrix of the γ_{qs}, then by (9.27.12)

$$\Gamma = \boldsymbol{T}'\boldsymbol{T} \tag{9.27.13}$$

$$\therefore \quad |\Gamma| = |\boldsymbol{T}|^2 = t_{11}^2 t_{22}^2 t_{33}^2. \tag{9.27.14}$$

Let $\boldsymbol{B}$ denote the 2×2 matrix of the β_{pr}, then it follows from (9.27.12) that

$$|\boldsymbol{A}_1| = \begin{vmatrix} \beta_{11}\gamma_{11} & \beta_{11}\gamma_{12} & \beta_{11}\gamma_{13} & \beta_{21}\gamma_{11} & \beta_{21}\gamma_{12} & \beta_{21}\gamma_{13} \\ \beta_{11}\gamma_{21} & \beta_{11}\gamma_{22} & \beta_{11}\gamma_{23} & \beta_{21}\gamma_{21} & \beta_{21}\gamma_{22} & \beta_{21}\gamma_{23} \\ \beta_{11}\gamma_{31} & \beta_{11}\gamma_{32} & \beta_{11}\gamma_{33} & \beta_{21}\gamma_{31} & \beta_{21}\gamma_{32} & \beta_{21}\gamma_{33} \\ \beta_{12}\gamma_{11} & \beta_{12}\gamma_{12} & \beta_{12}\gamma_{13} & \beta_{22}\gamma_{11} & \beta_{22}\gamma_{12} & \beta_{22}\gamma_{13} \\ \beta_{12}\gamma_{21} & \beta_{12}\gamma_{22} & \beta_{12}\gamma_{23} & \beta_{22}\gamma_{21} & \beta_{22}\gamma_{22} & \beta_{22}\gamma_{23} \\ \beta_{12}\gamma_{31} & \beta_{12}\gamma_{32} & \beta_{12}\gamma_{33} & \beta_{22}\gamma_{31} & \beta_{22}\gamma_{32} & \beta_{22}\gamma_{33} \end{vmatrix} \tag{9.27.15}$$

$$\therefore \quad |\boldsymbol{A}_1| = |\boldsymbol{B}||\Gamma| = |\boldsymbol{B}|\, t_{11}^2 t_{22}^2 t_{33}^2. \tag{9.27.16}$$

The general term of the 6×6 matrix $\boldsymbol{A}_2$ is of the form

$$a_{uv} = -E\left[\frac{\partial^2 \log L}{\partial t_{pq}\partial t_{rs}}\right]. \tag{9.27.17}$$

Except in the cases $p=q=r=s$,

$$-\frac{\partial^2 \log L}{\partial t_{pq}\partial t_{rs}} = \tfrac{1}{2}\sum_{i=1}^{n}\sum_{j=1}^{3}\left[\frac{\partial^2(\eta_{ij}^2)}{\partial t_{pq}\partial t_{rs}}\right]. \tag{9.27.18}$$

Now, unless $j=q$,
$$\frac{\partial \eta_{ij}}{\partial t_{pq}} = 0; \tag{9.27.19}$$

whereas
$$\frac{\partial \eta_{iq}}{\partial t_{pq}} = z_{qp}. \tag{9.27.20}$$

Hence, unless $q = s = j$,
$$\frac{\partial^2(\eta_{ij}^2)}{\partial t_{pq}\partial t_{rs}} = 0; \tag{9.27.21}$$

whilst
$$\tfrac{1}{2}\frac{\partial^2(\eta_{ij}^2)}{\partial t_{pq}\partial t_{rq}} = z_{ip}z_{ir}. \tag{9.27.22}$$

$$\therefore \quad \tfrac{1}{2}\sum_{i=1}^{n}\sum_{j=1}^{3}\left[\frac{\partial^2(\eta_{ij}^2)}{\partial t_{pq}\partial t_{rs}}\right] = 0, \text{ unless } q = s$$
$$= \sum_{i=1}^{n} z_{ip}z_{ir}$$
$$= \lambda_{pr} \text{ (say), if } q = s. \tag{9.27.23}$$

Hence the general term of the matrix A_2 may take any one of three forms:

$$(i) \quad a_{uv} = -E\left[\frac{\partial^2 \log L}{\partial t_{pq}\,\partial t_{rs}}\right] = 0, \text{ when } q \neq s$$

$$(ii) \quad a_{uv} = -E\left[\frac{\partial^2 \log L}{\partial t_{pq}\,\partial t_{rq}}\right] = E[\lambda_{pr}], \text{ when } p \neq r \text{ or } p \neq q$$

$$(iii) \quad a_{uv} = -E\left[\frac{\partial^2 \log L}{\partial t_{pp}^2}\right] = E[\lambda_{pp}] - n\frac{\partial^2 \log t_{pp}}{\partial t_{pp}^2}$$
$$= E[\lambda_{pp}] + \frac{n}{t_{pp}^2}. \tag{9.27.24}$$

We can now write out $|A_2|$

$$|A_2| = \begin{vmatrix} \frac{n}{t_{11}^2}+E[\lambda_{11}] & 0 & 0 & 0 & 0 & 0 \\ 0 & E[\lambda_{11}] & E[\lambda_{12}] & 0 & 0 & 0 \\ 0 & E[\lambda_{21}] & \frac{n}{t_{22}^2}+E[\lambda_{22}] & 0 & 0 & 0 \\ 0 & 0 & 0 & E[\lambda_{11}] & E[\lambda_{12}] & E[\lambda_{13}] \\ 0 & 0 & 0 & E[\lambda_{21}] & E[\lambda_{22}] & E[\lambda_{23}] \\ 0 & 0 & 0 & E[\lambda_{31}] & E[\lambda_{32}] & \frac{n}{t_{33}^2}+E[\lambda_{33}] \end{vmatrix} \tag{9.27.25}$$

Let L_3 denote the 3×3 matrix $\{\lambda_{rs}\}$, and let N denote the $n \times 3$ matrix $\{\eta_{ij}\}$. Then

$$N'N = T'L_3T$$

$$\therefore \quad |N'N| = |T|^2 |L_3|. \qquad (9.27.26)$$

Now $E(N'N)$ is the diagonal matrix with each of the 3 diagonal elements equal to n, so that its determinant is equal to n^3.

Hence

$$|E(L_3)| = n^3 / |T|^2 = n^3/(t_{11}^2 t_{22}^2 t_{33}^2). \qquad (9.27.27)$$

Similarly

$$\begin{vmatrix} E(\lambda_{11}) & E(\lambda_{12}) \\ E(\lambda_{21}) & E(\lambda_{22}) \end{vmatrix} = n^2/(t_{11}^2 t_{22}^2) \text{ and } E(\lambda_{11}) = n/t_{11}^2. \qquad (9.27.28)$$

Hence

$$\begin{vmatrix} E(\lambda_{11}) & E(\lambda_{12}) & E(\lambda_{13}) \\ E(\lambda_{21}) & E(\lambda_{22}) & E(\lambda_{23}) \\ E(\lambda_{31}) & E(\lambda_{32}) & \dfrac{n}{t_{33}^2}+E(\lambda_{33}) \end{vmatrix} = 2n^3/(t_{11}t_{22}t_{33})^2 \qquad (9.27.29)$$

$$\begin{vmatrix} E(\lambda_{11}) & E(\lambda_{12}) \\ E(\lambda_{21}) & \dfrac{n}{t_{22}^2}+E(\lambda_{22}) \end{vmatrix} = 2n^2/(t_{11}t_{22})^2$$

$$\text{and } \frac{n}{t_{11}^2}+E(\lambda_{11}) = 2n/t_{11}^2 \qquad (9.27.30)$$

$\therefore$ by (9.27.25), (9.27.29) and (9.27.30)

$$|A_2| = 8n^6/(t_{11}^3 t_{22}^2 t_{33})^2. \qquad (9.27.31)$$

Hence by (9.27.5), (9.27.16) and (9.27.31) I, the amount of information, is given by

$$I = |A| = 8n^6 \, |B| \, / \, (t_{11}^2 t_{22})^2. \qquad (9.27.32)$$

The numerator of (9.27.32) is independent of the parameters; hence, when we apply the procedure of making the prior probability density function for the 12 parameters proportional to the square root of the quantity of information, we obtain

$$\phi(c_{11}, c_{12}, c_{13}, c_{21}, c_{22}, c_{23}, t_{11}, t_{12}, t_{22}, t_{13}, t_{23}, t_{33}) \propto t_{33}/t_{11}. \qquad (9.27.33)$$

If we are provided with n independent sets of observations Y_{ir} of the three variates $\boldsymbol{y}_r$ under given values X_{i2} of the condition $\boldsymbol{x}_2$, we may use Bayes' Theorem to establish the posterior probability density function for the 12 parameters.

The likelihood of any set of values of the parameters is

$$L = (2\pi)^{-3n/2}(t_{11}t_{22}t_{33})^n \exp\{-\sum_{i=1}^{n}\sum_{j=1}^{3}\left[\sum_{r=1}^{3} t_{rj}(Y_{ir}-c_{1r}-c_{2r}X_{i2})\right]^2/2\} \tag{9.27.34}$$

and the posterior probability density function for the parameters is

$$\psi(c_{11}, c_{12}, \ldots, t_{23}, t_{33}) = Lt_{33}/t_{11}\int\int\cdots\int\int Lt_{33}t_{11}^{-1}dc_{11}, \ldots, dt_{33}, \tag{9.27.35}$$

where the integration extend over the whole space of the 12 parameters.

A brief consideration of the posterior probability density function (9.27.35) will be given in § **9.28**.

9.28. Interpretation of the Posterior Probability Density Function obtained in § 9.27

In the form (9.27.35), the posterior probability density function is not very informative, because it involves twelve variables simultaneously. The economist is likely to be interested primarily in the probability density function for the six regression coefficients, $c_{11}, c_{12}, c_{13}, c_{21}, c_{22}$ and c_{23}. The first step to be taken is to eliminate the other six parameters, t_{11}, t_{12} etc., by integration.

If we write $\alpha(c_{11}, c_{12}, c_{13}, c_{21}, c_{22}, c_{23})$ for the marginal posterior probability density function for the six regression parameters, then

$$\alpha(c_{11}, \ldots, c_{23}) \propto \int\int\int\int\int\int \psi\, dt_{11}dt_{12}dt_{22}dt_{13}dt_{23}dt_{33}, \tag{9.28.1}$$

where ψ is the posterior probability density function (9.27.35).

An easy way to perform the integration is to define a 3×3 upper triangular matrix, $\{p_{ij}\} = \boldsymbol{P}$, such that

$$\boldsymbol{P}'\boldsymbol{P} = \boldsymbol{Z}'\boldsymbol{Z}, \tag{9.28.2}$$

where $\boldsymbol{Z}$ is the $n\times 3$ matrix introduced in (9.27.1). In this case, if we further define a 3×3 matrix $s_{ij} = \boldsymbol{S}$, such that

$$\boldsymbol{S} = \boldsymbol{PT}, \tag{9.28.3}$$

then

$$\boldsymbol{S} = \begin{pmatrix} p_{11}t_{11} & p_{11}t_{12}+p_{12}t_{22} & p_{11}t_{13}+p_{12}t_{23}+p_{13}t_{33} \\ 0 & p_{22}t_{22} & p_{22}t_{23}+p_{23}t_{33} \\ 0 & 0 & p_{33}t_{33} \end{pmatrix} \quad (9.28.4)$$

so that

$$\frac{\partial(s_{11}, s_{12}, s_{22}, s_{13}, s_{23}, s_{33})}{\partial(t_{11}, t_{12}, t_{22}, t_{13}, t_{23}, t_{33})} = \begin{bmatrix} p_{11} & 0 & 0 & 0 & 0 & 0 \\ 0 & p_{11} & p_{12} & 0 & 0 & 0 \\ 0 & 0 & p_{22} & 0 & 0 & 0 \\ 0 & 0 & 0 & p_{11} & p_{12} & p_{13} \\ 0 & 0 & 0 & 0 & p_{22} & p_{23} \\ 0 & 0 & 0 & 0 & 0 & p_{33} \end{bmatrix} \quad (9.28.5)$$

Now by (9.28.1)

$$\alpha(c_{11}, \ldots, c_{23}) \propto$$
$$\int\int\int\int\int\int \left[\psi \middle/ \left|\frac{\partial(s_{11}, \ldots, s_{23}, s_{33})}{\partial(t_{11}, \ldots, t_{23}, t_{33})}\right|\right] ds_{11} \ldots ds_{23}ds_{33}. \quad (9.28.6)$$

Hence, by (9.28.5)

$$\alpha(c_{11}, \ldots, c_{23}) \propto \int\int\int\int\int\int \psi\, p_{11}^{-3}p_{22}^{-2}p_{33}^{-1}\, ds_{11} \ldots ds_{23}ds_{33}. \quad (9.28.7)$$

By (9.27.34), (9.27.35) and (9.28.4) and (9.28.7),

$$\alpha \propto \int\int\int\int\int\int (p_{11}p_{22}p_{33})^{-n-2} s_{11}^{n+1} s_{22}^{n+2} s_{33}^{n+3}$$
$$\exp\left\{-\sum_{i=1}^{3} \sum_{j=1}^{3} s_{ij}^{2} \middle/ s_{33}^{2}\right\} ds_{11} \ldots ds_{23}ds_{33} \quad (9.28.8)$$

$$\therefore \quad \alpha(c_{11}, \ldots, c_{23}) \propto (p_{11}p_{22}p_{33})^{-(n+2)}I, \quad (9.28.9)$$

where I is an integral whose value is independent of the six parameters $c_{11}, \ldots, c_{23}$.

Since $\boldsymbol{P}$ is upper-triangular

$$|\boldsymbol{P}| = p_{11}p_{22}p_{33}; \quad (9.28.10)$$

moreover

$$|\boldsymbol{P}|^2 = |\boldsymbol{P}'\boldsymbol{P}| = |\boldsymbol{Z}'\boldsymbol{Z}| \text{ by (9.28.2).} \quad (9.28.11)$$

Hence, by (9.28.9), (9.28.10) and (9.28.11) the posterior probability density function for the six regression parameters is

$$\alpha(c_{11}, c_{12}, \ldots, c_{23}) \propto |\boldsymbol{Z}'\boldsymbol{Z}|^{-(n+2)/2}. \quad (9.28.12)$$

In virtue of (9.27.1)

$$\alpha(c_{11}, \ldots, c_{23}) \propto |(\boldsymbol{Y}-\boldsymbol{XC})'(\boldsymbol{Y}-\boldsymbol{XC})|^{-(n+2)/2}. \quad (9.28.13)$$

Now denote by $c_{11}^*, \ldots, c_{23}^*$ the values of $c_{11}, \ldots, c_{23}$ for which the probability density (9.28.13) is a maximum, and let

$$\{c_{ij}^*\} = \boldsymbol{C}^*, \quad \boldsymbol{C}-\boldsymbol{C}^* = \delta\boldsymbol{C} \quad \text{and} \quad \boldsymbol{Y}-\boldsymbol{XC}^* = \delta\boldsymbol{Y}; \quad (9.28.14)$$

then

$$\alpha(c_{11}, \ldots, c_{23}) \propto |\delta\boldsymbol{Y}'\delta\boldsymbol{Y}+(\boldsymbol{X}\delta\boldsymbol{C})'(\boldsymbol{X}\delta\boldsymbol{C})|^{-(n+2)/2}. \quad (9.28.15)$$

The determinant in (9.28.15) is a 3×3 determinant and in principle one could integrate (9.28.15) over the five parameters c_{11}, c_{12}, c_{13}, c_{21} and c_{22} to find the marginal posterior probability density function for the sixth parameter c_{23}. One can save a lot of work by approximating to the value of the determinant in (9.28.15), proceeding as follows.

$$\alpha \propto |\hat{\mathbf{I}}+(\delta\mathrm{Y}'\delta\mathrm{Y})^{-1}(\delta\boldsymbol{C})'\boldsymbol{X}'\boldsymbol{X}(\delta\boldsymbol{C})|^{-(n+2)/2}. \quad (9.28.16)$$

We now proceed to ignore terms of order $(\delta c_{ij})^4$. On this basis (9.28.16) can be written in the form

$$\alpha(\theta_1, \theta_2, \theta_3, \theta_4, \theta_5, \theta_6) \propto (1+[\delta\boldsymbol{\theta}]'\boldsymbol{M}[\delta\boldsymbol{\theta}])^{-(n+2)/2}, \quad (9.28.17)$$

where the θ_i represent the c_{rs} as in (9.27.3) and where $\boldsymbol{M}$ is a 6×6 matrix. $\delta\boldsymbol{\theta}$ is defined as

$$\delta\boldsymbol{\theta} = \boldsymbol{\theta}-\boldsymbol{\theta}^* \quad (9.28.18)$$

where

$$\boldsymbol{\theta} = \boldsymbol{\theta}^* \text{ when } \boldsymbol{C} = \boldsymbol{C}^*. \quad (9.28.19)$$

To find the marginal posterior probability density function for any one of the coefficients, θ_6 say, we proceed as follows.

$$\pi(\theta_6) \propto \int\int\int\int\int (1+[\delta\boldsymbol{\theta}]'\boldsymbol{M}[\delta\boldsymbol{\theta}])^{-(n+2)/2} d\theta_1 d\theta_2 d\theta_3 d\theta_4 d\theta_5. \quad (9.28.20)$$

On integration, we obtain

$$\pi(\theta_6) \propto (1+K_6[\delta\theta_6]^2)^{-(n-3)/2}, \quad (9.28.21)$$

where K_6 is a constant depending on the observations.

On comparing (9.28.21) with (8.6.6), we find that the deviation of c_{32} (which equals θ_6) from its most probable value is a constant multiple of a t-variate with parameter $(n-4)$. The same is, of course, true of each of the other five regression parameters.

It will be recollected however that this result is not exact, since we introduced approximations between (9.28.16) and (9.28.17). The advantage of the form (9.28.21) is that, in any actual case, the value of the constant K_6 (and of the corresponding constants K_1 to K_5) can be quickly computed on an electronic computer, so that the marginal posterior probability density functions for all the regression parameters can be obtained in a familiar and readily understandable form.

Appendix to Chapter 9

'INVARIANT' RULE FOR CHOICE OF PRIOR PROBABILITY DENSITY FUNCTION

A9.1. The Amount of Information $I(t)$

$f(x, t)$ denotes the probability density function for a variate $\boldsymbol{x}$, given that a parameter θ takes the value t. The amount of information (per observation of $\boldsymbol{x}$) with respect to θ is defined as the following function of the value of t taken by θ:

$$I(t) = E\{u(x, t)\}^2 \mid (\theta = t), \tag{A9.1.1}$$

where

$$u(x, t) = \frac{\partial}{\partial t}\{\log f(x, t)\}. \tag{A9.1.2}$$

We shall see in the Appendix to Chapter 12 (in Volume 2) that under fairly general conditions governing $f(x, t)$ one can show that

$$E\{u(\boldsymbol{x}, t)\}^2 \mid (\theta = t) = -E\left\{\frac{\partial u(\boldsymbol{x}, T)}{\partial T}\right\}_{T'=t} \mid (\theta = t) \tag{A9.1.3}$$

so that we may also write

$$I(t) = -E\left\{\frac{\partial u(\boldsymbol{x}, T)}{\partial T}\right\}_{T=t} \mid (\theta = t) \tag{A9.1.4}$$

and we may probably abbreviate these equations, without serious risk of ambiguity, to the less pedantic form

$$I(t) = E\{u(\boldsymbol{x}, t)\}^2 = -E\left\{\frac{\partial u(\boldsymbol{x}, t)}{\partial t}\right\}. \tag{A9.1.5}$$

A9.2. The Asymptotic Behaviour of the Posterior Probability Density Function for θ as the Number of Observations of $\boldsymbol{x}$ Tends to Infinity

In § **9.20** it was asserted that the quantity $M\{\phi(t)\}$ can be made as large as possible by choosing $\phi(t)$, the prior probability density function for θ_3 to satisfy equation (9.21.4) which sets $\phi(t)$ proportional to the square root of $I(t)$, the quantity of information with respect to θ.

A rigorous proof of the validity of this assertion would involve far too detailed and tedious a discussion for it to be suitable for inclusion in a book of this kind. However, the difficult part of such a proof is concerned with establishing the asymptotic behaviour of the posterior probability density function for θ as the number of observations of x tends to infinity. The necessary features of this behaviour are stated, but without rigorous proof, on pages 129-130 of Volume 2 of Lindley's *Probability and Statistics*; moreover, a rigorous proof of this behaviour is to be published shortly in Series B of the *Journal of the Royal Statistical Society* in an article by Mr A. M. Walker entitled 'On the asymptotic behaviour of posterior distributions'. The features in question are stated as follows.

> If θ has an absolutely continuous prior probability density function $\phi(t)$, then, under suitable regularity conditions, the posterior probability density function for θ after N independent observations of $\boldsymbol{x}$ have been obtained, will, as $N \to \infty$, be asymptotically normal with mean equal to the maximum likelihood-estimator, θ_N^*, and with variance equal to the reciprocal of $NI(t)$, where t stands for the true value of θ. (A9.2.1)

We shall assume the truth of (A9.2.1), so that, since

$$\int_{-2\cdot576}^{2\cdot576} \frac{1}{\sqrt{2\pi}} e^{-v^2/2}\, dv = 0\cdot99, \tag{A9.2.2}$$

the narrowest range of values of θ which will enclose 99 per cent of the posterior probability will, as $N \to \infty$, behave asymptotically as

$$R = 5\cdot152\{NI(t)\}^{-1/2}. \tag{A9.2.3}$$

where t denotes the true value of θ.

It then follows that, as $N \to \infty$, the revealed inefficiency of the prior probability density function $\phi(t)$, at the value $\theta = t$, will be given asymptotically as

$$J(\phi, t) = \lim_{N \to \infty} \{i(N, \phi, t)\} = 5\cdot152\phi(t)\{I(t))\}^{-1/2}. \tag{A9.2.4}$$

It is now easy to show that, if t, the value of θ, is limited to lie in the range

$$t_2 < t < t_1, \tag{A9.2.5}$$

then the greatest lower bound $M\{\phi(t)\}$ of $J\{\phi, t)$, for t within this range, will be maximised if we set $\phi(t)$ to satisfy equation (9.21.4), namely so that

$$\phi(t) = \{I(t)\}^{1/2} \bigg/ \int_{t_0}^{t_1} \{I(v)\}^{1/2} dv. \qquad \text{(A9.2.6)}$$

For, suppose $\psi(t)$ to be any other prior probability density function for θ; then by equation (A9.2.4)

$$J(\psi, t) \,/\, J(\phi, t) = \psi(t) \,/\, \phi(t); \qquad \text{(A9.2.7)}$$

and, since throughout the interval $t_0 < v < t_1$

$$J(\phi, v) = M\{\phi(t)\} \qquad \text{(A9.2.8)}$$

$$M\{\psi(t) \,/\, M\{\phi(t)\} = \operatorname*{Min}_{v} \{\psi(v) \,/\, \phi(v)\}. \qquad \text{(A9.2.9)}$$

Since, however,

$$\int_{t_0}^{t_1} \phi(v)\, dv = 1 \geqslant \int_{t_0}^{t_1} \psi(v)\, dv, \qquad \text{(A9.2.10)}$$

equation (A9.2.9) ensures that

$$M\{\phi(t)\} \geqslant M\{\psi(t)\}, \qquad \text{(A9.2.11)}$$

and this establishes the result that the rule (A9.2.6) maximises $M\{\phi(t)\}$, as was asserted in the paragraph preceding equation (9.21.3).

A9.3. The Effect of Transformation from One Parameter ξ to Another Parameter θ which is a Monotonic Function of it

Suppose that the value t of one parameter θ is related to the value z of another parameter ζ by the monotonic functional relation $t = T(z)$, and let the derivative of this function be

$$\frac{dT}{dz} = T'(z). \qquad \text{(A9.3.1)}$$

Then by the usual formula (8.7.3) for transforming a probability density function, if $\phi(t)$ is the prior probability density function for θ, then $\psi(z)$, the prior probability density function for ζ must be

$$\psi(z) = |\, T'(z) \,|\, \phi\{T(z)\}. \qquad \text{(A9.3.2)}$$

Our rule (A9.2.6) will then involve us in a contradiction, unless when applied directly to the parameter ζ it leads us to choose the prior probability density function given by (A9.3.2). We shall now check that it does lead to this choice of a prior probability density function for ζ, so that no paradox or contradiction is involved.

Let $I(t)$ and $H(z)$ denote the amounts of information per observation with respect to the two parameters θ and ζ. When $f(x, t)$ and $g(x, z)$

are the probability density functions for $\boldsymbol{x}$ when $\theta = t$ and when $\zeta = z$,

$$g(x, z) = f\{x, T(z)\} \text{ for all admissible values } z. \qquad \text{(A9.3.3)}$$

Differentiating the logarithm of each side of (A9.3.3) partially with respect to z, we obtain

$$\frac{\partial}{\partial z}\{\log g(x, z)\} = T'(z).u(x, t), \text{ where } t = T(z); \qquad \text{(A9.3.4)}$$

and hence, remembering (A9.1.5)

$$H(z) = E\left\{\frac{\partial}{\partial z}[\log g(x, z)]\right\}^2 = \{T'(z)\}^2 I(t). \qquad \text{(A9.3.5)}$$

Hence our rule (A9.2.6) will lead us to choose the prior probability density function $\psi(z)$ for the value of ζ in such a way that

$$\psi(z) = | T'(z) | \, \psi(t), \text{ where } t = T(z); \qquad \text{(A9.3.6)}$$

and this agrees with equation (A9.3.2).

A9.4. The Effect of Allowing the Range of Values of θ to be Unlimited

The argument so far has assumed that the possible range of values for θ is limited to lie between two finite values t_0 and t_1. Now consider the effect of allowing t_0 and t_1 to move farther apart. Our rule (A9.2.6) is

$$\phi(t) = \{I(t)\}^{1/2} \Big/ \int_{t_0}^{t_1} \{I(v)\}^{1/2} dv. \qquad \text{(A9.4.1)}$$

The effect of moving t_0 and t_1 farther apart increases the denominator of this expression, but leaves the numerator unchanged for those values of t which already lie within the range. In the limit as $t_0 \to -\infty$ and $t_1 \to \infty$, the denominator may tend to infinity, in which case, the prior probability density function would tend everywhere to zero. If this happens, the prior probability density function (A9.4.1) should be replaced by the improper density function

$$\phi(t) = \{I(t)\}^{1/2}. \qquad \text{(A9.4.2)}$$

This choice of prior probability density function will ensure that the conditional measure $M\{\phi(t)\}$, given that $t_0 < t < t_1$, will be maximised for all pairs of limits (t_0, t_1) satisfying $t_0 < t_1$.

A9.5. Extension of the Results to the Case where there are Two or More Parameters

It we wish to choose the prior probability density function $\phi(t, g)$

for a pair of parameters (θ, γ) in the density function $f(x, \theta, \gamma)$ for the variate $\boldsymbol{x}$, we may still define the revealed inefficiency $i(\phi, \beta, N)$ as we did in § **9.21**, and denote the limiting value of the expected value of $i(\phi, \beta, N)$, when N tends to infinity, by $J(\theta, \gamma, \phi)$. We consider first the case where the possible region R of values of (θ, γ) is rectangular and finite.

In order to maximise $M\{\phi(t, g)\}$, the lower bound of $J(\theta, \gamma, \phi)$ with respect to (θ, γ), we should be led by the same argument as that of § **A9.2** to choose the rule

$$\phi(t, g) = \{I(t, g)\}^{1/2} \Big/ \int_R \{I(t, g)\}^{1/2} dt dg, \qquad \text{(A9.5.1)}$$

where integration is over the possible region R and where $I(t, g)$ is the amount of information per observation of $\boldsymbol{x}$ with respect to the pair of parameters when their values are t and g. This amount of information is defined by the equation

$$I(t, g) = \begin{vmatrix} -E\left\{\dfrac{\partial^2 \log f}{\partial t^2}\right\} & -E\left\{\dfrac{\partial^2 \log f}{\partial t \partial g}\right\} \\ -E\left\{\dfrac{\partial^2 \log f}{\partial g \partial t}\right\} & -E\left\{\dfrac{\partial^2 \log f}{\partial g^2}\right\} \end{vmatrix} \qquad \text{(A9.5.2)}$$

When the region R is allowed to include the whole range of values of θ and γ from minus to plus infinity, it may be necessary to suppress the denominator in (A9.5.1) and use the improper prior probability density function

$$\phi(t, g) = \{I(t, g)\}^{1/2} \qquad \text{(A9.5.3)}$$

which is analogous to that in equation (A9.4.2).

The above discussion of the case where there are two parameters may be easily generalised to the case where there are more than two parameters.

The case where observations are taken not of only one variate $\boldsymbol{x}$, but of two or more variates $(\boldsymbol{x}, \boldsymbol{w}, \ldots)$ introduces no new complication. The definition of the amount of information per set of observations is unaffected, and there is no change in the formulae for the choice of prior probability density function for the parameter or parameters.

A9.6. An Alternative Formulation of the Rule (A9.4.1) for choosing the Prior Probability Density Function

It was asserted in equation (9.21.5) that the rule

$$\phi(t) \propto \{I(t)\}^{1/2} \qquad \text{(A9.6.1)}$$

could also be expressed as

$$\phi(t) \propto N^{-1/2} g_N(t, t), \tag{A9.6.2}$$

where $g_N(t^*, t)$ denotes the probability density function (where t is the true value of θ) of t^*, the maximum-likelihood estimator of θ based on N independent observations of $\boldsymbol{x}$. A fuller discussion of maximum-likelihood estimators will be given in Chapter 12 (in volume 2), but at this stage we may quote the result that as $N \to \infty$, the density function for $N^{1/2}(t^* - t)$ tends under fairly general conditions to that of a normal variate with variance $\{I(t)\}^{-1}$ and of expected value zero. Hence the probability density at $\boldsymbol{t}^* = t$ for the variate $N^{1/2}(\boldsymbol{t}^* - t)$ tends to the value $\{I(t)/2\pi\}^{1/2}$, and so for sufficiently large N the probability density for $\boldsymbol{t}^*$ at $\boldsymbol{t}^* = t$ is almost exactly $\{NI(t)/2\pi\}^{1/2}$. Thus

$$g_N(t, t) = \{NI(t)/2\pi\}^{1/2}\{1 + 0(N^{-1/2})\} \tag{A9.6.3}$$

so that the rule (A9.6.1) may also be expressed in the form (A9.6.2).

The argument generalises with little modification to the case where a prior probability density function is required for 2 or more parameters. For example, if two parameters (θ, ω) are to be estimated, the rule (A9.6.2) becomes

$$\phi(t, w) \propto \{I(t, w)\}^{1/2} \tag{A9.6.4}$$

where

$$I(t, w) = \begin{vmatrix} -E\left(\dfrac{\partial^2 \log f}{\partial t^2}\right) & -E\left(\dfrac{\partial^2 \log f}{\partial t \partial w}\right) \\ -E\left(\dfrac{\partial^2 \log f}{\partial w \partial t}\right) & -E\left(\dfrac{\partial^2 \log f}{\partial w^2}\right) \end{vmatrix} \tag{A9.6.5}$$

and $f(x, t, w)$ is the density function for each observation of $\boldsymbol{x}$ when $\theta = t$ and $\omega = w$.

Let (t^*, w^*) denote the maximum-likelihood estimator for (θ, ω) and let $g(t^*, w^*; t, w)$ denote their probability density function when N observations are used and $\theta = t$ and $\omega = w$; then it is known (see Appendix to Chapter 12, in Volume 2) that the variate pair $N^{1/2}(t^* - t)$, $N^{1/2}(w^* - w)$ have a probability density function which, as N tends to infinity, tends to the bivariate normal form with zero expected values and with variance-covariance matrix $\boldsymbol{H}^{-1}$ where $\boldsymbol{H}$ is the matrix whose determinant is given above as the right hand side of equation (A9.6.5). This bivariate normal density function is

$$h(T, W) = |\,\boldsymbol{H}\,|^{1/2} e^{-Q}/(2\pi), \tag{A9.6.6}$$

where Q is a quadratic form in $(t^* - t, w^* - w)$ which takes the value zero at $t^* = t$, $w^* = w$. Hence

$$h(0, 0) = | H |^{1/2}/2\pi = \{I(t, w)^{1/2}\}/2\pi. \quad (A9.6.7)$$

It follows that

$$g_N(t, w; t, w) = N\{I(t, w)\}^{1/2}/(2\pi), \quad (A9.6.8)$$

and that the rule (A9.6.4) can also be expressed in the form

$$\phi(t, w) \propto N^{-1}g_N(t, w; t, w). \quad (A9.6.9)$$

Exercises 9

9.1. $\boldsymbol{x}$ has the probability function $f(0) = 1-\theta, f(1) = \theta$, where θ has the prior probability density $\phi(\theta) = 1/\pi\sqrt{\{\theta(1-\theta)\}}$. If n independent observations of $\boldsymbol{x}$ have yielded $(n-r)$ zeros and r ones, find the probability that the next observation will be a one.

9.2. A discrete parameter θ may take the values θ_r $(r = 1, 2, \ldots)$ and has the prior probability function $\phi(\theta_r)$. Evidence $\boldsymbol{X}$ is to be collected, and the posterior probability function $\phi[\theta_r | \boldsymbol{X}]$ will be calculated, by means of Bayes' Theorem.

Prove that where θ_T denotes the (unknown) true value of θ,

$$E\left[\frac{\phi(\theta_r \mid X)}{\phi(\theta_T \mid X)}\right] = \frac{\phi(\theta_r)}{\phi(\theta_T)}.$$

9.3. How would you explain matters to a critic who claimed that the result of question 2 revealed that according to Bayes' Theorem evidence would be expected not to shift the posterior probability function in favour of the true value of θ?

9.4. Distinguish between Bayes' Postulate and Bayes' Theorem. Explain the objections to their use which have been urged by adherents of the frequency theory of probability.

A bag contains 5 balls, and it is known that they are either (*i*) all white, or (*ii*) 3 white and 2 not white. Use Bayes' Postulate and Bayes' Theorem to find the posterior probability of alternative (*i*),

(a) if a ball drawn at random is white;

(b) if 2 balls drawn at random with replacement are white.

[Cambridge Economics Tripos Part 2, 1964]

9.5. The elasticity of demand η is to be estimated in a model

$$\log q_i = -\eta \log p_i + u_i,$$

where the u_i are independent observations of a normal variate of mean a and variance σ^2 and independent of p_i. Nothing whatever is known about a or σ, but it is considered probable (say $P = 2/3$) that η lies between 0·25 and 0·75, and very probable (say P is about 19/20) that η lies between 0 and 1.

What probability density function would you suggest as the prior for (a, σ^2, η)?

What functional form would the marginal posterior probability density function for η take if you used a set of n pairs of observations of (p_i, q_i)?

9.6. In order to estimate the parameters (α, β) in the model

$$y_t = \alpha + \beta x_t + u_t,$$

where the u_t are independent observations of a normal (0,1) variate and are independent of x_t, (α, β) are given the prior probability density function

$$\phi(a, b) = \frac{1}{20\pi} \exp\left[-\tfrac{65}{32}(\beta-1)^2 - \tfrac{1}{80}(\beta-1)(a-95) - \tfrac{1}{800}(a-95)^2\right]$$

and 10 observations (X_t, Y_t) $(t = 1, 2, \ldots, 10)$ are obtained.

It is found that

$$\sum_{t=1}^{10} X_t = 50 \quad \sum_{t=1}^{10} Y_t = 1110 \quad \sum_{t=1}^{10} X_t^2 = 260 \quad \sum_{t=1}^{10} X_t Y_t = 5553$$

$$\sum_{t=1}^{10} Y_t^2 = 12\,332.$$

Verify that the posterior probability density function for (α, β) is

$$\theta(a, b) = \frac{28\,007}{1600\pi} \exp\left[-5{\cdot}001\,25(a-118{\cdot}47)^2\right.$$

$$\left. -50{\cdot}0125\,(a-118{\cdot}47)(b-0{\cdot}5) - 132{\cdot}031\,25(b-0{\cdot}5)^2\right]$$

and that β has the marginal posterior probability density function of a normal variate with mean 1/2 and standard error $1/\sqrt{7}$.

9.7. Three bags A, B and C each contain 10 collar studs. Inside the bags 2, 4 and 6, respectively, of the studs in A, B and C are green. Joshua selects one bag at random and removes at random from it 2 studs, but neither of them is green. What are the probabilities that Joshua had selected bags A, B and C respectively? Give a frequency interpretation to your answer.

9.8. The variate $\boldsymbol{x}$ has the density function $f(x) = 1/\boldsymbol{\theta}$ for $0 < x < \theta$. Ten independent observations of $\boldsymbol{x}$ are found as 16, 12, 25, 8, 40, 17, 29, 15, 3 and 35. Find the posterior probability density function for $\boldsymbol{\theta}$ in the cases where its prior probability density function is taken as

(*i*) $\phi(t) = 0{\cdot}005 \quad (0 < t < 200)$;

(*ii*) $\phi(t) = 1/(2t \log_e N) \quad \left(\frac{1}{N} < t < N,\ N > 100\right)$;

(*iii*) $\phi(t) = 1/N \quad (0 < t < N,\ N > 100)$.

Write down the limiting forms of the posterior probability density functions as $N \to \infty$ in the cases (*ii*) and (*iii*). Find the expected values and variances for these posterior probability density functions.

9.9. $\boldsymbol{x}$ and $\boldsymbol{y}$ are Poisson variates with parameters λ and μ. The prior probability function for (λ, μ) is taken as $\phi(\lambda, \mu) = e^{-(\lambda+\mu)}$ $(\lambda, \mu > 0)$. Find the posterior probability function for (λ, μ) given that 10 observations of $\boldsymbol{x}$ total 20 and 20 independent observations of $\boldsymbol{y}$ total 10.

9.10. n observations of $\boldsymbol{\eta}$ have mean $\overline{Y}$ and variance s^2. $\boldsymbol{\eta}$ is a normal variate of unknown mean θ and unknown variance V. The prior probability density function for (θ, V) may be taken as proportional to $(1/V)$. Obtain the posterior probability density function for (θ, V).

9.11. In the notation of question 10, let $S_{\bar{x}}^2 = s^2/(n-1)$ and let $t = (\theta - \overline{Y})/S_{\bar{x}}$ $(S_{\bar{x}} > 0)$. Show that (provided $n > 1$) the posterior probability density function for t is

$$\phi(T) = C \bigg/ \left[1 + \frac{T^2}{n-1}\right]^{n/2},$$

where C is a constant depending only on the value of n. Find the posterior probability density functions for θ in the cases $n = 1$ and $n = 2$.

9.12. It is proposed to throw a die 120 times in order to estimate the probability of getting a six. Before the start of the experiment it is believed that 1/6 is the most likely value of this probability θ, and that θ very probably lies between 0·165 and 0·170. Discuss the choice of a prior probability density function for θ.

SELECTED BIBLIOGRAPHY

No attempt is made to provide a comprehensive bibliography. In the wide field of subjects touched on by this book, so many books have already been written that it is considered more useful to indicate a short list of some of the best books, than to provide a lengthy catalogue listing books, good, bad and indifferent. The author does not claim even to have glanced through *all* the many different books relating to this field; so it may be that the following list omits one or two excellent books. Many of the books included in the list, however, will be found to contain more comprehensive bibliographies of their own, to which the more voracious readers may refer in their search for yet more books to read on special subjects.

In Volume 1, three short lists are provided, relating respectively to the contents of Chapters 2–3, of Chapters 4–8 and of Chapter 9. If a book appears in more than one list, full details of it will be provided only in the earlier list.

Further such short lists will be provided in Volumes 2 and 3.

Chapters 2 and 3: Basic Probability Theory

BARKER, S. F. 1957. *Induction and Hypothesis*. Cornell University Press.

BOOLE, G. (died 1864), 1952. *Studies in Logic and Probability*. London: C. A. Watts.

BROAD, C. D. 1968. *Induction, Probability and Causation: Collected papers*; Synthese Library. The Humanities Press, New York.

CARNAP, R. 1950. *Logical Foundations of Probability*. London: Routledge and Kegan Paul Ltd.

DAY, J. P. 1961. *Inductive Probability*. London: Routledge and Kegan Paul. New York: The Humanities Press.

GOOD, I. J. 1950. *Probability and the Weighing of Evidence*. London: Charles Griffin and Co. Ltd.

HACKING, I. M. 1965. *The Logic of Statistical Inference*. Cambridge University Press.

KEYNES, J. M. 1921. *A Treatise on Probability*. London: Macmillan.

KNEALE, W. 1949. *Probability and Induction*. Oxford University Press.

KYBURG, H. E. 1961. *Probability and the Logic of Rational Belief*. Middletown, Connecticut: Wesleyan University Press.

LEVI, I. 1967. *Gambling with Truth*. London: Routledge and Kegan Paul.

RAMSEY, F. P. 1931. *The Foundations of Mathematics*. London: Routledge and Kegan Paul.

REICHENBACH, H. 1949. *The Theory of Probability* (English translation). University of California Press.

RUMSHISKII, L. Z. 1965. *Elements of Probability Theory* (translated by D. M. G. Wishart). Oxford: Pergamon Press.

WISDOM, J. O. 1952. *Foundations of Inference in Natural Science*. London: Methuen.

VON WRIGHT, G. 1955. *The Logical Problem of Induction*. Oxford: Blackwell.

Chapters 4, 5, 6, 7 and 8: Events, Variates and Measurements

ANDERSON, T. W. 1958. *An Introduction to Multivariate Statistical Analysis*. New York: John Wiley.

CRAMER, H. 1946. *Mathematical Methods of Statistics*. Princeton University Press.

CROXTON, F. E. and COWDEN, D. J. 1956. *Applied General Statistics.* London and U.S.A.: Prentice Hall.
EHRENFELD S. and LITTAUER, S. B. 1964. *Introduction to Statistical Method.* McGraw-Hill.
FELLER, W. 1957 & 1966. *An Introduction to Probability Theory and its Applications* (2 vols.). New York: John Wiley.
HOEL, P. G. 1960. *Introduction to Mathematical Statistics* (2nd. edn.) New York: John Wiley.
KARMEL, P. H. 1963. *Applied Statistics for Economists.* Melbourne: Pitman.
KEEPING, E. S. (1962). *Introduction to Statistical Inference.* Princeton; Van Nostrand.
KENDALL, M. G. and STUART, A. 1954. *The Advanced Theory of Statistics* (3 vols.). London: Griffin and Co.
MOOD, A. M. and GRAYBILL, F. A. 1963. *Introduction to the Theory of Statistics* (2nd edn.). McGraw-Hill.
RUMSHISKII, L. Z. 1965. See list for Chapters 2–3.
WEATHERBURN, C. E. 1949. *A First Course in Mathematical Statistics* (2nd edn.). Cambridge University Press.
WHITWORTH, W. A. 1901. *Choice and Chance* (5th edn.). Cambridge: Deighton Bell. 1959 ed. New York: Hafner.
WILKS, S. S. 1962. *Mathematical Statistics.* New York: John Wiley.

Chapter 9: Bayes' Procedures

WETHERILL, G. B. 1966. *Sequential Methods in Statistics.* London: Methuen.
GOOD, I. J. 1950. See list for Chapters 2–3.
JEFFREYS, H. 1961. *Theory of Probability* (3rd edn.). Oxford: Clarendon Press.
KEYNES, J. M. 1921. See list for Chapters 2–3.
LINDLEY, D. V. 1965. *Introduction to Probability and Statistics* (2 vols.). Cambridge University Press.
RAIFFA, H. and SCHLAIFER, R. 1961. *Applied Statistical Decision Theory.* Boston: Harvard Graduate School of Business Administration. Folkestone: Bailey Bros. & Swinfen.
SAVAGE, J. L. 1954. *The Foundations of Statistics.* New York: John Wiley.
SCHLAIFER, R. 1961. *Introduction to Statistics for Business Decisions.* New York: McGraw-Hill.

INDEX

The references are to pages.
Figures in **bold** *denote pages in which the entry is a principal theme.*